# Aristotle and The Philosophy of Law: Theory, Practice and Justice

# IUS GENTIUM

COMPARATIVE PERSPECTIVES ON LAW AND JUSTICE

VOLUME 23




For further volumes:
http://www.springer.com/series/7888

Liesbeth Huppes-Cluysenaer
Nuno M.M.S. Coelho
Editors

# Aristotle and The Philosophy of Law: Theory, Practice and Justice

Springer

*Editors*
Liesbeth Huppes-Cluysenaer
Department of General Jurisprudence
University of Amsterdam
Amsterdam, Netherlands

Nuno M.M.S. Coelho
Faculdade de Direito de Ribeirão Preto
University of São Paulo
Ribeirão Preto
São Paulo, Brazil

ISBN 978-94-007-6030-1 ISBN 978-94-007-6031-8 (eBook)
DOI 10.1007/978-94-007-6031-8
Springer Dordrecht Heidelberg New York London

Library of Congress Control Number: 2013932391



Printed on acid-free paper

Springer is part of Springer Science+Business Media (www.springer.com)

# Contents

1 **Virtue Jurisprudence: Towards an Aretaic Theory of Law** ............... 1
Lawrence B. Solum

2 **Reasoning Against a Deterministic Conception of the World** ............ 33
Liesbeth Huppes-Cluysenaer

3 **Law and the Rule of Law and Its Place Relative to *Politeia* in Aristotle's Politics** ............ 59
Clifford Angell Bates Jr.

4 **The Best Form of Government and Civic Friendship in Aristotle's Political Thought: A Discussion Note** ............ 77
Ki-Won Hong

5 **Controversy and Practical Reason in Aristotle** ............ 87
Nuno M.M.S. Coelho

6 **Aristotelian Ethics and Aristotelian Rhetoric** ............ 109
Marcel Becker

7 **Is There Any Theory of Value in Aristotle's Ethics?** ............ 123
António de Castro Caeiro

8 **Intellectual Excellences of the Judge** ............ 135
Tommi Ralli

9 **Justice *Kata Nomos* and Justice as *Epieikeia* (Legality and Equity)** ............ 149
Samuli Hurri

10 **Legality and Equity in the *Rhetoric*: The Smooth Transition** ............ 163
Miklós Könczöl

**11 Legal Rules and *Epieikeia* in Aristotle: Post-positivism Rediscovered** .......... 171
Jesús Vega

**12 Legal Vices and Civic Virtue: Vice Crimes, Republicanism and the Corruption of Lawfulness** .......... 203
Ekow N. Yankah

**13 A Neo-Aristotelian Notion of Reciprocity: About Civic Friendship and (the Troublesome Character of) Right Judicial Decisions** .......... 223
Iris van Domselaar

**14 Synallagma as a Paradigm of Exchange: Reciprocity of Contract in Aristotle and Game Theory** .......... 249
Mariusz Jerzy Golecki

**15 The General Principle of Proportionality and Aristotle** .......... 265
Eric Engle

# About the Authors

**Clifford Angell Bates Jr**. is a Professor at the American Studies Center of Warsaw University in Poland since 2004. He specialises in Political Philosophy. In 2003 he published at LSU Press his *Aristotle Best Regime: Kingship, Democracy, and the Rule of Law*. He has written several articles on Aristotle and is currently working on a second book on Aristotle dealing with the regime typology and its evolution.

**Marcel Becker** (1961) is assistant professor philosophical ethics at the Radboud University Nijmegen. He is specialized at virtue ethics and applied ethics, particularly ethics of public administration, business ethics and media ethics. He tries to make fruitful the Aristotelian heritage in each of these areas. Address: M.Becker@ftr.ru.nl

**Nuno M.M.S. Coelho** (Ph.D. in 2006, Habilitation in 2009) is Professor of Ethics and Philosophy of Law at USP (University of São Paulo) and Head of the Post Graduate Courses in Legal Philosophy in UNIPAC/Juiz de Fora. He published books and articles on Aristotle and Ancient Philosophy of Law, and on Hermeneutics and Legal Methodology.

**Antonio de Castro Caeiro** teaches Ancient and Contemporary Philosophy, Ancient Greek and Latin at the Faculty of Social Sciences and Humanities of the New University of Lisbon. He translated into Portuguese the Nicomachean Ethics of Aristotle and the Pythical Odes of Pindar. He recently translated and commented the Fragmenta Aristotelis collected by Valentin Rose and has been working on a project with many scholars, especially David Sedley, on the stoics. Contacts: acaeiro@mac.com, http://www.antoniodecastrocaeiro.com/.

**Eric Engle** is a Lehrbeauftragter at the Humbolt Universitaet, Berlin, Juristische Fakultaet. JD St. Louis, DEA Université Paris II (Mention), LL.M., Dr. Jur. Universität Bremen

**Mariusz Jerzy Golecki**: B.C.L. Warsaw (1998), LL.M. Cambridge (2002), Ph.D. Lodz (2004). Associate Professor- Department of Legal Theory and Philosophy of Law, Faculty of Law and Administration, University of Łódź. Visiting Scholar- the

Centre for European Legal Studies, the Faculty of Law, University of Cambridge (2008–2009). He specializes in Law and Economics, Jurisprudence and Comparative Law. The author of 2 monographs and about 40 articles.

**Ki-Won Hong** took his doctor's degree at the University of Aix-Marseille, France, with a dissertation on the political thought of Fr. Hotman, a sixteenth-century Aristotelian. His main interest consists in studying early modern political ideas in relation to classical republicanism. After several years of teaching experiences as assistant professor in Seoul, he is now enlarging his research area as visiting professor at the University of Panthéon-Assas (Paris).

**Liesbeth Huppes-Cluysenaer** specialises in the ontology and method of Law. On behalf of the Law faculty of the University of Amsterdam, she has developed a research website-blog dedicated to the English translation and comparative analysis of the main book by Paul Scholten, who defends an intuitive conception of the legal decision. (paulscholten.eu)

**Samuli Hurri** LL.D. works at the *Centre of Excellence in Foundations of European Law and Polity* at the University of Helsinki, Finland. He is the founder and former editor-in-chief of *No Foundations*, an open access journal in legal theory and philosophy.

**Miklós Könczöl** is an assistant lecturer at Pázmány Péter Catholic University, Budapest (Hungary) and currently Durham Doctoral Fellow at Durham University (UK). He holds degrees in Latin and Greek, Law, and Legal Theory. His research interests include Greek rhetoric, legal theory, the semiotics of law, and environmental philosophy. His latest publication is an edited volume, *Legal and Political Theory in the Post-National Age* (Peter Lang, 2011, co-edited with Péter Cserne).

**Tommi Ralli** obtained his Ph.D., on a thesis entitled 'Justice through Legal Dispute,' at the European University Institute, Florence. At Bremen University, he has been research assistant in the pan-European Reconstituting Democracy in Europe (RECON) project and lecturer at Hanse Law School. His research focuses on legal theory, social justice, the ethics of legal dispute resolution, and comparative law.

**Lawrence B. Solum** is John Carroll Research Professor at Georgetown University Law Center. He is author of many articles and books on a variety of topics in the philosophy of law, constitutional theory, and Internet governance; he edits Legal Theory Blog

**Iris van Domselaar**: Iris van Domselaar (LL.M., M. Phil.) is Assistant Professor in Legal Ethics at the Amsterdam Law School, the University of Amsterdam. She has written several articles on justice, law, legal ethics and adjudication. She is finishing her doctorate thesis in which she develops a tragic approach to adjudication. Currently she teaches the course The Legal Professions.

**Jesús Vega** is Professor of Philosophy of Law at the Universidad de Oviedo (Spain). Ph.D. in Law and B.A. in Philosophy. He has been visiting scholar in Germany and USA, as well as delivered lectures and doctorate courses in various Latin-American universities. Author of several contributions in the fields of legal epistemology, history of legal thought and legal methodology.

**Ekow N. Yankah** is a Professor of Law at Cardozo School of Law. His scholarship centers on analytical jurisprudence, criminal theory and political theory generally, particularly republican theories of political obligation and their application in criminal law. He holds degrees from the University of Michigan, Columbia Law School and Oxford University and his prior appointment was at University of Illinois School of Law.

# Introduction

Aristotle is a key figure of legal and philosophical theory. The Western academic tradition is founded on his open dialogue to Sophists, Socrates and Plato. His concept of prudence (*phronêsis*) had quintessential meaning for Roman legal culture, and Aristotle has been present in legal theory ever since. For Aquinas he was the paradigm of the Philosopher, currently he is seen as the thinker who was the main opponent to the Enlightenment, and his writings are used consequently to challenge Modern ways of thinking.

Most philosophical perspectives still assume Aristotle as their interlocutor. The new epistemological perspectives in legal theory, which arose during the twentieth century in connection to phenomenology and existentialism – for example philosophical hermeneutics, rhetoric, topics, theory of argumentation etc. – cannot be conceived without their reference to Aristotle. The same is true for jurisprudence, law and literature, contemporary natural law theories, legal pragmatism and virtue ethics. One way or another, most of us deal with the same problems as Aristotle faced and employ some of his theoretical tools to think these over.

In the field of practical philosophy and law, Aristotle's presence is even more significant. There are examples of practical legal problems in the texts of Aristotle, which are still used in exactly the same way in the major textbooks for students. It is a moving experience to read in texts of more than 2000 years ago such phrases on legal matters.

In legal theory, Aristotle is mostly discussed in the context of Legal Positivism versus Natural Law. In these debates the 'positivists' parade as the slightly cynical realists, who do not take refuge in vague and idealistic concepts, while the 'natural law' followers accuse the positivists of being formalists. One could easily assume that everything that could be said in this debate has been said already a hundred times. However, the debate is so fundamental to legal thought that every relevant change in society is reflected in a re-arrangement of the arguments of this debate.

At this moment in history, globalisation is clearly causing such a change in society. Ripple effects of actions spread nowadays extremely fast all over the world; sometimes it is merely the fear for effects which creates a worldwide chain of reactions. Is it possible to control these processes? It becomes clear how integrally positive law is tied to the nation state and how this holds an inherent weakness. Positivists have to face the question whether it is possible to prevent the nation state from fading away and whether the belief in a 'World Government' is not extremely unrealistic and dangerously idealistic.

The rapid spread of ripple effects of human interventions leads also to a re-evaluation of the technical predictive knowledge which until today has been pictured as the highest achievement of Western thinking. This type of knowledge is not helpful to cope with the occurrence of ripple effects on a world wide scale. A completely different type of rationality seems to be needed to keep people organised in a world full of uncertainty and ignorance. This means that the concept of nature, which was deeply transformed at the start of the Enlightenment, is being reconsidered today. And this in turn will affect the ideas of natural order and natural law.

How is it possible that Aristotle's views have been present in so many different conceptions of law and philosophy over more than 20 centuries? One of his discoveries gives us the clue to understanding this: his polyphonic conception of reason, which leads to a non-monolithic and pluralistic conception of rationality. For Aristotle there are different structures of reasoning existent, which accommodate the different ways intelligence is incited when experiencing the world – practical, technical, scientific or philosophical.

Within the Enlightenment concept of a rationality that strives for a unified science and a unified conception of the world, this discovery of Aristotle's was not appreciated. Aristotle's view was rejected by some as being inconsistent and making it possible to take from it whatever one liked, while others spent much effort in explaining the disorder away.

As globalisation unfolds, it becomes clear that it holds the germ for a turn in thinking that will lead to a new appreciation of Aristotle's polyphonic conception of reason. On a worldwide scale it will be unavoidable to find ways to accommodate pluralism, not as a kind of idealistic love for the exotic, but as a realistic condition for survival.

This book invites such a new reading of Aristotle.

The book presents a new focus on the legal philosophical texts of Aristotle, which offers a much richer frame for the understanding of practical thought, legal reasoning and political experience. It allows understanding how human beings interact in a complex world, and how extensive the complexity is which results from humans' own power of self-construction and autonomy. Unlike some Enlightenment perspectives and positivist theories of law, the Aristotelian approach makes it possible to think the task of justice from a non-linear and non-monological rationality. It recognizes the limits of rationality and the inevitable and constitutive contingency in Law. All this offers a helpful instrument to understand the changes globalisation imposes on legal experience today.

The contributions in this collection do not merely pay attention to private virtues, but focus primarily on public virtues. They deal with the fact that law is dependent on political power and that a person can never be sure about the facts of a case or about the right way to act. They explore the assumption that a detailed knowledge of Aristotle's epistemology is necessary, because of the direct connection between

Enlightened reasoning and legal positivism. They pay attention to the concept of proportionality, which can be seen as a precondition to discuss liberalism.

Most chapters of this book were presented in substance as papers at the IVR-conference 2011 in Frankfurt am Main, in the special workshop Aristotle and the Philosophy of Law. This workshop – which had its first meeting at the IVR-conference of 2007 in Krakow – aims at cooperation between specialists in the philosophy of Aristotle and legal theorists who use elements of Aristotle's philosophy in their theorizing. This leads to a very broad and general discussion in which nearly all aspects of legal theory are treated, sometimes deeply entrenched in specific interpretations of Aristotle's texts, sometimes focused on the practical use of Aristotle's ideas in current times.

Some of the chapters are written by authors who could not attend the IVR-conference and were invited to submit a contribution. Two chapters are discussion notes by authors who made interesting interventions about one of the papers during the workshop and were invited to develop these discussions into a chapter. The first chapter of the book was presented in Krakow and published together with all the other papers of that conference. We asked permission from Lawrence Solum to include his paper in this volume, because we think that it presents the state of the art of the debate in 2007 very well. It is in reference to this debate that we try to take a step further in this workshop.

The Editors would like to thank António Caeiro for his correction and unification of the transcription of Greek terminology and the International Association for Philosophy of Law and Social Philosophy (IVR) for offering the opportunity to organise Special Workshops on Aristotle in Krakow, Beijing and Frankfurt. They thank Springer, for taking up the challenge to publish this book and also all who participated in the project, especially the scholars who worked as reviewers of the submitted texts: Bruno Amaro Lacerda, Marcel Beckers, Annemarie Bos, Edith Brugmans, Marcelo Andrade Cattoni de Oliveira, Frans de Haas, Oliver Lembcke, Carlo Natali, Diego Poole, Jonathan Soeharno Lawrence Solum, Sebastião Trogo and Marco Zingano. The book would not have been possible however without the help of Carlo Natali and Marco Zingano, who as specialists in Ancient Philosophy showed the openness of mind to support and stimulate the making of a bridge to the Philosophy of Law.

# Chapter 1
# Virtue Jurisprudence: Towards an Aretaic Theory of Law

**Lawrence B. Solum**

## 1.1 Introduction: The Aretaic Turn in Legal Theory

*Virtue jurisprudence* is the name for a distinctive approach to normative legal theory. One way to introduce the core of virtue jurisprudence is by adumbrating its answers to three perennial questions about law. The first question is: "What is proper aim, goal, or *telos* of legislation?" Virtue jurisprudence answers that the fundamental purpose of law is to promote the flourishing of individual humans and their communities by creating the conditions that promote and sustain the development and exercise of the human excellence or virtue. The second question is: "What is the standard for a lawful or legally correct resolution of a dispute?" Virtue jurisprudence answers that the standard for lawfulness is best understood as congruence with a decision that would be rendered by an adjudicator who possesses the judicial virtues or excellences: such a judge is both *phronimos* (practically wise) and *nomimos* (lawful). The third question: "What is the nature of lawfulness?" Virtue jurisprudence answers that the virtue of lawfulness is best understood as the internalization of (and hence a disposition to comply with) the *nomoi*—the deeply held and widely shared social norms that function to enable flourishing in particular human communities. This essay makes the case for a particular version of virtue jurisprudence—one rooted in contemporary neoAristotelian virtue ethics.

How does virtue jurisprudence fit in the landscape of theorizing about law's nature and purposes? One approach to this question can begin with the observation that contemporary legal theory in the United States has been dominated by what

---

Published in Revista da Faculdade de Direito de Conselheiro Lafaiete, Nova fase, 2007, 3. See Introduction.

L.B. Solum (✉)
Georgetown University Law Center, Washington, DC, USA
e-mail: lsolum@gmail.com

L. Huppes-Cluysenaer and N.M.M.S. Coelho (eds.), *Aristotle and The Philosophy of Law: Theory, Practice and Justice*, Ius Gentium: Comparative Perspectives on Law and Justice 23,
DOI 10.1007/978-94-007-6031-8_1, 

might be called "the realist paradigm." The extreme version of realism is captured by the slogan of the critical legal studies movement: "Law is politics!" Other heirs to the realist tradition (including normative law and economics, the legal process school, legal pragmatism, and so forth) coalesce around what we might call *the instrumentalist thesis*—the point of legal institutions (especially courts) is to use the law as an *instrument* to achieve the goals of some normative theory (such as welfarism or deontology) or a political ideology (of the left, right, or center). There are, of course, opposing tendencies in contemporary legal theory. Some neoformalists emphasize the duty of adjudicators to follow the law and give the parties what they are due; in a rough and ready sort of way, these neoformalists adopt a deontological perspective on legal theory that competes with the consequentialism of contemporary neorealists.

For virtue jurisprudence, the end of law is not to maximize preference satisfaction or to protect some set of rights and privileges: the goal or *telos* of law is to promote human flourishing—to enable individuals and to lead lives that express human excellence or virtue.

For virtue jurisprudence, the best way to improve the ability of legal institutions to resolve disputes is not to populate the bench with economists or moral philosophers from either the left or the right; instead, achieving an excellent judiciary requires the selection of judges who possess the judicial virtues—civic courage, judicial temperament, judicial intelligence, wisdom, and, above all, justice. These answers to the big practical questions are unified by a central thesis: *the fundamental concepts of legal philosophy should* not *be welfare, efficiency, autonomy, or equality; the fundamental notions of legal theory should be flourishing, virtue, and excellence*. Thus, the proposal of the essay is that jurisprudence should turn from an emphasis on ideology, rights, and utility to a focus on virtue.

What is *aretaic* legal theory? The word for virtue or excellence in classical Greek was *arête*, from which we derive the English word "aretaic," of, or pertaining to, excellence or virtue. Virtue jurisprudence is one way that legal theory can execute a move already made by moral philosophy and epistemology—*the aretaic turn*. On one hand, the aretaic turn represents a renewed concern with human excellence as a unifying normative and explanatory concept. On the other hand, the turn towards human excellence is a turn away from the reductive programs of both consequentialist and deontological legal theory. Virtue jurisprudence offers a rich and fruitful account of the nature, means, and ends of law that simultaneously dissolves old problems and poses a new set of challenges for legal theorists. The aretaic turn in legal theory moves away from degenerating research programs that disconnect the academy from the bench and bar and moves towards the reintegration of legal theory and practice.

This essay has two aims. First, the essay unpacks these dense but nonetheless sketchy answers to the perennial questions about law by developing in detail the content of a virtue jurisprudence. Second, the essay develops the case for virtue jurisprudence as compared to some of its rivals.

## 1.2 Motivating the Aretaic Turn

*Why virtue jurisprudence?* The real answer to this question lies in the power of the theory itself; the proof is in the pudding. But before I put the pudding on the table, I sketch (in this Part of the Paper) the defects of the contemporary fusion of theory and practice. If all were well with modern jurisprudence, legal theorists would lack motivation to make the aretaic turn. But all is not well. Contemporary legal theory and practice are in serious trouble.

### *1.2.1 Mediocrity and Politicization*

Begin with the state of the law itself, focusing on the legal system of the United States as an illuminating (if somewhat atypical) example. No observer of contemporary American jurisprudence can feel satisfied with the current state of the judiciary. Although there are many judges who are both fine and fair, the overwhelming impression conveyed by a broad survey of the bench is that it faces two substantial and interrelated dangers: politicization and mediocrity.

The danger of politicization is a perennial one—shared to some extent by all legal systems. In the United States that danger is currently especially acute for a variety of reasons. Perhaps the chief among these is the American system of judicial review, which places the ultimate power of decision about almost any conceivable question in the hands of the United States Supreme Court.

The power of final decision creates temptation. On one hand, the political branches are tempted to use power of judicial selection to stack the bench with political hacks who will use their power to achieve the aims of high and low politics through judicial fiat. On the other hand, judges themselves are tempted to use their power to substitute their own judgment about how cases should come out and what the law should be, for decision according to the rules laid down. Each temptation reinforces the other. Politicians who see judges substituting their own political ideology for the rule of law (but who disagree with the results) are naturally tempted to balance the books by selecting new judges who will counteract the ideology of incumbents with ideological decisions of a different stripe. Judges who start on the bench without an ideological agenda may come to see their neutrality as self-defeating when ideological appointees do not share their restraint. If this cycle is not broken, it can become a downward spiral of politicization, with the political parties and judicial incumbents engaged in a race to the bottom.

Politicization breeds judicial mediocrity. If judges are selected for their loyalty to an ideological agenda, then they are not being selected for fidelity to the rule of law, for being learned in the law, for practical wisdom, or judicial integrity. Quite the contrary, judges who are learned, smart, wise, and independent are highly unlikely to be predictable votes in the contest for control of the judiciary. This is not to say that politicization makes every skill irrelevant. The most effective politicized judge may need rhetorical skill, the sophist's ability to make the better case appear the worse, to rationalize departure from the rules, and successfully to mask inconsistency. But these skills are worse than mere mediocrity—the Machiavellian version of the judicial "virtues" are not true excellences; rather, they are the enablers of the worst kind of judicial vices—the perversion of justice and degradation of the rule of law.

The downward spiral of politicization and its sibling the descent into mediocrity have not yet reached bottom. There are still many judges of integrity and intelligence, and even the most political judges may adhere to the rule of law if the political stakes are sufficiently low. But there is no guarantee that the uneasy balance of power between politics and the rule of law will long endure. Many would argue that one important line—between the so-called "high politics" of the New Deal and Warren Courts—and the polemically dubbed "low politics" of *Bush v. Gore* has already been breached. Balkin and Levinson (2001). If the judiciary is already moving from the politicized interpretation of equal protection and due process to the manipulation of election results, then the next steps are small ones. Every dispute is an opportunity for patronage and rent seeking—because in every case, either the parties or their lawyers are potentially the source of rents in the form of campaign contributions or other indirect payoffs.

The cost of a thoroughly politicized judiciary is very high indeed. Human flourishing is at risk in a society with a corrupt judiciary. The rule of law is a prerequisite for transparent markets and the protection of basic human rights. At the very bottom of a downward spiral of politicization, the rule of law is no more. At the bottom, the very great goods that the rule of law makes possible cannot long persist. Those goods are equality, prosperity, and liberty, the preconditions for human flourishing; their loss would be a heavy cost indeed.

### 1.2.2 Modern Moral Philosophy and Contemporary Legal Theory

There is a striking parallel between the state of contemporary legal theory after the turn of the millennium and the situation of modern moral philosophy in 1958, when Elizabeth Anscombe wrote her famous essay *Modern Moral Philosophy* (Anscombe 1958). Modern moral philosophy, Anscombe argued, has involved a competition between two great families of moral theories, consequentialism and deontology. Both views face severe difficulties and each provides a powerful critique of the other. Consequentialism has the advantage of providing a method that, in principle, is capable of resolving moral disputes, but purchases its discriminatory power by leaving no room for inviolable human rights and independent consideration of fairness. Deontology has the disadvantage of an uncertain method, and at least sometimes seems to exclude consideration of consequences that seem either relevant to or dispositive of the choice that must be made.

And what of the state of contemporary legal theory? Most readers will recognize the eerie parallels with Anscombe's sketch of the predicament of modern moral philosophy. Contemporary legal theory is characterized by two antinomies: *the antinomy of rights and consequences* and the *antinomy of realism and formalism*. Each antinomy captures a persistent controversy in contemporary legal theory that has proven resistant to resolution (or even clarification) through the practice of reasoned argument. In the less theoretical corners of the legal academy, many believe that legal scholars choose their position with respect to these antimonies on the basis of an *existential leap* as opposed to reasoned argument. In the pages of learned journals and in the introduction to learned monographs, readers may limn the contours of a struggle where rhetorical flourish and name calling take the place of careful scholarly analysis.

The antinomy of rights and consequences is the legal form of the modern philosophical debate between consequentialists and deontologists. In the legal academy, the flag of consequentialism is borne by the normative law and economics movement. An especially prominent and trenchant example is found in *Fairness versus Welfare* by Louis Kaplow and Steven Shavell of Harvard Law School (Kaplow and Shavell 2002).

If the flag of consequentialism is borne by normative law and economics, then surely the most prominent standard bearer for a rights-based approach to normative legal theory is Ronald Dworkin. Dworkin's theory, law as integrity, emphasizes the idea that the parties have preexisting rights that oblige judges to decide cases on the basis of principle rather than policy (Dworkin 1986). Of course, Dworkin is only one of many who carry the flag for deontology in the legal academy.

When I describe the lay of the jurisprudential landscape as characterized by an antinomy of rights and consequences, I mean to make a bold assertion about the state of debate between the partisans of consequence and the advocates of rights. This debate does not seem to be progressing towards a conclusion; instead we seem to be in a state of perpetual conflict (at best) or mutual disengagement (at worst).

Let me explain. On the one hand, there is considerable evidence for the proposition that normative legal theory is fragmenting. Normative law and economics has sufficient momentum so that it is institutionally feasible to proceed as if there were no deontological critique of the moral foundations of welfarism. Likewise, deontologists can debate among themselves, with egalitarians and libertarians arguing for the own preferred version of rights-based normative legal theory. Genuine dialog is rare.

If the *antinomy of rights and consequences* is characterized by perpetual warfare or mutual disengagement between two more or less equally matched forces, the *antinomy of realism and formalism* is reflected in a more fractured and less crystalline pattern of legal discourse. We can remind ourselves of the dialectic with a sweeping historical survey: the original legal realist movement of the 1920s and 1930s gave way to the law and process synthesis of the 1950s and 1960s, which in turn was challenged by the indeterminacy thesis advanced by the critical legal studies movement in the 1980s. CLS gave way to a blistering critique of implausible claims about radical indeterminacy in the 1990s, only to see realist cynicism reach a new zenith in the wake of the United States Supreme Court's decision in *Bush v. Gore* (2000).

Contemporary legal theory is of two minds about realism and formalism. The practitioners of legal theory have incorporated the standard realist moves into the conceptual toolbox. But legal formalism is surprisingly resilient to attempts to declare its demise. Once formalism is rescued from the realist caricature of a self-contained system of pure deduction, it is hard to deny that (1) there are easy cases and (2) while the law may underdetermine judicial decision making, it is rarely radically indeterminate. And neoformalism, in various forms, is on the rise. Originalism, textualism, and plain meaning—these are the watchwords of the neoformalists, a group which makes up in prominence and attention what it may lack in numbers.

The point of adumbrating the two antimonies—rights and consequences, realism and formalism—is to convey the sense that contemporary normative legal theory, despite its vibrancy and sophistication, is stuck in certain recurring patterns of irresolvable argument. One hesitates to say that contemporary legal theory is a degenerating research program, but there is surely reason for dissatisfaction. Things are not so hunky dory that contemporary legal theory should shut the door to alternative approaches.

### *1.2.3 Why Virtue Jurisprudence?*

Virtue ethics has hardly vanquished deontology or consequentialism, but there has been a flowering of aretaic approaches in moral philosophy and productive dialog between virtue ethicists, utilitarians, and deontologists. The best way to test the ability of virtue jurisprudence to contribute to legal theory is to build an aretaic legal theory and evaluate the insights that it generates.

## 1.3 Virtue Ethics

To get virtue jurisprudence off the ground, we need a basic understanding of its moral sibling—virtue ethics. Contemporary virtue ethics has deep roots—in the western philosophical tradition and elsewhere. In the west, the virtue ethics might be said to originate with Plato and Aristotle and the aretaic tradition includes the Stoics and Thomism. Virtue ethics has many different roots, reaching into a variety of intellectual traditions, but Aristotle's moral philosophy has played a key role in the development of aretaic moral philosophy and serves as a model for important contemporary versions of virtue ethics. For these reasons, Aristotle's moral theory provides an excellent starting point for an investigation of virtue ethics. Of course, even a brief exposition of Aristotle's moral theory is outside the scope of this Article; nonetheless, we can explore the broad outlines of his views and introduce the key terms in his conceptual vocabulary.

One place to start a survey of Aristotle's ethics is with Aristotle's investigation of the questions, "What ends or goals are most choice worthy for humans?" and "What is the highest humanly achievable good?" The answer to these questions, Aristotle argued, will possess three characteristics: first, the highest good will be desirable for itself, second, it will not be desirable for the sake of some other end, and third, every other goods will be desirable for its sake. The human good that meets these three criteria is *eudaimonia*—translated imperfectly as "happiness." It is important at the outset to distinguish the concept of *eudaimonia* from modern interpretations of the idea of happiness. *Eudaimonia* is not a pleasurable feeling or sense of wellbeing—that is, it is not merely a desirable psychological state. Rather, it is *eu zên* or "living well," an objective state of being and not only a subjective state of consciousness.

That happiness is good in itself seems obvious—the danger is not that happiness lacks intrinsic good, but rather that this conclusion is a mere tautology. Similarly, it seems clear that happiness is not pursued for instrumental reasons. Try completing the sentence, "I want to be happy in order to …" No other good seems appropriate as the further end for which the sake of which happiness is pursued. Finally, Aristotle argues that every other human end—wealth, health, and other resources—is pursued for the sake of happiness. I will not recapitulate Aristotle's argument for these conclusions here, but I do claim that Aristotle's position on the status of happiness is intuitively plausible and consistent with widely shared beliefs or intuitions.

If Aristotle's views about the choiceworthiness of happiness are intuitively plausible but abstract, his views about the nature of happiness are both more concrete and contestable. Aristotle develops his conception of happiness with one of the most famous arguments in all of moral philosophy—the function argument. That is, Aristotle answers the question, "What is happiness?" by posing another question, "What is the function *(ergon)* of a human being?" He argues that the characteristic function of humans is rational activity in accordance with the human excellences (or virtues). (*EN* 1097b22-1098a20) So happiness consists in using reason well over the course of a full life. Why "using reason"? Because humans are rational beings; rationality is part of human nature and it is what makes us distinctively human. Why "in accordance with the human excellences"? Because, in general, doing something well requires the appropriate excellences or virtues; hence doing human things well requires the human excellences. Why "over the course of a full life"? Because human lives can be spoiled by unredeemed tragedy; the appropriate vantage point for judging the happiness of a human life is from the end, looking backwards over the whole.

In order to live a happy life, however, possession of the virtues is a necessary but not a sufficient condition. Other goods are required to enable a life of excellent rational activity. Misfortune (a terrible accident) or extreme poverty (a lack of resources) can prevent humans from realizing their potential for living a life of rational activity in accord with the virtues. A life of exhausting physical toil and drudgery unrelieved by periods that offer the opportunity for higher pursuits does not offer opportunities for virtuous rational activity and hence cannot be a happy life.

The next question, then, is "What are the human excellences?" Aristotle contends that the virtues can be divided into two types, intellectual and moral. (*EN* 1103a1-10) The intellectual virtues are excellences of mind or intellect—what

Aristotle calls the rational part of the soul; the moral virtues pertain to character and emotion—the part of the soul that cannot itself reason but is nonetheless capable of following reason.

Of the intellectual virtues, two are important in the context of this discussion. The first is theoretical wisdom or *sophia*—think of the kind of excellence characteristic of a theoretical physicist, logician, or mathematician. The second intellectual virtue is practical wisdom or *phronêsis*—think of the quality that we describe as "good judgment" or "common sense." (*EN* 1139a3-8) Aristotle offers a long (and non-exhaustive list) of moral virtues—these include characteristics such as courage, temperance, and so on. He believed that each of the moral virtues was a mean with respect to a morally neutral emotion—although at least one of the moral virtues (justice) does not easily fit this pattern.

"A mean with respect to a morally neutral emotion"—that's quite a mouthful. We can unpack Aristotle's claim by looking at each of the component ideas in the context of an example. Let's use the virtue of courage. The emotion that is associated with courage is "fear." When Aristotle says that courage is a "mean" with respect to fear, he points to the relationship between fear and two opposing vices—which we might call cowardice and rashness. Cowardice is the vice that is associated with a disposition to excessive fear; humans with this vice will characteristically overreact to danger. Rashness is the vice with a disposition to insufficient fear; humans with this vice will characteristically be insufficiently alert to and evasive of various risks. Courage is the disposition to fear that is proportionate to the situation, neither too much nor too little but rather the mean that lies between excess and deficiency. Another point has been implicit in the discussion so far—the moral virtues are dispositions and not states. There is no particular amount of fear that is virtuous; the virtue of courage is the disposition to fear that is appropriate to the situation.

One more idea is important to Aristotle's understandings of the virtues. For Aristotle it is not sufficient that one's behavior is in accord with the virtues. Rather, the virtuous agent acts for the right reasons or from the right motives. The virtuous agent acts in conformity with courage and does the courageous act because it is courageous. But this does not mean that the virtuous agent must strain or act contrary to her emotions when she acts courageously. For the virtuous agent, virtuous action comes naturally—it does not require strength of will to overcome natural impulses to the contrary. For the fully virtuous agent, reason, emotion, and desire work together in harmony—they are not at war.

Aristotle viewed his ethical theory as continuous in an important way with his biology. Just as a biologist might ask what are the characteristics of a well-functioning antelope or lion, so Aristotle's ethics can be seen as asking the question, "What are the characteristics of a well-functioning human?" And his politics extends this question to, "What are the characteristics of a well-functioning community of humans?" Aristotle's naturalism poses many questions for our assessment of his theory, but one of those questions is this: since we now reject much of what Aristotle had to say about human biology and psychology, doesn't this undermine his account of the virtues? I am not going to answer that question, because contemporary virtue ethics provides a way for the project of virtue jurisprudence to avoid it. In a sense, the point of contemporary virtue ethics is to reconcile our understanding of the virtues with contemporary biology and psychology. This reconciliation may entail some important divergence between contemporary theories and Aristotle's account. In this essay, I will not attempt to give an account of contemporary virtue ethics.

## 1.4 A Virtue Jurisprudence

The next step is to offer an outline of virtue jurisprudence. A fully developed virtue jurisprudence would offer a complete theory of law. To be complete, an account of law must have three components: first, a theory of the proper function or end of law; second, a theory of adjudication or judging, and third, a theory of the nature of lawfulness.[1] No one will be surprised that the complete account of law offered here is described as an outline: the full statement of the theory could not be accomplished in a single article or monograph, or indeed, by a single author at an early stage in the development of the theory. But the goal is to offer an outline that is detailed and not sketchy. The aim is to put flesh on the bones—providing sufficient detail and supporting argument to give a lively sense of how the full statement might go. One drawback of this approach is that no single aspect of the theory is developed in full. Indeed, many components of the view are developed in just enough detail to expose controversial assumptions but without the full argument that might make those assumptions convincing (or at least reasonable). The advantage is that this breadth of coverage allows a view of virtue jurisprudence as a whole to emerge.

### *1.4.1 Legislating Virtue: The Aim of Law Is Human Flourishing*

The first element of a complete virtue jurisprudence would be a theory of legislation—a normative theory of the ends or proper function of law. The core of that theory may seem counterintuitive or implausible to some readers. At its core, virtue jurisprudence affirms that law aims at the inculcation of virtue and not only at right action or good states of affairs. This claim would not only be counterintuitive, it would also be wrong, if it meant that the law simply commanded that citizens be virtuous—that they perform the right actions *for the right reasons*. But virtue jurisprudence does not make that implausible claim. Law aims at human flourishing, but it does not command humans to flourish or prohibit the possession of a corrupt character.

An aretaic theory of legislation can begin with ideas that resemble some of Aristotle's thoughts. We start with the idea of human flourishing, faring well and doing well. Happiness, the highest humanly achievable good, consists in a life of

[1] The third topic, a theory of the nature of law, will not be addressed in this essay.

rational and social activity in accord with the human excellences or virtues. Roughly, this means that for law to serve its end—human flourishing—it must accomplish two tasks. First, law must foster the conditions for "faring well"—that is, the law must create and maintain the material and social conditions for human flourishing. These include sufficient material abundance, security of persons and property, opportunities for meaningful work, a healthy environment, and so forth. Second, the law must foster the conditions for "doing well"—that is, the law must create and maintain the conditions for humans to develop their capacities—in Aristotelian terms, the human excellences or virtues.

How can law create the conditions for human flourishing? Much or most of the answer to this question surely depends on myriad empirical inquiries. What forms of economic organization promote both material prosperity and the acquisition of good character? Which constitutional designs facilitate legislation that promotes human flourishing and which ones lead to corruption? How is compliance with the law best obtained, through optimal deterrence or legitimation? Our goal will not be to answer these complex and perennial questions—the domain of economics, sociology, political science, and other disciplines. Rather, we will attempt to see how an aretaic theory of legislation provides a distinctive approach—highlighting the ways in which these questions are cast in a new light by focusing on human flourishing as the end of law.

The first way that law can promote human flourishing is not unique to virtue jurisprudence. Almost any plausible theory of the ends of law aims at enabling humans to fare well—to live under conditions of peace and prosperity. Consequentialist theories of law share this aim, although most contemporary consequentialists affirm subjective theories of value (for example, the maximization of preference satisfaction): virtue jurisprudence differs on this point, because the aretaic theory of value is concerned with the objective conditions for human flourishing and not with the satisfaction of arbitrary preferences. Even deontological theories are likely to posit peace and prosperity as legitimate ends for law—although the deontologist is likely to see rights as trumps that may not be sacrificed for good consequences.

An aretaic theory of legislation value peace and prosperity as constitutive elements of human flourishing, but it also values them for another reason. It seems reasonable to assume that the human excellences are more likely to be acquired and maintained in conditions of peace and prosperity than in conditions of pervasive violence and deprivation. So peace and prosperity play an instrumental function in an aretaic theory of legislation: they facilitate human excellence.

For the most part, virtue jurisprudence will not have anything distinctive to say about *how* law can foster peace and prosperity. The *how* question is an empirical one, to be answered by the social and policy sciences. Nonetheless, an aretaic theory of legislation is likely to take a distinctive stand on some aspects of the *how* question. For example, virtue theories emphasize the role of *phronêsis* or practical wisdom in the solution of complex problems. This emphasis might lead to distinctive views about institutional arrangements. Virtue-centered theories might suggest that institutions should be designed so that key decisions are made by the *phronomoi*—persons of practical wisdom.

The second way in which law can promote human flourishing is by promoting the acquisition and maintenance of the human virtues. Once again, the key questions are empirical; they will draw on psychology, sociology, and the rest of the social sciences. It seems likely that the human excellences emerge in childhood and young adulthood. So an aretaic theory of legislation is likely to emphasize to promotion of families and educational institutions that provide an environment that is conducive to the development of the virtues. For similar reasons, we would expect an aretaic theory of legislation to endorse laws that prohibit conduct that interferes with the development of the human excellences: such laws would almost surely include prohibitions against child abuse and neglect.

What about so-called "vice laws"? Let us use the phrase "vice laws" to designate legal norms that prohibit activities that are characteristically the product of the vices. We need to be careful about terminology here. In ordinary usage, the phrase "vice law" has come to refer to laws that prohibit gambling, intoxicants, and prostitution. These activities can be the product of the human vices, and it is possible that engaging in these activities could interfere with the development of the human excellences. But from the perspective of virtue jurisprudence, the category of vice is much broader than gambling, intoxication, and some forms of sexual activity. Vice produces crimes of violence and crimes against property, but laws against these crimes are not considered "vice laws" as that phrase is conventionally understood.

One possible objection to an aretaic theory of legislation is that such a theory might lead to the prohibition of a special category of action produced by vice, the so-called "victimless crimes." Of course, victimless actions should not be crimes if the only purpose of the criminal law is to punish conduct that is harmful or rights violating—as some deontological theories of legislation may hold. But if the end of law of is human flourishing, then it is possible that flourishing can be promoted by prohibiting vicious action even when harm is inflicted on the character of the perpetrator. An aretaic theory of legislation *might* support the criminalization of gambling, recreational intoxicants, and prostitution on the ground that these activities interfere with the development and maintenance of the virtues.

But this is only a possibility—not an inevitable consequence of a virtue-centered theory of the ends of law. First, it is not clear that the activities prohibited by "vice laws" are inherently vicious. Gambling and consumption of intoxicants in moderate amounts might be recreational activities that form part of a flourishing life. Of course, excessive indulgence in such recreation might be harmful, but so can excess in any number of activities. Excessive work, and excessive exercise might be harmful to the development of character, but it surely does not follow from that fact that an aretaic theory of legislation must support criminal laws prohibiting work and exercise. Second, it is not clear that prohibition is the best strategy for dealing with gambling, intoxicants, and prostitution. It is possible that the harms generated by prohibition are very serious. For example, the criminalization of drugs might lead to violence, corruption, and the incarceration of large numbers of young people who become involved the lucrative trade in illegal substances. It is at least possible that the promotion of human flourishing would require the legalization of intoxicating substances. Again, the questions are empirical in nature. An aretaic theory of legislation is committed to the laws that will best promote human flourishing and not to a categorical rule that requires all vicious activity to be criminalized.

## *1.4.2 Virtuous Judging: An Aretaic Theory of Adjudication*

The second dimension of virtue jurisprudence is an aretaic theory of adjudication (or judging). The general idea of such a theory is that good judging is a function of the virtue of the judge. And hence, the content of an aretaic theory of judging is provided by an account of the judicial virtues. We can begin to develop such an account by focusing on those judicial excellences upon which there is likely to be widespread agreement. From there, we can turn to a more difficult topic—the development of a theory of the virtue of justice.

### 1.4.2.1 The Judicial Virtues

There is quite a lot of disagreement about the qualities that make for good judging. That disagreement is reflected in controversies about the selection of federal judges in the United States. Because judicial selection has largely been driven by the preference of political actors for certain outcomes on key issues (abortion, affirmative action, and so forth), political ideology has played a major role in the judicial selection process. This practical disagreement is reflected in legal theory as well. American legal scholars disagree on the criteria for a good legal decision, and hence they are likely to disagree about which judges are excellent as well. Nonetheless, it may be possible to identify a set of judicial excellences on which there is likely to be widespread agreement.

Some judicial virtues are uncontested. "Uncontested" in this context reflects the notion that these virtues are based on uncontroversial assumptions about what counts as good judging and on widely accepted beliefs about human nature and social reality. Just as some judicial virtues are uncontested, so are some judicial vices. For example, one judicial vice on which there is likely to be near universal agreement is "corruption." Judges who sell their votes undermine the substantive goals of the law, because corrupt decisions are at least as likely to be wrong as they are to be substantively correct. Moreover, corrupt decisions undermine the rule of law values of productivity and uniformity of legal decisions and likewise undermine public respect for the law and public acceptance of the law as legitimate.

Even the most zealous advocate of ideological judicial selection is likely to accept the conclusion that judicial corruption is a vice. Of course, it is possible that some judges might accept bribes or political favors in a way that systematically favors a particular political ideology or that advances the interests of a particular political party. But the partisans of ideological judging would have good reason to think that even a corrupt judge who "votes the right way" is a bad judge. *Why?* Two reasons. First, corrupt judges are not reliable ideological allies—precisely because their decisions are for sale, they can be lured to the side of one's ideological opponents by a "better offer" or more attractive bribe. Second, there is always a risk that corruption will be exposed, and when exposed, a corrupt decision that favors one's cause may actually be counterproductive—delegitimizing the ideology that it seemed to serve. Of course these two prudential reasons are likely to be supplemented by the obvious reason of principle—corrupt decisions are morally wrong. We have no reason to believe that the partisans of ideological judging would not condemn corruption on moral grounds.

If we accept the conclusion that judicial corruption is a vice, then what is the corresponding virtue? This question could become complex—because there are a variety of character flaws that might lead to corruption. One such flaw is greed (or *pleonexia*) may be an underlying cause of corruption—because a desire for more than one's share (or entitlement) could lead a judge to accept bribes. All humans are at risk of mistaking wealth (which can only be a means) for a final end (something worth pursuing for its own sake). Some judges may resent the fact that they receive compensation that is sometimes only a fraction of that provided their peers in private legal practice—some of whom may be less talented.

We do not need to identify all of the possible vices that could lead to corruption in order to see that *incorruptibility* is an uncontested judicial virtue. There is no real controversy over the proposition that judges should be disposed to resist the temptations that lead to corruption. We call this disposition the "judicial virtue of incorruptibility," even if it turns out that this virtue encompasses a variety of particular virtues each of which corresponds to a particular human vice that could lead to corruption.

There is another vice that is closely related to corruption but is distinct from greed. Judges can become corrupted because their desires are not in order—because they crave pleasure or the status (and corresponding envy) conferred by the possession of fine things. Judges, like the rest of us, can be corrupted by a taste for designer shoes, fast cars, loose companions, or intoxicating substances. More subtly, a judge could be corrupted by a desire for the finer things of life, for example, a magnificent home, the ability to confer lavish gifts upon one's children, or the opportunity for luxurious travel.

Let us use some old fashioned terminology and call the vice of disorderly desire "intemperance"—recognizing that modern ears may not be able to hear that word without summoning up an image of drunkenness caused by a craving for the pleasures of strong drink. Can a case be made that intemperance is not a judicial vice? One might argue that intemperance is a purely private vice—that a judge's preference for a third cosmopolitan, the latest from Jimmy Choo or Manolo Blahnik, or the company of good looking youthful companions is her own business and hence irrelevant to the question whether she is an excellent judge. Of course, a proportionate and well-ordered desire for such things is simply not a vice—or at least not an *uncontested* vice. But a disposition to disproportionate desires for such pleasures can lead to more than corruption. Most obviously, a judge who is intoxicated (or high) on the bench is likely to be prone to error, for obvious reasons. The inordinate pursuit of less intoxicating pleasures can also impair judicial performance—by focusing a judge's attention and energy away from judicial tasks.

There is a counter argument. It is a common human experience to have a friend, colleague, or acquaintance who is intemperate, but nonetheless "gets the job done"—even performs brilliantly at times. Who hasn't encountered the lawyer who is a star by day, but a lush in the wee hours or the friend whose life at work still holds together despite a drug problem? So, the argument does, intemperance is not a *judicial vice*—at least not until it interferes with the performance of judicial duty. Even if the intemperate judicial candidate is a disaster at home, her intemperance should not disqualify her from judicial office if she performs at the office. This counter argument is ultimately unpersuasive. Of course, an intemperate judge can get lucky and "get away with it," either appearing to do well or even actually doing well despite disordered desires. But in such cases "getting away with it" is a matter of luck; an intemperate judge is simply not reliable. A really damaging misstep is always just one cosmopolitan away.

The virtue that corresponds to the vice of intemperance could be called "temperance" in the classical sense that encompasses the ordering of all the natural desires. But I propose that we use another term to refer to the judicial form of temperance. We have a saying that captures the intuitive sense that judges must have their desires in order: we say of a temperate human that she or he is "sober as a judge," and this suggests that we name this virtue "judicial sobriety."

Consider another example of a judicial virtue and corresponding vices that are uncontested. Fear is one of the most powerful and familiar of the emotions. For Aristotle, the virtue of courage relates to the morally neutral emotion of fear. Following the pattern of the moral virtues, courage represents a mean between a vice of excess—cowardice—and a vice of deficiency, which we might call "rashness" or "recklessness." We can agree that cowardice is a judicial vice, and judicial courage is a virtue.

We might usefully subdivide the virtue courage into two parts—which I shall call "physical courage" and "civic courage." That judges need physical courage in order to be excellent *as judges* is a lamentable fact in many societies. A judge who could be intimidated by threats of physical violence could not reliably do justice in our society—much less under conditions where violence (or threats of violence) was even more prevalent—as may be the case where narcoterrorism or violent ethnic conflict is pervasive.

Judicial courage has a second dimension. Judges, like most humans, care about their reputations and social standing. Like the rest of us, judges seek the approval and companionship of their fellows. So in addition to physical danger, judges may fear consequences of their actions that involve threats to status and social approval. This is because the law may require judges to make unpopular decisions. A judge who ordered school integration in the South might be shunned socially. In societies where the judicial branch wields significant power in cases involving hot button issues (abortion, end of life disputes, and so forth), there will be occasions where doing what the law requires may be profoundly unpopular. For this reason, judges need the virtue of civic courage—the disposition to put the regard of one's fellows in proper place and to take it into account in the right way on the right occasions for the right reasons. A judge with this virtue will note be tempted to sacrifice justice on the altar of public opinion. A civically courageous judge understands that the good opinion of others is worth having if it flows from having done justice and that social approval for injustice is an impermissible motive for judicial action.

Like fear, anger is an emotion both familiar and powerful. Judges like the rest of us may be hot tempered or cool and collected. And like the rest of us, judges are likely to find themselves in situations where a hot temper could produce intemperate actions. This is especially true of trial judges, who are given the task of maintaining order in what may become emotionally charged circumstances. Litigants may ignore judicial authority or act with disrespect. Some lawyers may deliberately attempt to provoke the judge in order to elicit legal mistakes or "on the record" behavior that displays animus towards a party and serve as the basis for an appeal. In the face of such provocations, a judge with an "anger management problem" may "fly off the handle." Intemperate judicial behavior may lead the judge to misapply the law—misinterpreting the applicable legal standards in "the heat of anger." Moreover, a hot-headed judge may become partial—pulling against the party who is the object of anger and displaying favoritism to that party's opponent.

The corrective virtue for the vice bad temper can be called *proates* or "good temper." In the judicial context, this virtue is so important that we have a phrase that expresses the virtue as a distinctively judicial form of excellence—"judicial temperament." This phrase reflects our sense that the virtue of "good temper" is essential for good judging.

Is judicial temperament also required for judges who do not supervise trials? Appellate judges work in a cooler environment—provocative behavior by appellate lawyers is rare although not unknown. The parties to an appellate proceeding frequently do not appear, and if they do, they sit in the audience without any formal participation in the appellate process itself. Some appellate courts proceed almost entirely on the basis of the briefs, dispensing with oral argument and hence with the opportunity for "live and in person" provocations. Nonetheless, good temper is essential for excellence in appellate judging. Appellate judges hear cases in panels or en banc—creating opportunities for friction among the judges themselves. Hot tempers can destroy collegiality and with it the opportunity for compromise and mutual understanding. Moreover, even a brief can elicit anger, and if anger becomes rage, it can have a blinding effect, depriving the judge of the ability to recognize the merits of an argument or a weakness in the judge's own conception of the legal issues in a case.

If excessive anger is a vice, then what about its opposite? Is there a vice of deficiency with respect to anger? The notion that proportionate anger is virtuous rests on the premise that anger serves a valuable function—alerting us to wrongs and motivating us to respond to them. A simple way of framing the issue is to ask which character from the 1960s American television series *Star Trek* would make the best judge—Captain Kirk, Dr. McCoy, or Mr. Spock. Mr. Spock resembles the Stoic sage—he feels no anger and acts only on the basis of logic; we imagine Judge Spock reacting with equanimity to even the most severe courtroom provocations. Dr. McCoy is hot tempered; we imagine him flying off the handle in response to outrageous behavior by the lawyer for a greedy corporation. Captain Kirk represents a mean between these two extremes; we imagine Judge Kirk as appropriately outraged by bad behavior and injustice, but nonetheless remaining "in control," angered by the right things and responding with in an appropriate manner. The virtue of judicial temperament consists in having appropriate anger—anger for the right reasons on the right occasions with a clear understanding of the consequences of its expression.

More concretely, when a party flouts the law or disrespects the participants in a legal proceeding, anger may be appropriate. Such appropriate anger alerts the judge to the existence of a "situation that must be dealt with." In some circumstances, the judge will properly display such anger, giving a lawyer, party, or witness "a stern warning." When a lawyer, party, or witness persists in bad conduct, sanctions may be warranted; in such cases, an appropriate sanction is the right way to act on the basis of appropriate anger. But judges with the virtue of a judicial temperament will not display their anger by ruling against an offending party on issues that are close or exercising discretion on incidental matters so as to disfavor the anger-provoking party.

One reason that anger is an especially dangerous vice for judges is that anger can produce bias. For this reason, the virtue of judicial temperament is closely related to another judicial virtue, "judicial impartiality." This virtue is a familiar feature of our conception of good judging. We want judges to be neutral arbitrators. A judge should be open to the law and evidence and not biased in favor of one side or another. Such impartiality should extend not just to the parties but should also encompass the causes, movements, special interests, and ideologies that may be associated with those parties. When a judge takes the bench or lifts her pen to write an opinion, she should put aside her allegiance to left or right, liberal or conservative, religiosity or secularism.

It is a mistake, however, to view impartiality as synonymous with disinterest. The virtue of impartiality is not cold-blooded. This is because the role of judge requires insight and understanding into the human condition. A good judge perceives the law and facts from a human perspective. Some facts are hot—charged with emotional salience. Some legal rules are righteous—engaging our sense of moral indignation when juxtaposed with violative behavior. So the impartial judge is not cold blooded; she is not indifferent to the parties that come before her. Rather, the judge with the virtue of judicial impartiality has even-handed sympathy for all the parties to a dispute. When we say, "Impartiality is not indifference," we mean that the virtue of impartiality requires both sympathy and empathy without taking sides or favoring the legitimate interests of one side over those of the other.

Judging is hard work, involving its share of drudgery. Some trials are long and boring. Some opinions require long hours of research and even longer hours of careful drafting. The temptation to shirk this work is accentuated by the fact that judges are not (and should not be) closely supervised. And the lack of supervision is compounded in jurisdictions that grant judges life tenure or long terms in office. It is hard enough to remove a judge for outright corruption; one doubts that any American judge has been removed on the basis of sloth alone. But slothful or lazy judges can do real harm. They are tempted to delegate too much responsibility to judicial clerks, substituting the judgment of the clerk for the judge's own intellectual engagement with the case. Another temptation is to shape one's decision in order to minimize one's own workload. If granting the summary judgment motion takes a case off one's docket, the slothful judge might grant the motion for that reason alone, sacrificing justice on the altar of expediency.

What is the virtue that corresponds to the vice of sloth? We might call it diligence. The diligent judge has the right attitude towards judicial work, finding judicial tasks engaging and rewarding. But more than a good attitude is required. An excellent judge must have an appropriate "energy level"—a product of both physical and mental health. The combination of these traits should translate into a judge who is capable of hard work when hard work is required. Such a judge will put in the required hours and sweat out the difficult tasks. Such a judge will not hesitate to make the right decision, even if that makes more work for the judge. Nowadays, encouraging settlements may be an appropriate activity for judges, but a diligent judge will aim for just and efficient settlements and not for resolutions that serve the judge's own convenience.

Carefulness is closely related to diligence. No one can sensibly doubt that judicial carelessness is a vice. Careless decisions, careless drafting, careless research—any of these can lead to substantive injustice. Carefulness is especially important in the context of judging, because excellent judging frequently requires meticulous attention to details. The lazy judge may shirk the unpleasant task of mastering the structure of a complex statute or avoid the painstaking task of making sense of tangled body of precedent. Likewise, it requires diligence and care to draft an opinion in which each and every sentence is worded with careful appreciation of the importance of precision and accuracy. An excellent judge has an eye for detail and a devotion to precision.

Can anyone doubt that stupidity is a judicial vice? All humans need intelligence to function well—but some tasks require more intelligence on more occasions. Judging is the kind of task that sometimes requires extraordinary intelligence. Both law and facts can be complex. Only a judge with intelligence will be able to sort out the complexities of the rule against perpetuities or penetrate the mysteries of a complex statute. But more than intelligence is required. A truly excellent judge must also be learned in the law, because one cannot start from scratch in each and every case and because there is at least some truth to the notion that the law is a seamless web. To put these same points the other way round: stupid and ignorant judges will be error prone, likely to misunderstand and misstate the law and unlikely to make findings of fact that are correct.

The need for judicial intelligence and learnedness is accentuated rather than diminished in an adversary system. It is true that good lawyering makes a judge's job easier; the lawyers can identify the relevant issues and call the judge's attention to the best arguments on each side of those issues that are in dispute. But in an adversary system, successful advocates will try to make "worse case appear the better," by deploying sophistry and rhetoric. Intelligent and learned judges can "see through" the obfuscation and look past the appeals to prejudice and preconception.

So far, our investigation has focused on what Aristotle called the moral and intellectual virtues. These are dispositions of character and mind that make for human excellence. Good judging requires more than good character and intellectual ability. That is because judging includes elements of craft, and therefore a good judge must possess a skill set—the particular learned abilities that are to good judging what good bowing technique is to archery or good draftsmanship is to architecture. A full account of judicial craft is far beyond the scope of this Essay, but one particular aspect of judicial craft and skill cries out for attention. Excellence in judging (especially good appellate judging) requires particular skill in the use of language. Good judges must be good communicators. This aspect of judicial skill includes at least two parts—oral and written. It is obvious that trial judges need good oral communication skills; they must deliver a variety of oral instructions to the various participants in both trial and pre-trial proceedings. Among these, jury instructions are particularly important. Written communication skills are especially important for appellate judges in a common law system, because of the doctrine of *stare decisis*. Because appellate opinions set precedent, a badly written opinion can misstate the law or state the law in a misleading way. A really well drafted opinion, on the other hand, can clarify the obscure and illuminate the meaning of murky legal texts.

Good communication skills are also important to judges when they mediate between the parties to a dispute. A skilled judge can gain the trust and cooperation of the parties—resorting to the threat of sanctions only in those rare cases when force is truly necessary. In this way, good communication skills can increase the efficiency of judicial proceedings, allowing the judge to focus her attention on those issues and cases where settlement and cooperative processes are unavailing.

One advantage of a theory of judicial excellence is that it reveals a large zone of agreement. For all practical purposes, we can agree that judges should be incorruptible, courageous, good tempered, diligent, skilled, and smart. But these (mostly uncontested) virtues do not tell the whole story about judicial excellence. Even if we agree in our judgments about who the very worst judges are—the corrupt, ill-tempered, cowardly, lazy, incompetent, and stupid ones—there are strong and persistent disagreements about who the best judges are. The partisans of Lord Coke may deride the accomplishments of Lord Mansfield; the admirers of Justice Brennan may be among the critics of Justice Scalia. This section investigates the source of these disagreements about judicial excellence.

Once again, my strategy is to examine the judicial virtues. In particular, I shall argue that disagreements about judicial excellence are typically rooted in two disagreements about the nature of judicial virtue. The first disagreement is about the nature of the virtue of justice. The second disagreement concerns the role of equity and practical wisdom. On the one hand, some disagreements about judicial excellence turn out to be disagreements about and within conceptions of the virtue of justice—what some call "justice," others see as "unjust." On the other hand, other controversies hang on differences in the understanding of the role of practical wisdom in judging: some believe that wise judges will range far from the rules in the name of equity, while others believe that equity should be tightly constrained by the rule of law.

Although there are important disagreements about the virtues of justice and practical wisdom, there are certainly agreements as well. When stated at a high level of generality and abstraction, these virtues will command near universal assent. Almost everyone will agree that an excellent judge must be just (rather than unjust) and wise (rather than foolish). Let's borrow the concept/conception distinction (Rawls 2001, 4–5). We might say that there is agreement on the proposition that the concept of the virtue of justice is required for judicial excellence, but that there is disagreement about which conception of the virtue of justice is best (or correct or most adequate). And likewise, with the virtue of practical wisdom—we agree on the concept, but disagree about which conception of the virtue and its relationship to justice is the best one.

An excellent judge is just; a judge who lacks the virtue of justice has a serious defect. At this level of abstraction, the virtue of justice is likely to be the object of widespread agreement. But what does the virtue of justice require? In this section, I will examine two different conceptions of the virtue of justice: call these conceptions "justice as lawfulness" and "justice as fairness." (For short, I will use the phrases "the fairness conception" and "the lawfulness conception" to refer to these ideas.) I shall argue that conceptualizing the virtue of justice as fairness necessitates intractable disagreements about which judges are excellent, and that the competing conception, emphasizing the idea that excellent judges are lawful opens the door to agreement in judgments about who is just. My investigation begins with the fairness conception of the virtue of justice and then moves to the notion of justice as lawfulness.

One influential conception of the virtue of justice is based begins with the premise that the just and the lawful are separate and distinct. Of course, the view is not that all laws or unjust or that no just norms are law. Rather, the idea is that there is no necessary connection between legality and justice. If this were so, then the most plausible conception of the virtue of justice might be articulated as follows:

> The Virtue of Justice as Fairness: A judge, J, has the virtue of justice as fairness, V(j-f), if and only if P is disposed to act in accord with the best conception of fairness, F, in situations, S, where fairness provides salient reasons for action.

One might think that a judge who possessed *V(j-f)* would act solely on the basis of fairness with reference to the law, but this is not the case. If this were true, it would provide the basis for a devastating objection to the fairness conception—because it would require each judge to substitute her private judgments about what fairness requires for the duly enacted constitutions, statutes, and rules. Although I shall not provide the argument here, it seems plain that this would be a recipe for chaos.

But a defender of the fairness conception need not admit that a judge who acted on the basis of fairness would disregard the law entirely. *Why not?* Because the existence of legal norms will frequently give rise to considerations of fairness that will transform the moral landscape, creating salient reasons of fairness that motivate a judge who has *V(j-f)* to act in accord with the law. An example may help to clarify and illustrate this point. Suppose there is a dispute between Ben and Alice over Greenacre—a vacant and unimproved parcel of land. The law gives Ben title to Greenacre, which he has purchased, but Alice has begun to use Greenacre by planting a garden. In the absence of the institution of property law, it might be the case that Ben would have no claim on Greenacre—how would he acquire such a claim without some use or improvement of the land—but that given the existence of property law, Ben would have a claim of fairness, because he has paid for Greenacre and has reasonably relied on the legal institution of property. If this is so, then the law has created a claim of fairness that otherwise would not exist and a judge with *V(j-f)* would decide in favor of Ben—assuming, of course, that there were no other circumstances that created an overriding reason of fairness to decide in favor of Alice.

Nonetheless, the fairness conception faces a formidable objection because of the role that private judgment plays for judges with *V(j-f)*. To articulate this objection, we need to highlight the distinction between two questions about fairness—which I shall call "first order" and "second order" questions of fairness. A first order question of fairness is simply the question, "Which action is fair given the circumstances?" A second order question of fairness concerns whose judgment about first order questions will be taken as authoritative. Thus, the question, "Given the fact of disagreement about the correct answer to a first-order question of fairness, whose judgment should be taken as authoritative?," is a second order question of fairness. One possible answer to a second-order question of fairness is that one ought to rely on one's own *private* judgment about what action is fair. A quite different answer is that one should rely on some source of *public* judgment. For example, one might rely on duly-enacted and public laws.

The fairness conception implicitly requires judge's to exercise private judgment about first-order questions of fairness. In exercising that judgment, the judge may conclude that expectations generated by reasonable reliance on the law provide reasons of fairness—as in the case of Greenacre—but this is a conclusion of private judgment. One judge might conclude that Ben's reliance on property law was reasonable, and hence that fairness required a decision for Ben. A different judge might conclude that no one could reasonably rely on property law in cases in which they were allowing valuable land to lie fallow when others could make productive use of the land—and therefore decide for Alice. Yet a third judge might conclude that because of pervasive economic inequalities, the whole institution of property is unjust and award the land to a third-party, Carla, who was in greater need than either Ben or Alice. Because each judge makes a private judgment about the all-things-considered fairness of following the law in each case, these judgments can (and we expect will) differ with the political, moral, religious, and ideological views of the particular judge.

The objection to the fairness conception of the virtue of justice is that disagreements in private judgments about fairness would undermine the very great values that we associate with the rule of law. Because the fairness conception requires each judge to exercise her own private judgment about what fairness requires—all things considered—and because such judgments will frequently differ, the outcome of disputes adjudicated by judges with *V(j-f)* will be systematically unpredictable. If this were the case, then the law would be unable to perform the function of coordinating behavior, creating stable expectations, and constraining arbitrary or self-interested actions by officials. How bad this would be is a matter of dispute. A Hobbesian answer to this question is *very bad indeed*—in the absence of coordinating authority, life would be "solitary, poore, nasty, brutish, and short" (Hobbes 1994, 76). A Lockean answer is that reliance on private judgment leads to "inconveniences" (Locke 1988, 276), but even an optimistic realist would surely concede that the inconvenience of a society that cannot secure the rule of law would be serious.

We are now in a position to apply what we have learned about the fairness conception to judicial selection. If the fairness conception were correct, then the excellent judges are those who have the right beliefs about fairness and who are disposed to act on those beliefs. If we agreed on the content of the right beliefs about fairness, this would not be a problem, but we do not agree. So the fairness conception leads to disagreement about who has the virtue of justice. We can provide a crude translation of this point into the language of political ideologies of the left and right. For the left, only left-wing judges are just; because only left-wing judges have what the left considers true beliefs about what fairness requires. And of course, whereas for the right, the left-wing judges are unjust precisely because they have what the right considers false beliefs about fairness. Even the uncontested virtues—such as incorruptibility or courage—become problematic once the fairness conception has been accepted. For the left, an intelligent, diligent, and courageous right wing judge may be worse than one who lacks a keen intellect, is somewhat lazy, and who will succumb to the pressures of public opinion. And vice versa for the right.

It gets worse for the fairness conception. Anyone who holds the fairness conception is naturally tempted to apply a double standard of judicial excellence. The double standard works like this:

> For judges with whom I agree, the fairness conception supplies the content of the virtue of justice. Right-thinking judges are excellent when they act on the basis of their convictions about what is fair. But when it comes to judges with whom I disagree, a different standard applies. Wrong-thinking judges are excellent when they stick to the rules. For them, the lawfulness conception provides the standard for the virtue of justice.

You may say, "That's ludicrous, no one could hold such a blatantly inconsistent set of positions about the meaning of justice." In reply, I suggest that you pay careful attention to the political rhetoric that attends debates about judicial role and judicial selection.

If the fairness conception of the virtue of justice is unsatisfactory, is there an alternative? In the *Nicomachean Ethics*, Aristotle (1955) suggests an alternative understanding of justice as lawfulness, but to understand Aristotle's view, we need to take a look at the Greek word *nomos* which is usually translated as "law." For the ancient Greeks, *nomos* had a broader meaning that does "law" in contemporary English. Richard Kraut, the distinguished Aristotle scholar, explained the difference as follows:

> [W]hen [Aristotle] says that a just person, speaking in the broadest sense is *nominos*, he is attributing to such a person a certain relationship to the laws, norms, and customs generally accepted by some existing community. Justice has to do not merely with the written enactments of a community's lawmakers, but with the wider set of norms that govern the members of that community. Similarly, the unjust person's character is expressed not only in his

> violations of the written code of laws, but more broadly in his transgression of the rules accepted by the society in which he lives.
>
> There is another important way in which Aristotle's use of the term *nomos* differs from our word 'law': he makes a distinction between *nomoi* and what the Greeks of his time called *psêphismata*—conventionally translated as 'decrees'. A decree is a legal enactment addressed solely to present circumstances, and sets no precedent that applies to similar cases in the future. By contrast a *nomos* is meant to have general scope: it applies not only to cases at hand but to a general category of cases that can be expected to occur in the future (Kraut 2002, 105–106).

We can restate this last point by using our distinction between types of judgments (first and second order, private and public). If judges rely on their own private, first-order judgments of fairness as the basis for the resolution of disputes, then it follows inexorably that their judgments will be decrees (*psêphismata*) and not decisions on the basis of a second order, public judgment—in other words, not on the basis of a *nomos*. In other words, a judge who decides on the basis of her own private judgments about which outcome is fair—all things considered—is making decisions that are tyrannical in Aristotle's sense.

"How can this be?," you may ask. "Aren't decisions that are motivated by fairness the very opposite of tyranny?" But framing the question in this way obscures rather than illuminates the point. Of course, if there were universal agreement (or even a strong consensus) of first-order private judgments about fairness, then decision on the basis of such judgments would be *nomoi* and not *psêphismata*. But our private, first-order judgments about the all-things-considered requirements of fairness do not agree. So in any given case, a decision that the judge believes is required by fairness will be seen by others quite differently. At best, the decision will be viewed as a good faith error of private judgment about fairness. More likely, those who disagree will describe the decision as a product of ideology, personal preference, or bias. At worst, the decision will be perceived as the product of arbitrary will or self-interest. In no event, will a decision based on a controversial first order private judgment of fairness be viewed as outcome of a *nomos*—a publicly available legal norm.

We are now in a better position to appreciate why rule by decree (*psêphismata*) is typical of tyranny. Decision on the basis of private, first-order judgments about fairness is the rule of individuals and not of law. A regime that rules by decree does not provide the stability and certainty that is required for human communities to flourish. Kraut continues:

> We can now see why Aristotle thinks that justice in its broadest sense can be defined as lawfulness, and why he has such high regard for a lawful person. His definition embodies the assumption that every community requires the high degree of order that that comes from having a stable body of customs and norms, and a coherent legal code that is not altered frivolously and unpredictably. Justice in its broadest sense is the intellectual and emotional skill one needs in order to do one's part in bringing it about that one's community possesses this stable system of rules and laws (Kraut 2002, 106).

And with that point in place, we can now formulate the lawfulness conception of the virtue of justice:

The Virtue of Justice as Lawfulness: A judge, J, has the virtue of justice as lawfulness, V(j-l), if and only if J is disposed to act in accord with the nomoi (positive laws and stable customs and norms), N, in situations, S, where the nomoi provide salient reasons for action.

On the lawfulness conception, the virtue of justice does not require action in conformity with one's private, first-order judgments of fairness. Justice as lawfulness is based on a second order judgment that judges (or more generally, citizens) should rely on public judgments. The content of the public judgments are the *nomoi*—the positive laws and shared norms of a given community. Someone with the virtue of justice is disposed to act on the basis of the *nomoi*. In other words, the lawfulness conception holds that the excellent judge is a *nominos*, someone who grasps the importance of lawfulness and is disposed to act on the basis of the laws and norms of her community. A judge who is *nominos* cares about the laws and norms of her community. She is disposed to do that which is lawful, because she respects and internalize the *nomoi* of her community.

Finally, we are now in a position to compare the fairness conception and the lawfulness conception. Which of these offers a more satisfactory conception of the virtue of justice? On the surface, it might appear that the fairness conception is more satisfactory—after all, who can deny that we ought to do what fairness requires—all things considered? Although there is much more to be said in a full account of these matters, the argument advanced here provides good reasons to doubt that the fairness conception can offer a satisfying account of the virtue of justice. A view of justice must take into account the distinctions between first and second order judgments and between public and private judgments. Once these distinctions are introduced, the need for second order agreement on a public standard of judgment becomes clear. The lawfulness conception of the virtue of justice answers to this need; the fairness conception does not.

But the virtue of justice may not be exhausted by the lawfulness conception. Even if we concede that in ordinary cases justice requires adherence to the law, there question remains whether there are extraordinary cases—cases in which excellent judges would depart from the law (or, to put it differently, decide that the law does not really apply). Even if first order private judgment cannot do the work of filling in the content of a general conception of the virtue of justice, that does not necessarily imply that the judge's sense of fairness has no role to play. One reason we might doubt the adequacy of the lawfulness conception as the whole story about the virtue of justice flows from the fact that the positive law is cast in the form of abstract and general rules; such rules may lead to results that are unfair in those particular cases that do not fit the pattern contemplated by the formulation of the rule. If lawfulness were the whole story about the virtue of justice, then an excellent judge would apply the rule "come hell and high water" even if the rule led to consequences that were absurd or manifestly unjust. But this implication of the lawfulness conception seems odd and unsatisfactory. Another way of putting this concern is to distinguish between two styles of rule application, which I shall call "mechanical" and "sensitive."

Does the excellent judge apply the rules in a rigid and mechanical way? Or does a virtuous judge correct the rigidity of the lawfulness conception with equity? The classic discussion of these question provided by Aristotle in Book V, Chapter 10 of the *Nicomachean Ethics*:

> What causes the difficulty is the fact that equity is just, but not what is legally just: it is a rectification of legal justice. The explanation of this is that all law is universal, and there are some things about which it is not possible to pronounce rightly in general terms; therefore in cases where it is necessary to make a general pronouncement, but impossible to do so rightly, the law takes account of the majority of cases, though not unaware that in this way errors are made. And the law is nonetheless right; because the error lies not in the law nor in the legislator but in the nature of the case; for the raw material of human behavior is essentially of this kind. (*EN* 1137$^{b}$9–1137$^{b}$24, trans. Thomson)

This is the *locus classicus* for Aristotle's view of *epieikeia*, which is usually translated as "equity," but can also be translated as "fair-mindedness." As Roger Shiner puts it: "Equity is the virtue shown by one particular kind of agent—a judge—when making practical judgments in the face of the limitations of one particular kind of practical rule—those hardened customs and written laws that constitute for some societies the institutionalized system of norms that is its legal system" (Shiner 1994, 1260–1261).

But there is a problem with supplementing the lawfulness conception of the virtue of justice with the notion of equity. Understanding the problem begins with the fact that the virtue of equity seems to require the exercise of first-order private judgments of fairness. Once such judgments are admitted to have trumping force—to have the power to override the second order judgment that one should rely on the public judgments embodied in the law, the question becomes how the role of private judgment can be constrained. Without constraint, private judgment threatens to swallow public judgment and we are on a slippery slope that threatens to transform the lawfulness conception into the fairness conception.

The trick is to constrain equity while preserving its corrective role. To put the point metaphorically, we need an account of equity that provides enables us to navigate the slope while providing sufficient traction to avoid slipping or sliding. An Aristotelian account of the virtue of equity gives us three points of traction. The first point of traction is provided by the distinction between the equitable correction of law's generality and the substitution of private first order judgments for the *nomoi*. Equity is not doing what the judge believes is fair when that conflicts with the law; rather, equity is doing what the spirit of the law requires, when the expression of the role fails to capture its point or purpose in a particular factual context. The second point of traction is provided by the virtue of justice itself. A judge who is *nominos* simply isn't tempted to use equity to avoid the constraining force of the law. A *nominos* has internalized the normative force of the law; such a judge wants to do as the law requires.

The third point of traction is provided by Aristotle's understanding of the intellectual virtue of practical wisdom or *phronêsis*—think of the quality that we describe as "good judgment" or "common sense." A judge with virtue of practical wisdom (a *phronimos*) has the ability to perceive the salient features of particular situations.

We can say that a sense of justice requires "legal vision," the ability to size up a case and discern which aspects are legally important. The *phronimos* can do equity because she grasps the point of legal rules and discerns the legally and morally salient features of particular fact situations.

This account of equity can be contrasted with two rivals. On the one hand, we can imagine a conception of judging as pure equity—the idea that the judge would simply do the right thing in each particular fact situation. This conception of equity is simply a more particularistic version of the fairness conception of the virtue of justice. On the other hand, we can imagine a conception of judging that limits equity to the vanishing point—perhaps to those cases where the application of the rule is truly absurd. Neither of these two alternatives offers a fully satisfactory account of the virtue of equity. The first alternatives sacrifices the very great goods created by the rule of law. The second alternative pays too a high price for those goods, require more rigidity than is necessary. A constrained practice of equity done by judges who are both *nominoi* and *phronimoi* combines the values of the rule of law with the flexibility to bend the rules to fit the facts when that is required by the purposes of the rules themselves.

We now have an account of judicial virtue on the table. My next step is simply to transpose that account into a theory of judging. I do this by borrowing the approach adopted by Rosalind Hursthouse (Hursthouse 1999, 25–42). For the sake of simplicity and clarity, I shall formulate a virtue-centered theory of judging in the form of five definitions:

- *A judicial virtue* is a naturally possible disposition of mind or will that when present with the other judicial virtues reliably disposes its possessor to make just decisions. The judicial virtues include but are not limited to temperance, courage, good temper, intelligence, wisdom, and justice.
- *A virtuous judge* is a judge who possesses the judicial virtues.
- *A virtuous decision* is the decision (or one of the decisions) that would characteristically be made by a virtuous judge acting from the judicial virtues in the circumstances that are relevant to the decision.
- *A lawful decision* is a decision (or one of the decisions) that would be characteristically made by a virtuous judge in the circumstances that are relevant to the decision. The phrase "legally correct" is synonymous with the phrase "lawful" in this context.
- *A just decision* is a *virtuous decision.*

The central normative thesis of a virtue-centered theory of judging is that judges ought to be virtuous and to make virtuous decisions. Judges who lack the virtues should aim to make lawful or legally correct decisions, although they may not be able to do this reliably given that they lack the virtues. Judges who lack the judicial virtues ought to develop them. Judges ought to be selected on the basis of their possession of (or potential for the acquisition of) the judicial virtues.

How does this abstract theory work out in practice? One way to approach this question is to examine how a virtue-centered theory of judging would handle simple cases and complex cases.

Let's begin with *simple cases*. Some decisions will obviously be just. Even persons who have incomplete legal knowledge or who have obtained only an incomplete degree of virtue will be able to recognize the justice of the decision. Such cases involve legal rules that are easy to grasp and fact situations in which the salience and application of the rule can be comprehended even by judges who are not especially wise or learned. Of course, even in simple cases, someone who is thoroughly blinded by self-interest might not concur in a widely shared judgment about what outcome is just.

There are cases where the justice of a decision is not so obvious as in easy cases. The second context might be called *complex cases*. When the law is complex, a high degree of legal intelligence may be required to recognize the legally correct result. When the facts are complex, other intellectual skills, e.g., a highly developed situation sense, may be required to see what even relatively simple legal rules require. Thus, in complex cases, it may be the case that only someone with sufficient legal knowledge and in possession of a high degree of judicial virtue will be able to fully grasp which outcome is just and why this is so. Although we might say that a just decision is independent of the virtue of the particular judge who made the decision, it is not the case that the justice of the decision is independent of judicial virtue. There are cases in which the just outcome can only be recognized by a virtuous judge.

#### 1.4.2.2 Equity and the Rule of Law

Does virtue jurisprudence offer an adequate theory of judging? I shall answer this question with respect to two contexts, illustrating both the way in which a virtue-centered theory of judging can capture the insights of its rivals and the way in which it might differ from them. The first context, I shall call cases of "justice as lawfulness." These are cases in which the outcome required by the legal rules is in full accord with our sense of fairness. The second sort of case, I shall call "justice as fairness." Cases of justice as fairness involve the situation in which the outcome dictated by the rules of law alone is not consistent with our understanding of what is fair in a wider sense.

Can a virtue-centered theory of judging offer an adequate normative account of cases, either *easy* or *complex*, in which the legal rules determine the lawful outcome? The answer is surely yes. For the most part, a virtue-centered theory of judging will be in accord both with common sense and with other normative theories of judging with respect to the question as to what constitutes the just outcome in such cases. The virtue of justice ordinarily requires decision in accord with the letter of the law. Of course, the reasons offered by various normative theories of judging are likely to differ even in easy cases. Utilitarians will emphasize the good consequences that justify the rules and the bad consequences that would result if judges undermined the predictability and certainty created by the laws by failing to adhere to them. Deontologists might emphasize the rights that legal rules protect and the unfairness of failing to follow legal rules once they become a source of legitimate expectations.

The rivals of a virtue-centered theory of judging can agree on the idea that judges ought to possess the judicial virtues insofar as these are required for judges to reliably follow the law. No plausible normative theory of judging is inconsistent with an uncontested theory of the judicial virtues. No sensible theory would be indifferent to judges who are avaricious, cowardly, bad tempered, stupid or foolish, and no sensible theory would claim we should not prefer temperate, courageous, good tempered, intelligent, and wise judges. How then does a virtue-centered account differ from accounts that do not focus on the virtues?

Unlike other theories of judging, a virtue-centered theory makes the claim that virtue is an ineliminable part of the explanation for and justification of the practice of judging. According to a virtue-centered theory, the whole story about what the rules of law require in particular cases includes the virtues. If they were to be left out, the story would be incomplete. Moreover, a virtue-centered theory suggests that it may require judicial virtue to recognize the legally correct result. The rules do not apply themselves; judgment is always required for a general rule to be applied to a particular case. Practical wisdom or good judgment is required to insure that the rules are applied correctly.

The necessity for practical wisdom in rule application can be discerned by imagining an appellate judge and her interlocutor discussing the appellate review of a trial judge's finding of fact. "Why was the trial judge's finding of fact clearly erroneous?" the interlocutor asks. "Because it was not sufficiently supported by evidence on the record," answers the judge. "Why do you conclude that the support was insufficient?" asks the interlocutor. "Because a reasonable finder of fact could not move from that evidence to the conclusions that judge drew," answers the judge. "But why couldn't a reasonable finder of fact make the necessary inferences?" asks the interlocutor. Imagine that the interlocutor responds to each explanation with a demand for definite criteria for application of the clearly erroneous standard. At some point, the answers must stop. If the questioner were still unsatisfied, the judge would be forced to explain her lack of further justifications by saying, "because that's the way I see it, and I am a competent judge. I cannot say any more than that." Explanations must come to an end somewhere. The clearly-erroneous rule provides a particularly perspicuous example of the bottom-line role of practical judgment in rule application, because it is widely acknowledged that no criteria can be provided for sorting errors that are clear from those that are not.[2]

In the end, agreement and disagreement about what rules mean and how they are applied are rooted in practical judgments. Even with respect to some easy cases and more frequently with respect to complex cases, articulated reasons will not suffice to explain why, in cases of bottom-line disagreement about the application of a rule to the facts, one judgment is legally correct and competing judgments are not.

---

[2] See United States v. Aluminum Co. of America., 148 F.2d 416, 433 (2d Cir. 1945) (opinion by Learned Hand, J., stating, "It is idle to try to define the meaning of the phrase 'clearly erroneous'; all that can be profitably said is that an appellate court, though it will hesitate less to reverse the finding of a judge than that of an administrative tribunal or of a jury, will nevertheless reverse it most reluctantly and only when well persuaded.").

Indeed, a virtue-centered account allows us to appreciate the fact that explanations or justifications of legal decisions play more than one role. In some cases, when a judge explains a decision, the intention is to lay bare the premises and reasoning that moved the judge from accepted premises about the law and the facts to some conclusion about what result is legally correct. There are other cases, however, where explanations play a different role. When the decision of a case is based on legal vision or situation sense—that is, when the decision is based on the virtue of judicial wisdom of *phronêsis*—then the point of an explanation is to enable others to come to see the relevant features of the case. Such explanations do not recreate a decision procedure; rather, they are aimed at enabling others to acquire practical wisdom.

#### 1.4.2.3 A Virtue-Centered Account of Lawful Judicial Disagreement

At this point, one could object that a virtue-centered account fails for a different reason. It might be argued that a virtue-centered account requires that two inconsistent outcomes in the very same case could both be legally correct. As we shall see, this apparent objection to a virtue-centered theory of judging actually illuminates one of its greatest strengths. A virtue-centered theory allows us to account for the fact that there are frequently cases in which more than one outcome would count as legally correct.

The objection begins with a premise that we shall call *the multiplicity of virtuous decisions*. The core idea of this premise is quite simple: there are cases in which different virtuous judges would make different decisions with respect to a given issue and a given set of facts. The second premise shall be called *the uniqueness of legally correct decisions*. The idea of this premise is that given a particular issue and a given set of facts, only one decision of the issue can be legally correct. Call the claim that a decision is legally correct if and only if it is the decision that would be made by a virtuous judge under the relevant circumstances, *the identity of virtue and legality*. The shape of the argument should now be clear. From *the uniqueness of legally correct decisions* and *the multiplicity of virtuous decisions*, it would seem to follow that some virtuous decisions are incorrect. If these premises are true, it follows that *the identity of virtue and legality* is false.

The first premise, *the multiplicity of virtuous decisions*, asserts that two different virtuous judges could reach different decisions in the same case. This claim seems plausible. Different virtuous judges are likely to differ in ways that might affect their decisions. They will have different experiences and beliefs, and those differences could easily affect the decision on a variety of legal issues. The multiplicity of virtuous decisions seems especially likely in so-called hard cases, in which there are good legal arguments on both sides of the issue.

However, the second premise, *the uniqueness of legally correct decisions*, is false. There are a variety of situations in which more than one outcome is legally correct. This is true for a variety of reasons. First, it is sometimes the case that the preexisting legal rules underdetermine the outcome of a particular case. In the United States, a frequent pattern involves the situation where an issue of law has been resolved differently by different circuits of the United States Court of Appeal. This phenomenon is called a "circuit split." Unless the Supreme Court resolves the split, inconsistent results can be correct in different circuits. In a circuit that has not decided the issue, two different trial judges can reach different outcomes and neither judge has rendered a decision that is legally incorrect. In this first sort of case, however, one might argue that there is a sense in which the inconsistent decisions are only correct provisionally or temporarily. If the Supreme Court resolves the split, then one line of cases is approved and the inconsistent line becomes "bad law."

Of course, there are times when a circuit split is best explained as a competition between a correct line of authority and another position that is badly reasoned or that ignores relevant authority. But there are other times when both results are plausible. Because the Supreme Court leaves many circuit splits unresolved (for years, decades, or even permanently), the best description of the situation is that two inconsistent positions on the same issue of law are both correct, neither line of authority can be said to be bad law.

Second, it is sometimes the case that the law itself commits a decision to the discretion of the judge. A paradigm case of such discretion can be found in the power of trial judges to manage the mechanics of a trial. Trial court judges have discretion to decide how long a trial will last, how many witnesses each side can present, and how long the examination of a witness will be permitted to take. If a virtuous judge makes such a decision, then it is legally correct, even though another virtuous judge would have made a different decision. If, however, the decision was the product of judicial vice, e.g. it was a product of corruption, then the decision is in error—even though the very same decision would have been legally correct if it had been the product of virtue rather than vice. The law of procedure captures this phenomenon in the standard of appellate review for discretionary decisions. The relevant standard is called "abuse of discretion," and given an abuse-of-discretion standard is settled law that inconsistent decisions of the same issue on identical legally relevant facts can both be legally correct.[3]

Moreover, some legal standards sanction more than one legally correct outcome on a particular set of facts. A clear example of this is the "best interests of the child" standard in child custody disputes. Although formulated as a rule of law, this legal standard requires the application of practical judgment to a particular fact situation. As a consequence, an appellate court will affirm a trial court's decision to award custody, even when the appellate judges would have made a different decision.[4] In such a case, each of two inconsistent decisions (awarding primary custody to one parent versus the other) can be legally correct. Although "best" is superlative and therefore suggests a unique outcome, the best-interests-of-the-child standard is

[3] *See, e.g.*, Jones v. Strayhorn, 159 TEX. 421, 321 S.W.2d 290 (1959) ("The mere fact or circumstance that a trial judge may decide a matter within his discretionary authority in a manner different from what an appellate judge would decide if placed in a similar circumstance does not demonstrate that an abuse of discretion has occurred.").

[4] See Ford v. Ford, 68 Conn. App. 173, 187, 789 A.2d 1104, 1112 (2002) (indicating that a difference of judgment does not justify reversal of a child custody decision absent "abuse of discretion.").

understood by courts to permit a multiplicity of outcomes in the large range of cases in which both parents have good claims that they would provide the best for the child.

A virtue-centered theory of judging explains and justifies this feature of our judicial practice. There are circumstances in which two or more different (and in one sense "inconsistent") outcomes are legally correct. A virtue-centered theory explains this on the ground that two different virtuous judges could each make different decisions, even though each was acting from the virtues. In cases in which the judge was not acting from virtue, but was acting from vicious motives, such as corruption, willful disregard of the law, or bias, then a discretionary decision may be legally incorrect—even though the very same outcome would have been acceptable if it had been made by a virtuous judge.

#### 1.4.2.4 The Virtue of Equity

The distinctive contribution of a virtue-centered theory is even clearer in the second category of cases, those in which the result required by the legal rule is inconsistent with our notion of what is fair. In these cases, a virtue-centered theory suggests that the virtuous decision is guided by the virtue of equity, or justice as fairness, distinguished from justice as lawfulness. As we have already seen, the key to a virtue-centered account of equity is the virtue of practical wisdom or *phronêsis*.

A virtue-centered theory of judging offers a distinctive approach to cases that involve considerations of equity. Here is one way to put it. Other normative theories of judging have difficulty explaining why there should be a distinctive practice of equity. If an exception ought to be made to a legal rule, then amend the rule. (This is the approach favored by theories of statutory interpretation that require strict adherence to plain meaning.) Of course, sometimes rules should be amended, but a virtue-centered theory of judging stakes out the claim that there will be always be cases in which the problem is not that the rule was not given its optimal formulation. Rather, the problem is that the infinite variety and complexity of particular fact situations outruns our capacity to formulate general rules. The solution is not to attempt to write the ultimate code, with particular provisions to handle every possible factual variation. No matter how long and detailed, no matter how many exceptions, and exceptions to exceptions, the code could not be long enough. Rather, the solution is to entrust decision to virtuous judges who can craft a decision to fit the particular case.

No theory limited to the uncontested judicial virtues can incorporate the virtue of equity. Indeed, I shall stake the claim that only a virtue-centered theory offers a fully adequate explanation of equity. But no set of rules can do justice in every case. Thus, virtue jurisprudence offers a normative and explanatory theory of judging that explains and justifies the practice of equity, but many other theories of law stumble at precisely this point.

## 1.5 Conclusion: Towards an Aretaic Theory of Law

This essay has taken a few steps towards the development of an aretaic theory of law. A first step is to sketch a virtue-centered theory of legislation. The central principle of such a theory is the idea that human flourishing is the central aim of law: because human flourishing is (in part) constituted by the exercise of the human excellences, the law should aim at the development of the virtues and the creation of conditions favorable to their exercise. A second step is to sketch an aretaic theory of adjudication. The core idea of such a theory is that just and lawful decisions are best understood as virtuous decisions. Among the judicial virtues are justice (understood as lawfulness) and practical wisdom.

The completion of these two preliminary steps leaves much to be done. A more complete aretaic theory of legislation would offer an account of each of the major branches of law. A fuller statement of an aretaic theory of adjudication would offer a more complete account of the judicial virtues and provide an account of their relevance to particular problems of adjudication (e.g., the interpretation of legal texts, the evaluation of evidence, and so forth). Moreover, this preliminary essay leaves an important topic (the nature of law) mostly unexplored. Nonetheless, the account offered here does suggest a research program for virtue jurisprudence.

## Bibliography


Anscombe, G.E.M. 1958. Modern moral philosophy. *Philosophy* 33(124): 1–19.

Aristotle. 1955. *Nicomachean ethics.* Trans. J.A.K. Thomson and H. Tredennick, ed. J. Barnes. Harmondsworth: Penguin Books.

Balkin, Jack M., and Sanford Levinson. 2001. Understanding the constitutional revolution. *Virginia Law Review* 87(6): 1045–1109.

*Bush v. Gore*, 531 U.S. 98 (2000).

Dworkin, Ronald. 1986. *Law's empire*. Cambridge, MA: The Belknap Press of the Harvard University Press.

*Ford v. Ford*, 68 Conn. App. 173, 789 A.2d 1104 (2002).

Hobbes, Thomas. 1994. *Leviathan*. Ed. Edwin Curley. Indianapolis: Hackett Publishing Company.

Hursthouse, Rosalind. 1999. *On virtue ethics*. New York: Oxford University Press.

*Jones v. Strayhorn*, 159 Tex. 421, 321 S.W.2d 290 (1959).

Kaplow, Louis, and Steven Shavell. 2002. *Fairness versus welfare*. Cambridge, MA: Harvard University Press.

Kraut, Richard. 2002. *Aristotle: Political philosophy*. New York: Oxford University Press.

Locke, John. 1988. *Two treatises of government*. Ed. Peter Laslett. Cambridge: Cambridge University Press.

Rawls, John. 2001. *A theory of justice*. Cambridge, MA: The Belknap Press of the Harvard University Press.

Shiner, Roger A. 1994. Aristotle's theory of equity. *Loyola of Los Angeles Law Review* 27(4): 1245–1264.

*United States v. Aluminum Co. of Am.*, 148 F.2d 416 (2d Cir. 1945).

# Chapter 2
# Reasoning Against a Deterministic Conception of the World

**Liesbeth Huppes-Cluysenaer**

> The nations inhabiting the cold places and those of Europe are full of spirit but somewhat deficient in intelligence and skill, so that they continue comparatively free, but lacking in political organization and capacity to rule their neighbours. The peoples of Asia on the other hand are intelligent and skilful in temperament, but lack spirit, so that they are in continuous subjection and slavery. But the Greek race participates in both characters (…) It is clear therefore that people that are to be easily guided to virtue by the lawgiver must be both intellectual and spirited in their nature. (…) it is spirit that causes affectionateness, for spirit is the capacity of the soul whereby we love. A sign of this is that spirit is more roused against associates and friends than against strangers, when it thinks itself slighted. (…) Moreover it is from this faculty that power to command and love of freedom are in all cases derived; for spirit is a commanding and indomitable element. (…) Hence the sayings "For brothers' wars are cruel" and "They that too deeply loved too deeply hate." (*Pol.* VII. v.1327b20–1328a19, trans. Rackham)

## 2.1 Introduction

In the humanities the deterministic conception of the world has been primarily discussed in relation to the justification of punishment: can a person be held responsible for acts if there is no free will and the action had already been determined before the person was born? (Nagel 1987, 37–44). Time and again this argument is used by behavioural psychologists who insist that people should not be punished by lawyers, but rather mentally treated by psychiatrists and psychologists (Kenny 1978, 1–13). Recent insights derived from brain-research are used by psychologists to revitalize this debate. The deterministic conception itself, however, has

L. Huppes-Cluysenaer (✉)
Department of General Jurisprudence, Faculty of Law, Unviersity of Amsterdam,
Oude Manhuispoort 4, 1012 CN Amsterdam, The Netherlands
e-mail: huppes.uva@gmail.com

L. Huppes-Cluysenaer and N.M.M.S. Coelho (eds.), *Aristotle and The Philosophy of Law: Theory, Practice and Justice*, Ius Gentium: Comparative Perspectives on Law and Justice 23,
DOI 10.1007/978-94-007-6031-8_2, 

never been seriously challenged in this debate on justifying punishment, although determinism is currently rejected in science.

The introduction of determinism was one of the main issues of the Enlightenment. Determinism refuted the Aristotelian view on the world, which assumed an intelligent force in nature. According to Aristotle there was a double impulse working in every living creature, one of which was intelligent. He described the intelligent impulse as a free and indomitable element in the natural make-up of human beings, while the capacity to reason tends to subject and enslave human beings to rules. Determinism attacked this enchanted Aristotelian picture of the world and rejected the idea of an intelligent – spirited – force in nature. By this determinism, a reversal of the Aristotelian view was brought about: to be driven by natural force meant now to be enslaved, while the capacity to reason could now free people of this enslavement.

It seems relevant to ask whether the rejection of determinism by current science could lead to a return of the Aristotelian view on nature. Popper is one of the very few authors who asked this question.[1] Most authors do not treat this issue seriously.

Solum (this volume, Sect. 1.3), for example, explicitly announces that he will develop an account of the virtues that is consistent with current science, even though this could entail some important divergences between contemporary theories and Aristotle's account. He does not specify these divergences. Therefore the question remains open what exactly is the divergence between Aristotle's account and current science.

Kenny (1978, 13–26) stated that the philosopher can and should remain agnostic on these questions. The consequence of this was that he did not acknowledge the fact that physical determinism had been rejected by current science.

Nagel (1987, 37–44) argued that the rejection of determinism by current science had not solved the problem of the justification of punishment. This meant that human action was the result of pure chance and that therefore nothing was identifiable as the cause. He thus did not seriously take the Aristotelian view into account.

Strawson (1973, 1–25) has given the most elaborate and original analysis of the possibility of a justification of punishment. He stated that people adopt punitive or tolerant reactive attitudes towards wrongdoers because of situational facts and *not because* they believe in the general truth of determinism or indeterminism. By this argument Strawson denied that the development of a moral community could be influenced by the choice for or against a deterministic view on the world.

The arguments of Strawson, Kenny and Nagel seem to be designed to divert attention from the way determinism leads to a connection between rationality and domestication, resulting in discipline and slavery. This connection was not only pointed out in classical times by Aristotle, but was extensively discussed by authors

[1] See Briegel (2012) for a current attempt to combine an indeterminist physical approach with the possibility of creative machines.

such as Foucault, Marcuse, Habermas and Berger[2] in the period in which the essays of Strawson, Nagel and Kenny were written.

Foucault (1979) showed how the Enlightenment had created a society of discipline and punishment. Not only Foucault, but also, for example, the Frankfürters[3] tackled the same issue contemporaneously with their theories about the Military Industrial Complex[4] and the exchangeable subject, while Berger (1966) wrote about the imprisonment of the individual by society and the relevance of a destabilization of routines as defended by critical sociologists.

Foucault (1988) turned to the Ancient 'technologies of the self', but most other participants in this debate simply tried to save the project of Enlightenment. As a remedy they propagated a severe form of democratization of authority.[5] In doing so, they negated the inherent connection between Enlightenment and determinism, and the possible consequences of a rejection of determinism for the project of Enlightenment.

This chapter will describe the indeterminist view of Aristotle and will explore the inherent connection between determinism, Enlightenment and enslavement. It will analyse the attempt of Popper to save the project of Enlightenment within the context of indeterminism and use his analysis to explain how the Aristotelian view could agree with the indeterminism of current science. It will conclude with the consequences of a choice for determinism or indeterminism for politics and law.

## 2.2 The Greek Concept of Free Spirit: Desire

According to Plato, the human soul has three parts. In the dialogue *Phaedrus* he used his famous metaphor of a chariot with a black and a white horse to explain these three parts and their interplay (245–247). The driver of the chariot is the intellect, the black horse is lust and the white horse is an element described by Plato in terms of ardency, rage and scolding oneself. To explain the working of these three elements Plato described how the soul acts when somebody suddenly falls in love. The black horse of his soul wants to run immediately to this person, while the white horse halts, befallen by a sudden shame. The driver has the greatest difficulty to keep the chariot on the road. So the black and the white horse represent a double impulse of attraction and restraint, which simultaneously stimulate a human being to act. The intellect can steer between these two impulses, but does not itself develop any power to act.

Aristotle (*EN* I.xiii.1102a17–1103a10) also recognized three parts of the soul: the vegetative, the appetitive/desiderative and the rational. The vegetative part is

---

[2] This discussion referred to by earlier authors such as Weber.

[3] For example Marcuse (1955, chpt. IV).

[4] The term gained popularity after its use in the farewell address of President Eisenhower in 1961.

[5] Leading in this respect Habermas (1981).

purely irrational and doesn't act, the appetitive part acts and is partly irrational and partly rational, while the intellect doesn't act, but thinks and is purely rational. In this respect there seems to be a great continuity between the conception of Plato and Aristotle, although there are also some differences. Aristotle did not for example believe in an immortal soul.[6] He rejected the idea that the intellect has control over the desiderative element, like a driver over horses[7] (*EN* VII.ii.1145b21–1146a). Moreover, he considered the idea of a low passion to be not very important or attractive [8] (*EN* VII.xiv.1154a22–1154b11). In principle every desire is good. The important point for Aristotle was the development of the strength of will to follow the lead of the rational element of the desire. Aristotle describes the desiderative element in the following way:

> But there seems to be another element of the soul (desire LHC), which while irrational, is in a sense receptive of reason. Take the types of man which we call continent and incontinent. They have a principle – a rational element in their souls – which we commend, because it urges them in the right direction and encourages them to take the best course; but there is also observable in them another element, by nature irrational, which struggles and strains against the rational. Just as in the case of the body paralysed limbs, when the subject chooses to move them to the right, swing away in the contrary direction to the left, so exactly the same happens in the case of the soul. (*EN* I.xiii.1102b6–1102b28, trans. Thomson/ Tredennick)

Many authors think that merely the lust can be perceived as a bodily desire. They do not situate the root of moral behaviour in the body, but in the intellect.[9] According to Werner Jaeger (1945, 186–187), Plato's dialogue *Phaedrus* had no relevant connection with the concept of love as elaborated upon in his dialogue *Symposium*. Maybe Jaeger shrank back from the fact that in *Phaedrus* the homo-erotic love between elder men and younger boys is in fact treated. One should however realise that in Plato's days this kind of love was part of public life – while the love between men and women obviously was not. Further there is no type of relationship which is so gratifying when care and teaching dominate it, but which is at the same time so susceptible to turn into corruption by sexual desire. It is this counterbalance between wrongfulness and goodness on which Plato focused and to which he also turned in the final passages of *Symposium*.

Plato described the white and the black horse as follows:

> The right-hand horse is upright and cleanly made; he has a lofty neck and an aquiline nose; his colour is white, and his eyes dark; he is a lover of honour and modesty and temperance, and the follower of true glory; he needs no touch of the whip, but is guided by word and

[6] "...not the least absurdity is the doctrine that there are certain entities apart from those in the sensible universe, and that these are the same as sensible things except in that the former are eternal and the latter perishable" (*Met*. III. ii. 997b5–10, trans. Tredennick).

[7] See however (*Phdr.* 253–257) where there is hardly any control of the driver.

[8] See however also (*EN* I.xiii.1102b6–28, trans. J.A.K. Thomson/H. Tredennick): "Probably we should believe nevertheless that the soul too contains an irrational element which opposes and runs counter to reason – in what sense it is a separate element does not matter at all."

[9] Barnes (1955) however, is very explicit about this root of moral behaviour in the body, which makes the behaviour of animals and human beings comparable.

> admonition only. The other is a crooked lumbering animal, put together anyhow; he has a short thick neck; he is flat-faced and of a dark colour, with grey eyes and blood-red complexion; the mate of insolence and pride, shag-eared and deaf, hardly yielding to whip and spur. (*Phdr*. 253–257, trans. Jowett)

The fact that the condition of the will – the indomitable, lion-like element in Aristotle's terms – is shown in bodily posture and reactive attitude is an important fact, which underscores the bodily interpretation of the Greek concept of desire.

Aristotle described desire as a general force, which is active in human beings and animals. He referred in this respect to spiders, which show some kind of intelligence in weaving their webs (*Phys*. II.viii.199a20–25). At the same time desire is a specific force in each individual, formed by the experiences which it has had. In the last part of *Politics,* Aristotle explains how the state should take care of the development of a right sense of pain and pleasure by promoting literature in the education of children: "habituation in feeling pain and delight at representations of reality is close to feeling them towards actual reality" (*Pol*. VIII.v.1340a20–30, trans. Rackham).

## 2.3 Desire and Habituation

For Aristotle, reality meant endless variability and continuous growth. His biology is a "natural history" (*Hist. An.*) in which scientific study is primarily focused on descriptive taxonomies.[10] Aristotle believed that at a genus level, formative principles are induced in matter and that reality exemplifies the complete gradation of full presence and complete absence of such a principle. A full description of this continuously moving and changing endless gradation would ask for a description in terms of potentiality and would be completely inarticulate. Without specifying different species and subspecies, indicating different ranges within this endless gradation, without stating that something belongs either to this or to that type and that a third possibility is not open, there would be no science possible, nor any reflection on actions. The art of definition and the related principle of contradiction are thus for Aristotle "Strukturformelle der Realität" as Segalerba (2011) formulates it.

To understand the art of definition means to be able to understand why Aristotle's works on logic are certainly not inconsistent with his metaphysics, but continuous, specifying respectively the scientific method and the ontology.[11] To understand the art of definition it is important to realize that taxonomic guides – to birds for example – often work with pictures, which emphasize the differences between species. These pictures are ideal types. As is further explained in Sect. 2.6, Aristotle

[10] Only as late as the mid-eighteenth century were systematic taxonomies made. In earlier times people lacked a belief in one true criterion, which is necessary for such an enterprise and which was indeed introduced by Linnaeus. The first edition of Systema Naturae was published in 1735. See for a more detailed discussion of the subject Huppes (2005, or 2008).

[11] See for a different opinion D.W. Graham (1987) and Klaus Brinkmann (1996).

deems it one of the largest mistakes of Plato to have reified these ideal types. For Aristotle circles and other pictures are simply means by which people reason about the world (*Met.* VII, x, 1036a). Therefore it is in a certain sense arbitrary how these pictures are made, as they hold no truth value.

However, when somebody is able to make a guide which helps many people to memorise detailed differences between phenomena and to think about these things in an orderly way, the maker of such a guide is a very good scientist. In this sense – when not taken as a mirror of reality, but as a frame of reference – such a guide exemplifies an insight in the order of things: in the nature of the completely abstract (imageless) formative principles as well as in the details of the variety and growth these principles induce in the actual world.

One cannot acquire knowledge of the actual world by simply reading the guide. One has to study hard cases and discuss them against the background of the frame of reference offered by the guide, in cooperation with people who have done so already for many years. This way the ideal types of the guide are 'filled' with experiences.[12] The guide is nothing else than a help to organize experiences and to be able to communicate these experiences with others. This art of definition and the intellectual order and communication it constitutes has been the great discovery of the Greek Academicians.

Concerning action Aristotle does acknowledge another way of perceiving reality. When one aims for a result which can be clearly specified, for example in medicine the decrease of fever, it is not only possible to acquire knowledge about the effects of certain actions, but also to gather knowledge about the reasons why a specific action generates such an effect[13] (*Met.* I.i. 981a–981b15). But in the field of law and morals people strive for happiness and there can be no such learning of general rules and remedies. In this field a type of learning is possible which is called habituation.

As also acknowledged by Jaeger (1945, 215–227), there is a great continuity between Plato's *Laws* and Aristotle's thought in this respect.[14] In *Laws* Plato explains what habituation means. The training of continence is discussed in the first book by way of a comparison of the laws in Crete, Sparta and Athens on drinking behaviour. While in the other places drinking is strictly regulated, the public meetings for drinking are deemed important in Athens for their educational effect on the youth. Only by really experiencing the dangers of incontinence one can be trained to fight ones weaknesses. The state has to create the institutions – a more or less safe environment – in which the youth can be confronted with its weaknesses and can cross boundaries.

---

[12] See Robert R Sokal (1974) for research in which imaginary animals served to illustrate how different individuals of the same specialisation take different classification decisions. Classification is never evident.

[13] To treat a patient, however, the physician has to adjust the general remedy to the particular situation of the individual patient.

[14] See for Aristotle especially *EN* book I.

For example, to internalise the rule that one should not drink more than one glass an hour, does not mean for Plato and Aristotle to put oneself in the place of the father/mother, or any other authority, but rather to associate such a rule with personal experiences. Habituation is about acquiring strength and not about obedience or conscience. Although the drinking-rule may function as a warning to be careful with alcohol and may specify a safe guideline, it is, according to Plato and Aristotle, certainly not good to simply follow guidelines blindly. This would lead to a mediocre practice, to the enslavement by a rationality, which is poorly instructed by experiences and simply follows the lead of others.

Habituation leads to a type of knowledge about human affairs which is closely connected to the descriptive taxonomies in biology. The formulation of rules is the formulation of ideal types. At the one hand rules are arbitrary, as many legal theorists have argued, while at the same time an ingenious composed set of rules can exemplify – in the same way as an taxonomic guide – an insight in the natural order of human affairs. Such a set of rules constructs a conceptual world, which can be used as a frame of reference to articulate and memorise experiences. The formulation of common rules institutionalises a practice of reflection by which experiences can be ordered and communicated. This reflection and communication constitute an insight in what belongs to human affairs.

Rules have to be connected continuously with actual experiences. Only by experiencing situations in which it is hard to discern the good and the bad, can the individual develop personal knowledge about what it means to act well and how the rule has to be interpreted. The fact that rules are arbitrary and represent at the same time an insight in formative principles cannot be explained to scientists who believe in determinism. As will be explained in Sect. 2.6, determinism is founded on the reification of ideal types. The incapacity of determinism to understand the constitutive features of the legal academic tradition has caused the slow decay of this tradition.

Although strength is developed by habituation, for humans there are, like for every other animal, differences in the strength and eagerness with which different people are born (*EN* III.1114a8–1114b15). It is remarkable how lenient and tolerant Aristotle is concerning incontinence (*EN* V.1134a6–23). For him, evidently, everything is better than to be domesticated.

## 2.4 The Indeterminism of Aristotle

Barnes (1955, 20–21) states that Aristotle's notion of 'akribeia' has aroused little scholarly comment and remains somewhat obscure. 'Akribeia' means that a subject matter only allows for a modest amount of precision. Barnes describes how Aristotle accepted precision in mathematical science, which consists of analytical truths, but not in biological science. He indicates how Aristotle was "[i]mpressed by the seemingly infinite variety of human circumstances and situations" and thinks it worth to underline "[t]he fact that Aristotle is here adopting an extreme position, not unlike the one taken up by some existentialist thinkers" and to point out that Aristotle thought that:

> The most we can hope for is a group of roughly accurate generalizations – principles which will meet most ordinary situations, but which are always liable to come unstuck.[15]

By rejecting the possibility of precise general propositions in biological science, Aristotle revealed his indeterminist view. He stated (*Met.* VI. ii.1027a-1–20) that the accidental exists and that matter allows for deviation from the usual.

Aristotle acknowledged only two ways to talk about the real world: on the one hand in terms of the laws which govern it (the formulas) and on the other hand in terms of the particulars which are partly governed by laws, and partly by the variation of which matter admits (combination of formula and matter). While the laws just 'are', being independent of generation and destruction, particulars are perishable. People only can obtain empirical knowledge by studying particulars. However, particulars cannot be defined; they only can be grasped intuitively (*Met.* VII. xiv–xv.1039b20–1040a20). Therefore, according to Aristotle, empirical knowledge can only consist of rough generalisations. Knowledge of the laws which govern the world is partly mathematical and can be stated with precision, and partly empirical, consisting of speculative, rough generalisations.

When Aristotle's indeterminism is applied to his ethics, the following account of his ideas is possible. The double impulse involves a moment of choice. However, there is not a strict dichotomy of choice or no choice at all, but rather somewhat more or less room for choice: some creatures being more capable of adaptive behaviour than others. Since every individual being or concrete event is a very complex composite, it is impossible to state anything about these things with any precision. How free a person was in his or her choice in a concrete case cannot be a scientific statement of fact, but only a personal – speculative – judgement about his or her character.

The more strength a person has, the more he is free. According to Aristotle some people are born as slaves and others as masters. This is one of the most contested elements of his theory and can easily lead to the conclusion that after all Aristotle defends a deterministic view.[16] To understand this part of his theory, one has to place it against the background of his indeterminist view on nature: the endless variety of the world and the fact that this variability is the essence of being. This means that people are not born either as a slave or as a master, but with more or less talent for being strong. Aristotle distinguishes between slaves by law and slaves by nature. People who are masters by nature can be enslaved by law, while people who have the position of a master, can be slaves by nature.

The indeterminist view of Aristotle has been completely lost in the reception and tradition of his ideas. This can be explained by the fact that Aristotle's texts gained

---

[15] See also Kenny (1978, 13–26) who uses the term defeasible rules for this phenomenon, and who clearly reveals his own belief in physical determinism by reserving this notion for the field of ethics alone.

[16] In this volume see the essays by Engle en Yankah.

a renewed impact on the Western world through the way they were re-interpreted by Thomas Aquinas in accordance with the development of a very strong, religious, deterministic outlook on the world. Because today very few people have read Aristotle's original *Metaphysics* thoroughly, Thomas' interpretation is still generally accepted. Barnes' remark that the notion of 'akribeia' remains somewhat obscure has to be read against this background.

## 2.5 Introduction to the Concept of Truth

> The study of Truth is in one sense difficult, in another easy. This is shown by the fact that whereas no one person can obtain an adequate grasp of it, we cannot *all* fail in the attempt; each thinker makes some statement about the natural world, and as an individual contributes little or nothing to the inquiry; but a combination of all conjectures results in something considerable. Thus in so far as it seems that Truth is like the proverbial door which no one can miss, in this sense our study will be easy; but the fact that we cannot, although having some grasp of the whole, grasp a particular part, shows its difficulty. (*Met.* II. i. 993a30–993b10, trans. Tredennick)

Only knowledge of the world of particulars is knowledge of the world in its full sense, according to Aristotle. At the same time he acknowledged that knowledge of particulars is impossible, because to know means to be able to define something in general terms (*Met.* VII.xiv–xv.1039b20–1040a20). This insight however did not lead to cynicism or relativism, but was celebrated by the Greeks and the Romans as the key notion of ethical theory: the complexity of the world makes it impossible to grasp it or to be enslaved by it. It means that nature is a free place.

Both the Greeks and the Romans favoured the idea that the more one knows, the more one understands that it is impossible to know anything with certainty about the world of particulars. Especially in legal theory and in the field of ethics this was expanded upon as an insight to promote moderateness and temperance and to avoid absolutism. Because truth was unattainable, the full focus was given to institutional devices for the division of power and the protection of the independence of authority.[17]

There was, however, a difference between the Greeks and the Romans in the way they valued scientific study. Both shared the idea of habituation and the value of a broad and reflective knowledge of the feelings aroused by practical experience. For Plato and Aristotle however, the study of science added a dimension over and above the political level, in which the human being situates himself in an order in which not only humans partake, but all living beings. Having a broad practical experience makes a person prudent, but the rough knowledge of eternal things makes a person wise (*EN* VI.vii.1141a19–b8). A wise person is not very interested

[17] See not only the discussions on money and age for those who held jurisdiction, on the education of them and on the function of philosophy to gather strength, but specifically also all the institutional arrangements for unpopular offices (*Pol.* VI.v.1321b40–1322a30).

in money or applause. This is important for a politician, in order to be impartial and have integrity.

The Romans on the other hand did not value science as highly as the Greeks did. Cicero (1928) described the point of view of the Greeks in the following way:

> What power, moreover, what office, what kingdom can be preferable to the state of one who despises all human possessions, considers them inferior to wisdom, and never meditates on any subject that is not eternal and divine; who believes that, though others may be called men, only those are men who are perfected in the arts appropriate to humanity.

He then confronts this position with the question

> For why is it that the grandson of Lucius Paulus, the nephew of our friend here, a scion of a most worthy family and of this most glorious republic, is asking how two suns could have been seen, instead of asking why, in one State, we have almost reached the point where there are two senates and two separate peoples? (*Rep.* xvii28–xix31, trans. Keyes)

That Cicero did not value science in the way the Greeks did, doesn't mean that he was not very well acquainted with Aristotle's view. For the Romans the philosophy of the Greeks was still common knowledge and authors did not need to refer to their Greek source. This may be the reason why in later days the indeterminist view of the Romans is generally ascribed to Stoicism, indicating the mechanistic interpretation of it by Epicurus and Lucretius (Pullman 1998).

The indicated lack of depth in Roman culture – as far as wisdom is concerned – is filled but at the same time altered considerably by the introduction of a religious conception of truth. Augustine introduced the concept of truth – explicitly in discussion with the Ancient understanding – in his essay *Contra Academicos.* The concept rests upon the assumption that there is a place – a Panopticum or Archimedical point, currently indicated by the more prosaic 'helicopter view'- from which everything can be seen and known. It is a religious concept. God occupies the panoptical point and is the all-knowing and all-seeing creator and governor of the world.

This religious concept of truth certainly brought back the dimension of the eternal, with its specific function for integrity in political life, but it did so in opposition to the great achievement of the Greeks, namely their scientific research attitude. This meant that the reintroduction of Greek philosophy in the Renaissance had to face the integration of science and religious belief.

The God, introduced by Aristotle was the unmovable Mover, a Mind which had no grip on matter, which only was the seat of the eternal presence of formative principles and the source of the attraction which these exercise on living things. The God, introduced in philosophy by Augustine, was a Mind with creative power, determining everything, the eternal presence of the formulae as well as the details of the life-cycle of particulars.

Notwithstanding this enormous difference, there still is a great similarity between Augustine's theory and Ancient Greek theory, because Augustine defended the belief that it was only possible for somebody to act as a good person, when he had been selected by God. This way Augustine saved uncertainty as a key notion in ethics, although it gained a completely different connotation: as an act of faith rather than an epistemological insight.[18]

Science and religion both have a function for political life, creating a level at which political power can be embedded and moderated. The pragmatic attitude of the Romans led to the rejection of Greek science. Religion was introduced to fill this gap. Through the fusion which resulted from the reintroduction of Greek philosophy, religion became a pragmatic source for absolute power. This power-claim was then successfully attacked by science. Through the claim of truth, however, science brought the secularisation of the panoptical point, which represents an even more dangerous absolutism. A conscious return to indeterminism could break this absolutism.

## 2.6 Determinism and Enlightenment

When Greek philosophy was reintroduced in the Western world, it brought about the integration of science and religion. The belief in a natural language played a decisive role in this integration. Natural language is the key-assumption for a deterministic view of the world. Aristotle rejected the existence of a natural language and attacked Plato, for having introduced this belief. Aristotle thought that it had been one of the biggest mistakes of Plato that he assumed:

> [t]hat the problem of definition is concerned not with any sensible thing but with entities of another kind; for the reason that there can be no general definition of sensible things which are always changing. These entities he called 'Ideas', and held that all sensible things are named after them and in virtue of their relation to them; for the plurality of things which bear the same name as the Forms exist by participation in them. (*Met.* I. v. 987b3–11, trans. Tredennick)

Between these Ideas and sensible things Plato introduced mathematical symbols as intermediates, according to Aristotle. Aristotle rejected this Platonian view in which theoretical notions such as the "circle" have an independent existence (the absolute circle) and are represented by a symbol (a drawing).

> For Platonists say nothing more or less than that there is an absolute Man, and Horse and Health, in which they closely resemble those who state that there are Gods, but of human form; (...) Again, if anyone posits Intermediates distinct from Forms and sensible things, he will have many difficulties; because obviously not only will there be lines apart from both Ideal and sensible lines, but it will be the same with each of the classes. (*Met.* III. ii. 997b10–20, trans. Tredennick)

Firstly the position of Aristotle will be further clarified. Thereafter the different relevant variants of determinism will be treated generally.

---

[18] How the new religious outlook on the world could be brought in agreement with the Ancient philosophy by the epistemological notion of God's Providence was very well explained by Boethius' Consolation of Philosophy (1969, Book IV and V).

## *2.6.1 Aristotle*

It can be concluded that Aristotle rejected both the independent existence of an Absolute Idea (in religious terms, the image God had in mind when he created) and of an Intermediate (the image in the mind of human beings). For Aristotle there are only particulars and the independent existence of general laws.

The formulas – the general laws – have according to Aristotle an independent eternal existence by their seat in the mind of the unmoved Mover. They exercise an attraction to living beings, such that these beings will attain that which belongs to their form. This is their finality and explains the growth in particulars.

However, for Aristotle the formulae of man and horse are the same, because it is at the level of genus and not of species that the formula merges with matter. Species are the variations between total privation and total completion of matter by a formula (*Met.* X. iii. 1054b–1059a; *Phys.* I. vi–II. ii.188b and 194b). There is no Absolute Idea to which the names of species refer, neither do these names refer to concepts in the human mind which have an independent existence as Intermediates. This means that Aristotle rejects natural language.

Aristotle did acknowledge mathematical knowledge as precise and sure knowledge about necessity. But these mathematical propositions – determinate laws such as:

> What is drawn up must cool and what has been cooled must become water and descend. (*Phys.* II. Viii. 198b10–199a10)

– are only hypotheses. “In mathematics the principle is the principle of reasoning only, as there is no action” (*Phys.* II. ix. 200a1–30). In other words, the propositions of mathematics are analytic and not synthetic.

## *2.6.2 Different Relevant Variants of Determinism*

### 2.6.2.1 Fundamental Religious Determinism

Fundamental religious determinists stick to the Augustinian belief of Grace as a gift of God, which rests on Gods will and not on his reason. They believe that God determines the world at the level of each particular event, and they deny the possibility for a living being to know anything about Gods ways. This view comes in a certain sense very near to the Aristotelian view. Both views, determinism and indeterminism, lack of free will and free will, are extremes which converge in their conception about what a man can know. Both cherish the lack of certainty about the world of particulars as a key notion of ethics. The practical difference between both is a different attitude towards wisdom and a different outlook on life: for the religious determinist, faith and a life at the edge of despair; for the Aristotelian, speculative intelligence and the joy of a complete life. Both views share however the values of individualism, freedom and moderation, while both also dislike Absolutism.

#### 2.6.2.2 Cultivated Religious Determinism

The view of *cultivated religious determinists*, such as Thomas Aquinas, is often indicated as the belief in a free will, because it rejects a belief in predestination. As Thomas Aquinas played an important role in the reintroduction of the ideas of Aristotle, his view has mistakenly been portrayed as an Aristotelian view. It is quite important to avoid this mistake, as it conceals Aristotle's indeterminism. Thomas Aquinas accepted Absolute ideas for every class of things that indicate a fixed goal for every member of the class, which goal is the essence of their being and can be intuited by experience. This makes correspondence between knowledge and a world of natural classes possible. A person can understand Gods plan to a certain degree by reflecting on the concepts which are naturally acquired through experience.[19] One can choose to act accordingly and therefore one can influence one's final destiny. In this view, 'that which is called free will' is in fact to 'act according to Gods general plan'. It is here that the reversal of freedom and slavery starts.

#### 2.6.2.3 Scientific Determinism

*Scientific determinists* took the nominalist stance of Ockham. Ockham (1974, Summa Logicae, Part I, 3) accepted Absolute Ideas for every class, but not the natural development of insight into Gods ways through experience.[20] He rejected thus the possibility to know anything about the finality of things through conceptual analysis. However his belief in an independent status of concepts as intermediates made it possible for him to distinguish between concrete concepts (which indicate a natural class) and abstract concepts (which indicate the Absolute Idea).

The main argument of Ockham for the independent existence of Intermediates referred to the mysteries of faith: the fact that a man can also be the son of God and that bread can also be the body of Christ. He distinguished between abstracta such as 'Manhood' (being the son of God) and concreta such as 'Man' (Part I, 7, trans. Loux).

According to him it is possible to say that Man walks, while it is impossible to say that Manhood walks. He thus accepted a complete correspondence between the names of species and a world in which particulars can be recognized as members of

---

[19] See also *Averroes Middle Commentary,(Anatolio's Introduction)*, trans. H.A. Davidson (1969): "Speech designates the conceptions that are present in the human mind, those conceptions have their reference in things that exist outside the mind; and the totality of existent things provides knowledge of the Cause of their existence and confesses that He created them. Therefore everyone who truly desires to seek God stands in need of the science of logic."

[20] "Thus, suppose a spoken word is used to signify something signified by a particular concept of the mind. If that concept were to change ist signification, by that fact alone it would happen that the spoken word would change ist signification, even in the absence of any new linguistic convention.(…) For one thing the concept or impression of the soul siginifies naturally, whereas the spoken or written term signifies only conventionally. We can decide to alter the signification of a spoken or written term, but no decision or agreement on the part of anyone can hev the effect of altering the signification of a conceptual term."

a natural class. This made it possible to describe and gather further knowledge – not about essential nature – but about the appearance of particulars. This scientific determinism is further developed by Hobbes (1968):

> The first author of Speech was *God* himself, that instructed *Adam* how to name such creatures as he presented to his sight; For the Scripture goeth no further in this matter. But this was sufficient to direct him to add more names, as the experience and use of the creatures should give him occasion; and to join them in such manner by degrees, as to make himself understood; and so by succession of time, so much language might be gotten, as he had found use for; though not so copious, as an Orator of Philosopher has need of. For I do not find any thing in the Scripture, out of which directly or by consequence can be gathered that *Adam* was taught the names of all Figures, Numbers (...) and least of all, of *Entity, Intentionality, Quiddity*, and other insignificant words of the School. But all this language gotten, and augmented by *Adam* and his posterity, was again lost at the tower of *Babel*, when by the hand of God, every man was stricken for his rebellion, with an oblivion of his former language. And being hereby forced to disperse themselves into several parts of the world, it must needs be, that the diversity of Tongues that now is, proceeded by degrees from them, in such manner, as need (the mother of all inventions) taught them; and in tract of time grew everywhere more copious. (Leviathan I. IV.)

Although Hobbes thus rejected the idea that concepts such as 'intentionality' can be founded on the Scripture, and defended the view that language develops for pragmatic reasons, he nevertheless based the primordial correspondence between language and natural classes on the Scripture. A short way to express the same belief in the possibility of correspondence is the statement of Galilei that the world is written in mathematical terms.

#### 2.6.2.4 Sceptical Determinism

*Sceptical determinists*, such as Descartes and, much later, Hume, were impressed by the sobering thought that the new philosophy which was developing in Early Enlightenment and which attacked the Church, could not be founded rationally. Their scepticism can be summarised by the statement: it cannot be explained rationally, but has to be accepted. Both Descartes and Hume played a crucial role in the disenchantment of the scientific understanding of the world.

Descartes rejected the diversity of animals and plants – which had been all important in the Aristotelian view. Descartes stated that sensible qualities have no existence outside human awareness. These sensible qualities have to be reduced, according to him, to the primary 'mechanical' properties of insensible particles (van Ruler 1995, 118–120). Instead of the idea of individual living beings having a double stimulus to act *from within*- every movement was explained by causality as an omnipresent pressure *from without*[21] (van Ruler 1995, 129–131). Instead of the four causes of Aristotle, there was only one cause accepted for all natural mechanisms in Cartesian philosophy: God as an efficient cause.

[21] More extensively on this Huppes (2011).

Hume (Human Nature, II. 3. iii) fully and explicitly realised the consequences of this reduction of a double impulse to one mechanical impulse. He stated that a man desires what he desires. According to Hume, man doesn't have a natural instinct according to which it is irrational for him to desire the destruction of the whole word. His rationality has only an instrumental function to help him realise what he wants.

## 2.7 The Secularisation of the Panoptical View: The Rise of Pragmatism

Kant made an impressive attempt to reconcile the old and the new philosophies. He made a distinction between the phenomena with which science deals and the substances which are dealt with in ethics; he thus maintained the Aristotelian view that general knowledge can only be obtained about forms and not about substance; he envisaged the individual as an end in itself, by which he secured the Greek concept of desire; and he protected this individual finality by a formal conception of norms.

In respect of the further development of modernity, however, Kant contributed something else much more important. Kant was the first to understand the pragmatic and political turn which science had been making during the Enlightenment.

He stated in the preface of his Kritik der Reinen Vernunft that Gallilei, Torricelli and Stahl had:

> [l]earned that reason only perceives that which it produces after its own design; that it must not be content to follow, as it were, in the leading-strings of nature, but must proceed in advance with principles of judgement according to unvarying laws, and compel nature to reply its questions. (…) Reason must approach nature with the view, indeed, of receiving information from it, not, however, in the character of a pupil, who listens to all that his master chooses to tell him, but in that of a judge, who compels the witnesses to reply to those questions which he himself thinks fit to propose. To this single idea must the revolution be ascribed, by which, after groping in the dark for so many centuries, natural science was at length conducted into the path of certain progress. (trans. J.M.D. Meiklejohn)

This means that Kant understood how science used a hypothetic-deductive method for research. Later this method was further elaborated by scientists like Peirce, Dewey, Hempel, Nagel and Popper.

During the Enlightenment, belief in God aroused a lot of discussion. This did not mean that the panoptical view in itself was debated, but that there was a competition as to who was going to occupy the panoptical point: God or Man. To be a humanist, objective, or rational, to be a good politician, a good person, a good legislator and so on, this all meant to think in these terms: if I could govern this all, knew everything, and were not tied by personal interests, what would I ask, want, know, or do? It is this type of reasoning – exemplified by the Categorical Imperative – this type of control, this type of political goal to realize paradise on earth, which is the mark of determinism.

Reading the words of Kant, cited above, the relevant question to ask is: 'whose reason?' In reference to truth, the bearer of this reason was for Kant a cosmopolitan community; in reference to the determination of individual destiny it was the negative freedom of the individual; and in reference to collective action it was the executive power to represent a collective.

Ever since Kant, the whole of Western philosophy has occupied itself with the unification of these three requirements. It can therefore be concluded that – contrary to what Strawson (1973) stated – the humanities have developed their ideas and theories completely and solely in reference to the thesis of determinism, i.e. the thesis of a Collective Mind.[22] The idea of a historical growth in which the three contradictory requirements merge into the 'Enlightened' unity of a Collective Mind has been predominant in the humanities since the end of the eighteenth century. Individual freedom is primarily considered with respect to its function in the development of this collective enterprise.

Just like Kant's theory, the whole of modern Western philosophy reflects a new concept of freedom which was generated during the Enlightenment and came to full flush in the French Revolution: the freedom of the political individual to determine the destiny of the self and the world *as part of a collective*. Authors such as Arendt and Agamben have rooted this concept of the political individual in Aristotle's text[23] (*Pol.* I.i.1253a1–35 ). Indeed it is true that Aristotle stated that a man is a political animal and that the city is prior in nature, implying that there is in every man a natural impulse to form a political partnership. However, for Aristotle politics was not about realising paradise on earth. To him governing was, like any other art, situated in a context. Aristotle explicitly rejected pragmatic science, as this would involve the idea that the human being is the highest being (*EN* VI.vii.1141a19–b8). For Aristotle politics was about habituation. Although Aristotle shares with pragmatism the value that theory is the highest virtue attainable for people, and although both acknowledge the fact that theory can only exist as a collective enterprise, they have a different conception of science: respectively indeterminist/organic and deterministic/mechanistic.

When evolution theory developed in the second half of the nineteenth century, this was the first sign of a return of an indeterminist view in science. This biological indeterminism however was rapidly transformed into social determinism. Even Peirce (1960) defended such a social determinism:

> [i]deas are not all mere creation of this or that mind, but on the contrary have a power of finding or creating their vehicles, and having found them, of conferring upon them the ability to transform the face of the earth. (I, 95)

and

---

[22] Compare the first sentences of Foucault's Political Technology (1988, 145) "The general framework of what I call the 'Technologies of the self' is a question which appeared at the end of the eighteenth century. It was to become one of the poles of modern philosophy (...) The question (...) is: What are we in actuality? (...) Kant, Fichte, Hegel, Nietzsche, Max Weber, Husserl, Heidegger, the Frakfürterschule have tried to answer this question".

[23] More on this Huppes (2011).

> This activity of thought by which we are carried, not where we wish, but to a fore-ordained goal, is like the operation of destiny. No modification of the point of view taken, no selection of other facts for study, no natural bend of mind even, can enable a man to escape the predestinate opinion. (V, 268)

Popper (1972) defended in *Objective Knowledge* an evolutionary approach, which he presented as consistent with Darwinian biology. At first sight his main argument appears to be against determinism. At second sight however Popper is primarily interested in the political freedom of humans to change – as a collective – the face of the earth.

In the next section a comparison will be made between Popper and Aristotle, to clarify why Popper's fight *against* determinism and *for* freedom is, seen from an Aristotelian perspective, a fight *for* determinism and *against* freedom.

### 2.7.1 The Objective Knowledge of Popper versus the Subjective Knowledge of Aristotle

In his youth Popper had attacked the idea of Historicism. But, as Popper (1972, 241) confessed, he ended up more or less with this same Historicism. In his later work he defended his belief in interactionism, i.e.

> the belief that non-physical aspects (aims, purposes, traditions, tastes, ingenuity) play a role in the development of the physical world (footnote 35, 223).

In the Preface of *Objective Knowledge* Popper announces that he will reject the subjectivist tradition that could be traced back to Aristotle and will replace this old theory with an objective theory of essentially conjectural knowledge. In one of the chapters he calls his theory "epistemology without a knowing subject." (106)

His theory involves an evolutionary approach which can be summarised in three arguments: (1) growth of knowledge is a development through natural selection; (2) natural selection can be understood in terms of intelligent design by humanity; (3) intelligent design by humanity can be explained by the propensity structure of clouds. These three arguments will be elaborated below and compared with the subjectivist Aristotelian approach. Before doing so the indeterminist position of Aristotle will be compared to Popper's view.

### 2.7.2 Indeterminism of Popper and Aristotle

Popper (1972, 212) refers to Peirce to explain physical indeterminism:

> I may perhaps quote one of Peirce's brilliant comments: '... one, who is behind the scenes'(Peirce speaks here as an experimentalist) '... knows that the most refined comparisons [even] of masses [and] lengths, ... far surpassing in precision all other [physical] measurements, ... fall behind the accuracy of bank accounts, and that the .... Determinations of physical constants...are about on a par with an upholsterer's measurements of carpets

> and curtains....' From this Peirce concluded that we were free to conjecture that there was a certain looseness or imperfection in all clocks, and that this allowed an element of chance to enter. Thus Peirce conjectured that the world was not only ruled by the strict Newtonian laws, but that it was also at the same time ruled by laws of chance, or of randomness, or of disorder: by laws of statistical probability.

Compared with the explanation given by Aristotle it becomes clear that the indeterminism of Aristotle is much more radical as Popper speaks of imperfection while Aristotle focusses on the perfection that can be realized through individual adjustment. Aristotle (*Phys.* II.v.196b10–viii.199b120, trans. Ross) distinguishes between (1) movements which are '*always the same*', which means that they are '*of necessity*', which refers to mathematical knowledge and eternal things; (2) movements which are '*for the most part*', which means that they are '*by nature*' and '*for the sake of*' *i.e.* functional; and (3) movements which are '*at random*' such as the movements of seeds "among the seeds anything must come to be at random. But the person who asserts this entirely does away with nature and what exists by nature." Within that which occurs driven by functionality, some are '*according to intention*' and some not. This can be interpreted as referring to respectively learning processes and feed-back mechanisms. The way Aristotle describes it allows for the interpretation that he assumed that there is a sliding scale between these two. To speak of '*spontaneity and chance*' means to refer to accidents: *spontaneous* when relating to processes which are not intended and *by chance* when relating to processes in which intention plays a role.

### 2.7.3 Growth of Knowledge

The theory of knowledge which Popper proposes is, according to him, a largely Darwinian theory of the growth of knowledge. He describes (258–260) knowledge as the result of a process closely resembling what Darwin called 'natural selection'. Every animal is born with expectations, which can be framed as hypotheses, and when it is disappointed this inborn knowledge creates the first problems. Knowledge does not start from observations, but from problems, practical or theoretical. While animal knowledge grows mainly through the elimination of those animals that hold unfit hypotheses, human knowledge grows by eliminating hypotheses. Human beings will conjecture a solution for the problem they are confronted with, which they then will criticize:

> From the amoeba to Einstein, the growth of knowledge is always the same: we try to solve our problems, and obtain, by a process of elimination, something approaching adequacy in our tentative solutions. (261)

However at the level of applied knowledge there is a growth of differentiation and specialization,

> [p]ure knowledge grows in a very different way. It grows almost in the opposite direction to this increasing specialization and differentiation. (...) we should have to represent the tree of knowledge as springing from countless roots which grow up in the air rather than down, and which ultimately, high up, tend to unite into one common stem. (262/3)

Popper (263/4) explains this upside down tree as our wish to find *true theories which agree to the facts,* together with the fact that *our curiosity and passion to explain by means of unifying theories* is universal.

A comparison with Aristotle clarifies the difference between subjective and objective knowledge, which Popper pointed out. For Aristotle, knowledge about the world of particulars had primordial meaning. The particular, however, could not be described. The individual is affected by a particular in the form of a double impulse which leaves a range of reactions open to him or her. One can intuitively understand the quality of the individual configuration of one's environment by one's tendency to act in a certain way. Descriptive knowledge is about general things and generates from the comparison of personal experiences. This generality means a loss of 'truth value'. The meaning of general knowledge is that it prevents an indulgence in feelings. General knowledge creates distance and makes it possible to put the (always ambivalent) personal experiences into a broader perspective. To act solely on general knowledge is inadequate (*Met.* I.i.–ii.980a22–982a3).

From an Aristotelian perspective, Popper's idea 'that perception always happens in the light of expectations which can be framed in the general terms of hypotheses', has led to a kind of expropriation of the personal intuitive understanding of things. This expropriation has especially affected the 'arts' and craftsmanship.

### 2.7.4 Intelligent Design

> What Darwin showed us was that the mechanism of natural selection can, in principle, simulate the actions of the Creator, and His purpose and design, and that it can also simulate rational human action directed towards a purpose or aim. If this is correct, then we could say from the point of view of *biological method*: Darwin showed that we are all completely free to use teleological explanation in biology – even those of us who happen to believe that all explanation ought to be causal. For what he showed was, precisely, that *in principle* any particular teleological explanation may, one day, be reduced to, or further explained by, a causal explanation (…) we have to add that the phrase 'in principle' is a very important restriction. Neither Darwin nor any Darwinian has so far given an actual causal explanation of the adaptive evolution of any single organism or any single organ. All that has been shown – and this is very much – is that such explanations might exist (that is to say, they are not logically impossible). (Popper 1972, 267)

Popper more or less follows here a line of reasoning, which is also used by Hempel (1959, 122) in one of his essays in which he treated the question whether teleological accounts must be seen as pseudo-explanations.[24] Hempel explained that

[24] "A magnetic field is not directly observable any more than an entelechy; but the concept is governed by strictly specifiable laws concerning the strength and direction, at any point, of the magnetic field by a current flowing through a given wire and by other laws determining the effect of such a field upon a magnetic needle in the magnetic field on the earth. And it is these laws which, by their predictive and retrodictive import, confer explanatory power upon the concept of a magnetic field"(122).

the kind of phenomenon which people want to explain by a teleological – functional – account is

> [s]ome recurrent activity or some behavior pattern in an individual or a group; it may be a physiological mechanism, a neurotic trait, a culture pattern, or a social institution, for example. And the principal objective of the analysis is to exhibit the contribution which the behavior pattern makes to the preservation or the development of the individual or the group in which it occurs. (123)

Hempel explained that the functional account involves the fallacy of affirming the consequence in regard to the premise (127) or is trivial. Only with additional knowledge (130) or further specification of the statements (134) can it get explanatory import. Hempel concluded that

> The preceding considerations suggest that what is often called 'functionalism' is best viewed (…) as a program for research guided by certain heuristic maxims or "working hypotheses." (142)

In an attempt to exhibit the value of science for the preservation and development of human society, Popper refers to the argument – also used by Hempel – that only a causal explanation can establish truth value. However he ads something to this account and is quite aware that this would be highly objectionable to many biologists (and certainly also to Hempel, LHC). He states that Darwin's idea of natural selection suggests – and thus explains – the existence of a 'strong tendency or disposition or propensity to struggle for survival', which becomes part of the genetic structure of all organisms and which shows in their behavior and in much if not all of their organization (268).

A comparison with Aristotle highlights the following points. Aristotle was not acquainted with paleontology, which forms the hard backbone of Darwinian analyses. He did not think in terms of the struggle of species (according to Aristotle these are speculative rational reconstructions) or the history of the world. He did think however about the conceptual difficulties of theories about the generation of particulars (*Met.* II.i.–ii. 993b–994b33). These difficulties were the main reason for Aristotle to turn from efficient causation to final causation. The difference between Darwin and Aristotle is primarily that for Darwin the moving force is 'need', a one-sided impulse of sexual desire and wish for food, while for Aristotle the moving force is 'love', a double-sided impulse leaving a wide range of possible reactions open.[25] The difference between Popper and Aristotle is greater because it involves

[25] Darwin stated in the preface of his Origin of Species that Aristotle's Physics foreshadowed the principle of natural selection, but that his remarks on the formation of teeth show how little Aristotle fully comprehended the principle. Indeed Aristotle's account on the formation of teeth doesn't say anything about the competition of species or about the origin of species, it gives an account of the fact that the natural growth in particulars presupposes a kind of tendency or orientation: "Similarly if a man's crop is spoiled on the threshing-floor, the rain did not fall for the sake of this – in order that the crop might be spoiled – but that the result just followed. Why then should it not be the same with the parts in nature, e.g. that our teeth should come up of necessity – the front teeth sharp, fitted for tearing, the molars broad and useful for grinding down the food – since they did not arise for this end, but it was merely a coincident result; and so with all other parts in which

the replacement of individual feed-back and learning mechanisms by collective feed-back and learning.

### 2.7.5 Central Propensity Structure

Popper wondered how the small deviations which accidentally occur come to be used by an organism. Most deviations will be lethal. He concluded that only a central propensity structure – a (collective, LHC) Mind – which includes an aim-structure and a skill-structure, can guarantee the use of such deviations (275). He explained this central propensity structure initially by way of the metaphor of an 'automatic pilot'.

In later work he connected this idea with the distinction between determinism and indeterminism. For him, determinism was a nightmare because it destroys the idea of creativity, the idea that the brain can create something new (222). He envisaged determinism in terms of the metaphor of a clock (physical systems which are highly predictable in their behavior) and indeterminism in terms of clouds (disorderly and more or less unpredictable physical systems): the solar system is closest to the clock metaphor and a cloud of small gnats closest to the other extreme (207).

Popper used the concept of a cloud to elaborate the idea of a central propensity structure, which exercises a plastic control (not by force, but by feedback or learning LHC):

> Like the individual molecules in a gas, the individual gnats which together form a cluster of gnats move in an astonishingly irregular way. It is almost impossible to follow the flight of any individual gnat, even though each of them may be quite big enough to be clearly visible. (...) Their keeping together can be easily explained if we assume that, although they fly quite irregularly in all directions, those that find that they are getting away from the cloud turn back towards that part which is densest. (...) This assumption explains how the cluster keeps together even though it has no leader, and no structure – only a random statistical distribution resulting from the fact that each gnat does exactly what he likes, in a lawless or random manner, together with the fact that he does not like to stray too far from his comrades. (...) Yet the cluster of gnats is an example of a whole that is indeed nothing but the sum of its parts – and even in a very precise sense; (...) the movement of the whole is, in this case, precisely the (vectorial) sum of the movements of its constituent members, divided by the number of the members. (208–210)

Before turning to a comparison with Aristotle, some remarks have to be made about the concept of a cloud as developed by Popper. Animation techniques have resulted in a different view on vector movement: not the (common) wish to be near

---

we suppose that there is purpose? Wherever then all the parts came about just what they would have been if they had come to be for an end, such things survived, being organized spontaneously in a fitting way; whereas those which grew otherwise perished and continue to perish, as Empedocles says his 'man-faced oxprogeny' did. Such are the arguments (and others of the kind) which may cause difficulty on this point. Yet it is impossible that this should be the true view" (*Phys*. II. vii–viii 198b1–199a1, trans. Ross).

to the densest part is the essential characteristic but the fact that every single gnat wants to stay near to the gnats nearest to it. When some start to move for some reason, the rest follows as a consequence of this characteristic. Although Popper stated that his concept of a cloud is an example of a whole that is nothing but the sum of its parts and not the kind of holistic whole which he had attacked in his studies on historicism, his belief in a kind of control issued by the densest part constitutes just this kind of holistic whole and is explicitly used by him as a model for human government.

Compared to Popper, Aristotle had quite another view on creativity, since for him human action and human freedom were part of nature.

> Thus if a house, e.g. had been a thing made by nature, it would have been made in the same way as it is now by art; and if things made by nature were made not only by nature but also by art, they would come to be in the same way as by nature. The one then is for the sake of the other; and generally art in some cases completes what nature cannot bring to a finish, and in others imitates nature. (*Phys*. II.viii.199a1–20, trans. Ross)

The influences of human action will therefore be of a kind which can also be issued in other ways by other natural phenomena.

Aristotle's view on desire and love easily fits in with the interpretation of vector movement, as proposed by animation techniques: as far as there is social cohesion in a group, this will be the result of the quality of the relations between people who interact, and not of a collective identity or intentionality. For Aristotle the simple formula 'Love your neighbour' will be counterbalanced by a contradictory impulse, which can cause the group to fall apart. Habituation creates stability as it constitutes a common frame of reference, which enables communication about personal experiences.

## 2.8 Determinism and the Concept of Law, a Few Conclusive Considerations

Traditionally there are two concepts of law: law as the codified ruling of an authority, which claims obedience, and law as a written academic tradition, which is instructive as a conceptual frame of reference for anyone who has to make agreements with others or has to account for decisions in conflicts.

As a result of the introduction of the concept of truth during the Enlightenment, the academic tradition of law has been incorporated and usurpated in the codified ruling which claims obedience. The legal study has gradually lost its status of being academic, of being a (practical) science. The search for and production of scientific truth replaced the academic tradition of law as the main source of instruction for Governments. The scientific discussion annexed the themes and problems of the academic tradition of law, while vice versa the political discussion annexed the academic discussion of science. It has become impossible to discuss human affairs without ending up with conflicting claims about "what is shown by empirical research." The scientific discussion in turn has become completely contaminated by vested interests.

Although it was already clear from the start that determinism was untenable,[26] even today people normally talk about scientific truth and the difference between norms and facts as if they believe in determinism. The main reason may be that they are afraid to promote relativism and irrationalism.

Aristotle would agree with them. In the quote with which this chapter starts Aristotle describes the inhabitants of the cold places of Europe as free. They are free but without any political organisation, like the gnats in a cloud. The Asians however are cultured and have political organisation, but they are like slaves. It is important, according to Aristotle, to make a mix of these two forms of being: free and/or rational.

Aristotle distinguished between the art of definition (method) and metaphysics (ontology). When people do not practice the art of definition, relativism and irrationalism will reign. However, when people reify their ideal types or definitions they will lose contact with reality, they will be enslaved by their own definitions.

Determinists think that legal norms are either describing regularities in behaviour or prescribing behaviour. They thus reify the definitions of law. From the perspective of law as an academic tradition, norms are ideal types, just like circles for mathematicians. They create a conceptual world. The legal ideal types do not correspond with reality, but they are necessary to be able to talk and think about right and wrong. To create political order people will have to practice the art of definition.

Within the context of a set of definitions, one can speak of truth or justice being established. But to keep contact with the real world, people will have to study ontology. They will have to try to understand the world at large, with all the variability and complexity it contains. At this level there is no truth or certainty to be obtained. At this level the arbitrariness of all definitions hits the mind. This is why wisdom is so important for Aristotle and why the exercise of power has to be embedded in the study of ontology.

Aristotle valued science highly. He realised that science can only proceed when clear definitions are established, but that at the same time these definitions endanger freedom and contact with reality. This insight was generally shared by the Greeks and has resulted in the Antique world in the design of constitutions with an impressive array of measures and institutional arrangements to guarantee that the exercise of authority was continuously checked and counterbalanced. The Enlightened belief in (objective) truth has promoted the view that all these measures and institutional arrangements are hindrances to an effective and evidence based ruling.

The academic tradition of law has been the main moral force in Europe in old times and could be so in the future in a global society. To summarise the characteristics of this concept of law: Clear and definite definitions can only function when

[26] As Popper (1972, 212 ftnt 11) reveals "Newton himself may be counted among the few dissenters, for he regarded even the solar system as *imperfect*, and consequently as likely to perish. Because of these views he was accused of impiety, of 'casting a reflection upon the wisdom of the author of nature'."

they are not reified. Only then a common understanding and discussion can be free from vested interest. The exercise of power has to be divided, checked and balanced. Only those citizens are free, that take actively part in the many civil services that are needed in a society and exercise power accordingly. The rule of law reigns when the people who exercise power let themselves be instructed by the definitions of the law, but are at the same time conscious of the gap between these definitions and the real world. Only those, who struggle to bridge this gap and reflect about their actions, acquire a more general understanding of human life as such. This understanding is the essence of happiness.

Traditionally authoritative ruling has been part of the law. This is the public law, which is enforced by violence and administrative forms or procedures. Public law claims obedience, but this claim cannot be legitimated on a worldwide scale. This means that this type of law can only function on a global level as far as effective control is possible. Law as an academic tradition does not claim obedience. It creates a collective asset which people can use or not: a Collective Mind without aspirations to be a collective actor, who determines the world and changes the face of the earth.

The discovery of the societal function of a common conceptual order, has been the great invention of the Greeks, brought to theoretical understanding by Aristotle. This is the civil law, to which people voluntarily comply as far as they feel instructed by it, meaning that they experience these definitions as helpful to get a better understanding of reality. The creation of sets of definitions which have coherence and are used by generations of people, is the work of great scientists, who have a deep understanding of the formative principles in their field. Great legislators, such as Solon and Lycurgus, were such scientists.

To know a set of definitions is not enough. One cannot acquire knowledge of the actual world by simply learning the definitions, *i.e.* by reading the code. One has to study hard cases and discuss them against the background of the frame of reference offered by the definitions, in cooperation with people who have done so already for many years. This way the ideal types of the codes are 'filled' with experiences. It has been the great contribution of the Romans to have developed such a legal practice.

Legal theorists are amazed that Roman Law has been a great moral and political force in the mediaeval society which was so completely different from the Roman society. But the conceptual world created by Law as an academic tradition doesn't determine behaviour and doesn't create expectations, like the Collective Mind of determinists like Popper. According to Aristotle the abstract formative principles of human dealings can materialise in many ways. Behaviour is explained by the fact that feelings just happen to people and that their behaviour is stimulated by these feelings. The conceptual world makes it possible to talk and reason about what happens. It makes it possible for people to think their own existence and this again makes them happy and gives them a feeling of completeness which makes them calm, moderate and strong.

## Bibliography


Aristotle. 1933/1935. *Metaphysics*. Trans. H. Tredennick. Cambridge MA/London: Harvard University Press, Part I 1933, Part II 1935.

Aristotle. 1944. *Politics*. Trans. H. Rackham. Cambridge MA/London: Harvard University Press.

Aristotle. 1955. *Nicomachean ethics*. Trans. J.A.K.Thomson, and H. Tredennick, ed. J. Barnes. Harmondsworth, Middlesex: Penguin Books.

Aristotle. 1984. Physics. In *The complete works of Aristotle: The revised Oxford translation*, ed. J. Barnes and Trans. W.D. Ross. Princeton: Princeton University Press, 315–447.

Averroes. 1969. *Averroes middle commentary on Porphyry's Isagoge and on Aristotle's Categoriae*. Trans. H.A. Davidson. Cambridge, MA: The Mediaeval Academy of America.

Barnes, Jonathan. 1955. Introduction. In *Nicomachean ethics*. Harmondsworth, Middlesex: Penguin Books, 9–43.

Berger, Peter. 1966. *Invitation to sociology, a humanistic perspective*. Harmondsworth, Middlesex: Penguin.

Boethius. 1969. *The consolation of philosophy*. Trans. V.E. Watts. Harmondsworth, Middlesex: Penguin Book.

Briegel, Hans J. 2012. On creative machines and the physical origins of freedom. *Scientific Reports* 2: 522. doi:10,1038/srep 00522.

Brinkmann, Klaus. 1996. The consistency of Aristotle's thought on substance. In *Aristotle's philosophical development, problems and prospects*, ed. W. Wians. London: Rowman and Littlefield.

Cicero. 1928. *De Re Publica*. Trans. Clinton Walker Keyes. London/Cambridge, MA: Harvard University Press.

Foucault, Michel. 1979. *Discipline and punish, the birth of the prison*. New York: Vintage Books.

Foucault, Michel. 1988. The political technology of individuals. In *Technologies of the self*, ed. Luther H. Martin, Huck Gutman, and Patrick H. Hutton. London: Tavistock Publications.

Graham, D.W. 1987. *Aristotle's two systems*. New York: N.Y. Oxford University Press.

Habermas, Jürgen. 1981. *Theorie des Kommunikativen Handelns*. Suhrkamp Verlag: Frankfurt am Main. Band I and II.

Hart, H.L.A. 1961. *The concept of law*. New York/London: Oxford University Press.

Hempel, Carl G. 1959. The logic of functional analysis. In *Readings in the philosophy of science*, ed. Baruch A. Brody, 121–147. Englewood Cliffs: Prentice Hall.

Hobbes. 1968. *Hobbes Leviathan*, ed. C.B. Macpherson. Harmondsworth, Middlesex: Penguin Books.

Huppes-Cluysenaer, E.A. 2007. Towards a theory of new natural law as a basis for future legal positivism. In *Legal theory, legal positivism and conceptual analysis*, ed. José Juan Moreso, pp. 107–117. Proceedings of the 22nd IVR World Congress 2005. Stuttgart: Frans Steiner Verlag.

Huppes-Cluysenaer, E.A. 2008. Turning the state into a household. In *Legality, legitimacy & modernity, reconsidering Max Weber's concept of domination*, ed. E.A. Huppes-Cluysenaer et al., 125–156. 's-Gravenhage: Reeds Business BV.

Huppes-Cluysenaer, E.A. 2011. The fallacy of continuity. *Ethic@*, 223–253.

Jaeger, Werner. 1945. *Paideia: The ideals of Greek culture*, vol. III. Oxford: Basil Blackwell.

Kant, I. 2010. *Critique of pure reason*. Trans. J.M.D. Meiklejohn, Penn State Electronic Classics Series Publication. http://www2.hn.psu.edu/faculty/jmanis/kant/critique-pure-reason6x9.pdf.

Kenny, Anthony. 1978. *Freewill and responsibility; four lectures by Anthony Kenny*. London: Routledge & Kegan Paul Books.

Marcuse, Herbert. 1955. *Eros and civilisation*. Boston: Beacon.

Nagel, Thomas. 1987. *'Free Will': What does it all mean? A very short introduction to philosophy*. Oxford: Oxford University Press.

Ockham, G. 1974. *Ockham's theory of terms, Part 1 of the Summa Logica*. Trans. Michael J. Loux. Notre Dame/London: University of Notre Dame Press.

Peirce, C.S. 1960. *Collected papers of Charles Sanders Peirce. Part I Principles of Philosophy, Part V Pragmatism and Pragmaticism*, ed. Ch. Harthorne and P. Weiss. Cambridge, MA: Harvard University Press.

Popper, K.R. 1972. *Objective knowledge, an evolutionary approach.* 'Evolution and the tree of knowledge' (256–285) and 'Of clouds and clocks' (206–256). Oxford: Clarendon Press.
Pullman, Bernard. 1998. *The atom in the history of human thought.* Trans. A. Reisinger. Oxford/ New York: Oxford University Press.
Ruler, J.A. van. 1995. *The crisis of causality, voetius and descartes on God, nature and change.* Doctoral thesis, University of Groningen.
Segalerba, Gianluigi. 2011. Ohne Staz vom Widerspruch keine Entität – der Satz vom Widerspruch als Strukturformel der Realität. *Journal of Ancient Philosophy* V(2): 1–57.
Sokal, Robert R. 1974. Classification: Purposes, principles progress, prospects. *Science* 185(4157): 1115–1123.
Strawson, P.F. 1973. Freedom and resentment. In *Freedom and resentment and other essays,* 1–25. Oxford: Oxford University Press.

# Chapter 3
# Law and the Rule of Law and Its Place Relative to *Politeia* in Aristotle's Politics

**Clifford Angell Bates Jr.**

## 3.1 Introduction

When scholars and statesmen speak of the rule of law, they often talk about it as though it is a particular form of government. In fact Herodotus argued this being ruled by law rather than by the will of the ruler (*isonomacy*) is what distinguished the Greeks from the non-Greeks (Arendt 1958 or Finley 1985). Yet Aristotle in his *Politics*, which is the most complete discussion found in the Classical world of the forms of political rule (also known as *politeia* or regimes) make no special mention of the rule of law as a form of political rule (or type of regime). In fact there is a clear suggestion, that in the political thought of Aristotle, the rule of law is rather something common to all decent forms of regime.

This paper will focus in on Aristotle's treatment in his *Politics* of the question of law and the problem of the tension between the rule of law and the rule of rulers simply. This paper hopes to return to the original teaching of Aristotle's text and not hold on to the received interpretations that so powerfully shape our understanding of the question of law in the *Politics*. We children of the modern world see politics as a product of a unified whole, a process of the joining of a body politics out of voluntary monads who join together to form a unified community in order to escape the harsh realities of nature that the pre-political world leaves human beings within (contrast Holmes 1979; see Arendt 1958). This is the world that Bodin, Grotius, Hobbes, Locke, Rousseau, Kant, Hegel et al. give to us and our understanding of the modern state.

But for Aristotle the political community is not the state and when we apply to the polis the understanding we have of the state, we impose on this political community the impulse to unitary cohesion of a willing being, possessing the solitary

C.A. Bates Jr. (✉)
Institute of The Americas and Europe, Uniwersytet Warszawski,
Al. Niepodleglosci 22, Warsaw 02-653, Poland
e-mail: c.a.bates@uw.edu.pl

L. Huppes-Cluysenaer and N.M.M.S. Coelho (eds.), *Aristotle and The Philosophy of Law: Theory, Practice and Justice*, Ius Gentium: Comparative Perspectives on Law and Justice 23,
DOI 10.1007/978-94-007-6031-8_3, 

will of a mentally healthy creature that is whole. Rather the polis of Aristotle is a discrete community (*koinonia)* of other lower order communities (one of these lower order communities is the household (*oikos*) (see Nagel 2006)). Given the fact that the political community is this collection of interacting communities, the question then becomes which part of this collective is the ruling part and what justifies their rule over the whole community. This is the heart of what drives classical political order and indicates its different character from modern politics with the latter's central concept of the state that is a product of the willing of its originally autonomous parts that formed it and constituted its coming into being. Thus political community of this pre-modern classical political model constantly has the need to explain and legitimate its claim of why this given part rules over the other parts. And this is why the *politeia*, the regime, plays such a key role in helping to understand the character of any given political community. The *politeia,* the regime tells us which part of the community rules and why it rule.

## 3.2 The General Character of Regimes

In *Politics* 3, only after five chapters discussing the citizen (which in this context means one who shares in rule in the political community), which presupposed but did not discuss the regime, do we get a discussion of the concept of the regime, or *politeia*. Aristotle begins *Politics* 3.6 with an injunction:

> Since these things have been discussed, what comes after them must be investigated – whether we are to regard there being one regime or many, and if many, which and how many there are and what the differences are between them. (3.6.1278b6–8, trans. Carnes Lord[1])

A discussion of the regimes and their number necessarily follows a discussion of citizens. Why? As I have already argued, the discussion of citizens presupposed but did not spell out the importance of the regime in understanding fully what it means to be a citizen (Bates 2003). In other words, the discussion of the regime should have come before the discussion of the citizen so that that discussion would not have been as troublesome as it was.

The question of the regime concerns not only their number – i.e., how many regimes are there – but also the differences among them (3.6.1278b7–8). This is the question that Aristotle is pursuing in *Politics* 3.6. He appeals to the heart of the matter by describing what a *politeia* is: "an arrangement of a *polis* with respect to its offices, particularly the one that has authority over all [matters]" (3.6.1277b9–10). For what has authority in the *polis* is everywhere the governing body (*politeuma*), and the governing body is the regime. I mean, for example, that in democratic regimes the people have authority while, by contrast, it is the few in oligarchies. The regime, too, we say, is different in these cases; and we shall speak in the same way concerning the others, as well (3.6.1277b10–14).

[1] All Aristotle quotations in this chapter are generally from Carnes Lord (1987) with slight modifications based on differences in reading the Greek text of Aristotle 1957, edited by Ross.

Thus, to understand what type of regime one is dealing with one must first examine what type of governing body it has. But, for Aristotle, there are many types of governing bodies and, therefore, there are many types of regime. But how does the governing body – understood as the regime – form the *polis*, or political community. Leo Strauss one of the leading scholars dealing with the recovery of classical political philosophy in the early to mid-twentieth century, has said that the relationship between the *polis* and the regime is one of matter and form, where the *polis* is the matter and the regime shapes it (Strauss 1989, 32; 1978, 45–47). Thus to understand how the regime gives shape to the political community we need to address the question of form.

Again, the regime is that which gives form (*eidos*) to the particular political community. The particular form of a regime will by definition imply a different *telos*, or end, which that regime will hold as its authoritative way of life.[2] This is to say, the form a regime will have will structure the authoritative body within that given political community. In doing this, the way of life of those who have authority will become authoritative, or normative, within that political community. The structure, or form, of a regime allows one access to the *telos* of the regime. Therefore, different regimes will have different forms and, because of their different forms, they will have differing ends. These differing ends will lead to differing authoritative ways of living (see Strauss 1989, 32, 59–79; 1978, 31–33), which leads in turn to differing understandings of justice (see Strauss 1978, 47–51).

It is reasonable that the difference of form will lead to a difference in ends. Each regime, because it has a different form, will have a different end. Differences in ends are to be understood as differing conceptions of ways of life and so, by implication, differing conceptions of what is just. Because of this, there exists a great variety of possible regimes – one for each possible variation in form. But, in practice, the possibilities of the variety of forms which the regime may have are limited to the various social elements which exist within the political community.[3] So, for Aristotle, there exists, in understanding human political communities, a radical dependency upon form, which arises from the interdependency between form and end (see Zuckert 1983; Nichols 1991).

The regime is what gives any political community (and here for Aristotle the given political community is the polis, the city) its particular character or its true

[2] See Swazo (1991, 405–420) and Strauss (1978, 47–48). Also it is important to recall the connection between Aristotle's teleological treatment of the political community and his biology. Aristotle uses his biological teleology in his understanding of human political activity, see Arnhart (1990, 1994), Nussbaum (1994, 477–488) and Masters (1989). Both agree with Arnhart's claim regarding the biological take on Aristotle's use of teleology. Given this, one should refer to what role *eidos* plays in Aristotle's biology. For the best treatment of the role of *eidos* in Aristotle's biology, see Preus (1979).

[3] See Quinn (1990, 170–186) on the relation between the parts to the whole in *Politics* 3. Also, Saxonhouse 1992, 215–218 discusses how the parts relate to the whole. She argues that the parts are only meaningful in relation to the whole. The example she uses is that of the hand. When it is separated from the whole body, it is no longer truly a hand. The nature of the hand is seen in its functional relation to the whole.

shape. The regime is the form (*eidos*) that has authority over the whole community and gives the polis its true character. The specific character of a regime leads to the "true" differences between the different political communities. Thus it is right to say that the regime is truly the stuff that defines how politics works within the political community in Aristotle.[4]

## 3.3 The Place of Law in the Politeia

From this picture of the *politeia*/regime, we now come to ask – what role does law play here? The clear teaching of Aristotle on the nature of law is that law is something that is shaped by the regime (*politeia*) of the political community and not a form of political community. For Aristotle, and Plato and Socrates as well, laws are authoritative opinions about what is just and right of a given political community. Laws are the particular expression of what a given political community holds to be what is justice and the right way of living and shaping the life of the whole community. Now given that law would be the city's/political community's authoritative opinions (or orthodoxy – which in Greek means right/correct opinion) about what is just, the relationship between law and justice, is akin to the relationship between what is by nature (*phusis*) and convention/law (*nomos*), as well as the relationship between what is the true (*aletheia*) and what is opinion (*doxa*). While the just is always the target of what law aims at, law as law is more like *doxa* (an opinion) rather than the truth of what nature holds to be the just simply.[5]

Justice, as such, is thus found in nature, that is to say the nature of things, especially the nature of man. Thus justice is something that is there to be found within the nature and not something to be imposed upon from something outside or beyond. And being there in the nature of things as such, it is what is true simply (see Strauss

---

[4] As to the question of the mixed-regime, I have argued in Bates (2003, 102–121), that, given the logical character of Aristotle's teaching of the regime, the idea of a specific type of regime named 'regime' that is different from other regimes by its being a mixture of two different regime principles (views of justice) is rather unlikely and problematic. Now I hold that it is not possible for a regime to hold two usually logically opposite views of the just without having a kind of schizophrenic regime order. Also if we look carefully at Aristotle's teaching about the nature of regimes in *Politics* 4 we see that all regimes are in fact not pure but a mixture. Thus it the crux of my argument that the traditional view of polity as the mix-regime and a regime of the rule of law versus the rule of the people is one incorrectly attributed to Aristotle. For the traditional view see Mulgan (1977), Finley (1985), Johnson (1988), Stoker and Langtry (1986), and Bluhm (1962). Whereas Blythe (1992) and Fritz (1954) present alternative origins to the concept of the mixed regime or mixed constitution, found either in the political thought of the Middle Ages for the former and Polybius' teaching of the political regime of Rome for the latter. Also see Cherry (2009) and Ewbank (2005) who are responding to aspects of what I argued in Bates (2003) on the issue.

[5] See Strauss (1953, 1988, 1978) for a systematic presentation of the nature of the regime in relation to the political community, and the relationship between the law of the given regime and the just by nature.

1953; Voegelin 1978, 53–70). Now in that law is to be understood as an opinion about the just that a given political community holds to be true. Thus the political community holds that what it holds as law truly encompasses what nature holds to be true about the justice of the matter in question. Also we must note that law is something more powerful than mere opinion, in that it is something the political community holds to be true as what are the key beliefs about what the given political community is all about—the justification of its particular way of living and its claim about why its rule is the best type of rule for those of its subjects and citizens. Hence law is an authoritative opinion of a political community about the just and the unjust and about what the given political community [and/or ruling part of that community] believes true about itself.

Also for Aristotle law is not something trans-political, but rather sub-political. Yes laws can be found in most if not all regimes. Yes it's something common to all regimes. But all laws are not the same and how they will differ from one another is found in the differences among regime types. Thus laws are relative to the given type (or variety) of regime. This is to say that different regimes will give shape to different laws—the laws of a democracy are different from the laws of an oligarchy or a tyranny. Also, although it is true that not every democratic regime will have exactly the same laws, yet the laws of various democratic regimes will have much more in common with each other than to the laws of an oligarchy or a tyranny or any other different regime.[6]

So from this reading of the nature of law as something shaped by the regime (*politeia*) of a given political community, echoes the teaching that concludes *Politics* 1 is about the household (*oikos*) and its relationship to the political community. Aristotle at the end of the *Politics* 1 will make the case that although the household (*oikos*) is a common unit found in all political communities, he found that, contrary to what others have said, the rule of a household and political rule differ in kind rather than in degree and this points to the fact that the nature of the household in a political community will differ among the different political communities by the difference in regime type more than by anything else. Thus for Aristotle law has a similar relationship to the household (*oikos*), in that its character and nature will depend on the given regime type.

Now some might claim that this view is rather one-sided. They might argue that rather it is things from below like the household or the laws that truly shape and give character to the given political form (its regime/*politiea*) of a given political community. This problem leads to a classic "chicken or egg" problem. And yes

---

[6] Looking at the relative character of law to the given regime will let us see the problem of the differing character of the citizen among regimes. Since law will shape the character of the citizen and different regimes will have different laws (and laws that are at odds with what other regimes hold to be good or just) different regimes will lead to a different understanding of what is the good citizen. Thus a good citizen, who is one who not only obeys the law but is so shaped in his/her character that the laws are perfectly reflected in it, will differ from regime to regime, and a good citizen in one regime would not be a good citizen in another (the good Nazi may be a good citizen in 1930s Germany, but he would not be a good citizen in 2012 America or 2012 Germany).

alterations in household structures and character or changes in the law can and often do lead to changes in the type of regime the political community has. But this does not mean that either the household or the law is more authoritative than the regime. Rather it suggests that those changes in the household or law stem from the fact that the given actors responsible for those changes hold a view of justice and right that is at odds with the view held by the given political community that they seek to change. Aristotle teaches in *Politics* 5 that political change or revolution often arises from small things as much as larger ones. Thus alterations in the laws and the household that lead to regime change, large or small, come from the principle underlying new household habits or norms or the new given laws are different from the given principle that underlie the regime of the given political community.

## 3.4 Laws and the Question of the Best Man

From what we said above let us move first to an examination of the question of the rule of law in *Politics* 3 about the rule of law and whether it is superior to the rule of the best man or men. After that examination, we will turn to the way law shapes the character and variation within a given regime type as shown by the variation within regime types as pointed out in *Politics* 4 (see Quinn 1990).

In the last four chapters in Book 3, Aristotle presents a debate concerning whether law or best man should rule. A closer examination of this debate would be very useful for us to see a clear picture of how Aristotle understands the limits and true value of the rule of law as a political concept. At the start of the debate Aristotle opens up with the question, "is it more advantageous to be ruled by the best man or by the best laws?" (3.15.1286a8–9). This question begins with a dialogue between a partisan of the laws and a partisan of the best man, where one side argues for the rule of law and the other for the rule of the best man.

The partisan for the best man puts forth the argument against the laws: "the laws only speak of the universal and do not command with a view to circumstance" (3.15.1286a10–11). The laws cannot be superior since they only speak generally. Also, to rule in accordance with the written laws, argues the partisan for the best man, is foolish because it would be like requiring a doctor to treat sick people by a written set of instructions ("as it is done in Egypt") without regard to the individual circumstances of the patient in question (3.15.1286a12–15).

Another problem is that the laws cannot simply address problems that arise out of the consequences of implementation. That is, the laws cannot give order to what comes from the laws (3.15.1286b10). These objections suggest that something other than the laws needs to guide what the laws themselves cannot directly control or provide (see Bodeus 1993, 54–57, 1991; Yack 1993, 175–208).

Judith Swanson reads "law" in this section of the *Politics* text not to refer to that derived from custom and the political character of the regime, but from natural law (Swanson 1992, 98–101). Such a position seems a gross misrepresentation of the text. In Swanson's correct reading of Book 3.15–17 of the *Politics* – the debate between the rule of law and the rule of the wise king – Aristotle sides politically with the rule of law over that of human will, regardless of how wise or noble that ruler can be. In this she is correct. But she goes on to argue that the law being advocated in this debate is natural law rather than everyday, conventional law, *nomos* (Swanson 1992, 98–106). Putting aside the problems with ascribing a natural law teaching instead of a natural right teaching to Aristotle's Politics, the text in question – *Politics* 3.15–17 – uses law (*nomos*) in its conventional meaning, or customary law. Although conventional law may be in accord with natural law, or the principle of natural right, it is not natural law per se.[7]

If Aristotle were arguing for the rule of natural law in this text, he would not have allowed the rule of law argument to win, because the rule of law argument is in fact the continuation of the democratic argument earlier in Book 3. Instead, if he were supporting an argument for natural law/natural right, he would have let the argument for the absolute rule of the wise king defeat the rule of law argument. The absolute rule of the wise king would seem to be the perfect embodiment of the rule of natural law/natural right (see Strauss 1953; Voegelin 1978, 55–70; Rhodes 1991).

To restate the argument made against the laws by the partisan of the best man: first, the difficulty of the laws is that they speak generally and, second, because they speak generally, they do not attend to the particular circumstances. Hence, the partisan for the best man concludes that the "regime" of written laws cannot be best. In response to the attack on the rule of law, the partisan for the laws declares that "what is unaccompanied by the passionate element generally is superior to that in which it is innate" (3.15.1286a16–18). He argues that passion is not present in law but is necessarily possessed by every human soul.

The partisan for the best man interrupts, stating that such indeed is the case but that this problem is addressed "by the fact that he [the best man] will deliberate in a finer fashion concerning particulars" (3.15.1286a20–21). Also, as Aristotle clearly states that "the ruler must necessarily be a legislator, the laws must exist but they must not be authoritative" (3.15.1286a22–23). The laws cannot be authoritative because circumstances change. Issues of justice tend to admit of degrees of variation in circumstances that affect the outcome of the judgment. Laws cannot be authoritative because they are dependent on the particular type of regime a polis happens to have. The laws of a democracy are fundamentally different from those of an oligarchy, an aristocracy, or a kingdom. The same is true for the offices. The regime itself is prior to both the laws and the offices and is thus fundamentally more authoritative than either.

---

[7] See Strauss (1953) for a presentation of a teaching about what is just by nature in the history of political thought from the Ancients to the Moderns. The latter hold that the doctrine of natural rights is merely the modern variety of natural right. What is meant by natural right here is not a teaching about natural rights but about what is just or correct (right) by nature. Often there is an assumption that what is meant by natural right is akin to what Catholic political thinkers (especially St. Thomas) teach about Natural Law, but for the Greeks the very concept of natural law is a contradiction in terms in that nature (*physis*) and law (*nomos*) are opposites, as the latter concept is a product of human making or human willing, whereas the former is that which simply is, either in terms of the nature of things or the nature of a particular thing.

The partisan for the best man admits the need for the laws but claims that they ought to be subordinate to the best man because he is best able to deliberate about circumstances, whereas the laws cannot. The laws cannot change themselves. Because what is right and wrong is determined by the given circumstances, the possibility arises that the laws may be in contradiction to what is right. Once the laws deviate from what is right, they become unjust. Therefore, the possibility of the unjustness of the laws supports the claim for the rule of the best man.

The partisan of the laws then asks, "as regards the things that law is unable to judge either generally or well, should the one best person rule, or should all?" (3.15.1286a22–25). The partisan for the laws, noting that the laws can at times be unjust and may be unable to deal with specifics, changes the question. He asks who is a better judge, the best man or the many? (3.15.1286a25). In response to this question, we see that the partisan of the laws also reveals himself to be a partisan of the many.

The argument for the laws is in fact the justification for the rule of the many over the laws, regardless of the best man's character. The partisan of the laws notes that any single person taken separately, like most human beings, might (or even, will) be inferior to the best man (3.15.1286a27). But, he argues, "the city is made up of many persons, just as a feast to which many contribute is finer than a single and simple one, and on this account a crowd also judges many matters better than any single person" (3.15.1286a26–31). Here the partisan of the laws argues that the numerical strength of the many makes up for the defects of single individuals, and together the many's collective strengths will exceed even the best man's. This is similar to the argument made at *Politics* (3.11.1282a13–19).

The partisan of the laws goes further by arguing that the many are less corruptible than the one best man (3.15.1286a32). This is so, he argues, because they are like "water" and, as such, are "more incorruptible than the few." The judgments of a single person are necessarily corrupted when he is dominated by anger or some other passion of this sort, whereas it is hard for all to become angry and err at the same time (3.15.1286a33–35).

The partisan of the laws seems to make a comparison between the laws and the many. They (both the laws and the many) are said to be less corruptible by the passions than is the one best man. This is the case for the many because, to restate, it is harder to corrupt them than it is to corrupt one man. The partisan of the laws does not say that it is impossible for the many to be corrupt or to become angry, but that it is harder to make them corrupt or angry.

Experience tells us, however, that a corrupt people can be far worse than any single tyrant. Publius, in the *Federalist Papers*, clearly indicates this, by his concern about majority tyranny. But the partisan of the laws does not exaggerate the many's incorruptibility, so he limits the many: they "must be free people acting in no way against the law, except in those cases where [the law] necessarily falls short" (3.15.1286a36–37). The multitude ought to consist of the free men who do nothing against the law unless the law does not or cannot deal with the issue at hand (3.15.1286b35). The partisan of the laws also argues that the many are better able to judge well than the one best man simply because their number lessens the possibility for error due to mere passions.

So the partisan of the laws limits the many's judgment in that they must be, first, free men, second, obedient to the law, and third, careful to change the law only when it falls short. These three limits, or criteria for limiting the judgment of the many, point to the power of the many; if these are not present, the laws will be ignored and the many will rule according to their whims. Hence, the rule of the many is potentially worse than, or at least as bad as, the bad rule of one man. The tendency of this argument is to downgrade the superiority of the rule of law in favor of the rule of the best man.

The partisan of the best man argues that the limits placed on the many by the laws are easily evaded by them. The partisan of the laws then poses the question: "if there were a number who were both good men and good citizens," then "is the one ruler more incorruptible or rather the larger numbers who are all good?" (3.15.1286a38–39). The partisan of the best man answers that it is clearly not one, because the many good will have difficulty with factions, whereas the single good ruler will be without factions (3.15.1286b1–2).

The partisan of the laws at first seems to ignore the problem of faction raised by the rule of the many. He instead raises the question whether the good man or the good majority is less corruptible. If there can be a good multitude, argues the partisan of the laws, then to argue that the good one is better than the good many will create a situation in which the many will rise up in factions (3.15.1286b1). Aristotle here suggests that those who believe themselves to be good or as good as the good single ruler will view his absolute reign as a slight to their excellence. In this light, they will strive for equal status with him (*Politics* 3.15.1286b11–13). However this does not deal with the question of how to address problems that occur because of factions within the many. I contend that Aristotle deals with factions and their problem in his discussion of the so-called "mixed regime" in *Politics* 4.7–9 and 4.11–16. What is discussed there is not a particular form of regime but what elements constitute a regime and how they can be made harmonious. I argue that what Aristotle presents at *Politics* 4.7–9 and 4.11–16 is not a new regime form which is mixed (which is commonly believed by many interpreters of those parts of the *Politics* text) but that all regimes per se are inherently composed of parts that point to other regimes types and if a regime is to be stable it must address the good (or advantage or benefit) of all parts of its political community, else factional conflict will emerge.

Instead of addressing the problem of factions in order to show that the rule of the many good is superior to the rule of the single best ruler, Aristotle raises the question whether it is more likely that there could be one good man or a good multitude. To clarify, the question is which is more possible, an aristocracy – which at this point in the text he calls the rule of a good multitude – or kingship. If there could be a majority of good men, it would be better to be ruled by them than by one single good man. However, the principle that it is better to be ruled by the good simply, regardless of number, than to be ruled by the many is maintained, in light of the previous argument for the many's excellences – their excellences in judging, providing for the city's needs, and so on. This puts an interesting twist to the debate. The partisan of the laws has gotten the partisan of the best man to accept the premise that the rule of the many good men is better than the rule of the one good man (3.15.1286b5).

This establishes the direction for the argument that the rule of the many is simply better than the rule of one man. Since aristocracy is more choiceworthy than kingship – "provided it is possible to find a number of persons who are similar" (3.15.1286b7–8) – then the groundwork is laid for the rule of the many being better than the rule of a single ruler. Given this line of reasoning, the partisan of the best man has accepted the premise that the rule of the many good is better than the one good, which can be used to support a fundamentally democratic premise, that the many are simply better than either the one or the few. Thus, the possibility of a good multitude provides the basis to rescue the rule of the many – i.e., democracy – from its status as a merely base regime.

Yet at this point, to avoid the trap set by the partisan of the laws, the partisan of the best man argues that only if the majority is seriously good can it avoid the creation of factions (3.15.1286b1). (I argue that Aristotle sets aside the problem of how to resolve the problem of factions until Book 4.) The underlying argument is that the one good man can be seriously good, whereas it seems improbable that there can be a seriously good multitude.

However, if such a seriously good majority could exist, it would be an aristocracy. Yet, what is aristocracy? Is it merely the rule of the good men, as suggested above, or is it the rule of the few, who rule for the sake of the common good as suggested by *Politics* 3.1279a35? Recall that Aristotle earlier in Book 3 seems to reject the usefulness of the twofold typology of regimes – composed of the quantitative (e.g., one, few or many) and qualitative (i.e., its justice, or its rule for the common good) claims – established in *Politics* 3.7. His rejection takes the form of his making the case that what defines oligarchy is not that its rulers are few but that they are rich, and that they claim their rule is just simply because only the wealthy should rule (3.8.1279b11–80a6). Aristotle argues that the rule of the many rich is as much an oligarchy as the rule of the few rich. Therefore, the quantitative claim of the regime is not a basis to understand what type of regime one is examining. Instead, the qualitative claim of a regime, its claim about the best way of life, is the distinctive criterion for examining the varieties of regimes (3.10.1281a13–39).

Therefore, what defines aristocracy is its claim that it is the rule of the best men (*aristoi*). But this claim is too generic. Would not all regimes claim that their rule is the rule of the best men? In this light, aristocracy then becomes merely the name of whatever actual regime is best.

Earlier in *Politics* 3, the rule of the many was not justified as better than the best man in judging because they were the virtuous multitude. Rather, it was justified in spite of the fact they were far from virtuous. So if one accepts the earlier position as valid, then the standard needed to rescue democracy from becoming a base regime may not be the possibility of the good multitude but the collective judgment of the multitude that is not overly slavish. This should underlay any acknowledgement of the unlikelihood of a virtuous multitude.

To return, the partisan of the best man argues that the many good have to be excellent in soul, just like the single good man. But if one good man is hard to find, then a good multitude would be even harder to find. Also some even suggest that not only is a good multitude difficult to find, it is most likely that one could not even exist. It is stated that if the condition set forth is true, although the aristocracy of many good rulers is more choiceworthy in cities than kingship, its creation is highly improbable.

## 3.5 On Laws and the Rule of Law

The partisan of the laws takes the argument one level higher. He notes that, although the arguments for the written laws made previously are good, "laws based on unwritten customs are more authoritative" (3.16.1287b5). The partisan for the laws then admits that the rule of human beings might be safer than the rule of written laws. However, he continues to note that the rule of human beings is not as safe as the rule of unwritten laws, or custom (3.16.1287b6–7). To take it even further: if the rule of human beings is safer than the rule of written laws, why is not the rule of one man also safer? Perhaps written laws are not as good as the rule of one good man. Then is this also not true for unwritten laws? Is the rule of one good man truly better than unwritten laws?

Now, however, both the rule of one and of many seem to be clearly inferior to unwritten laws. Customs are more authoritative about more authoritative things, such as how to worship the gods, whom one should obey, and so on, than written laws (3.16.1287b5). This argument for custom shows that unwritten laws can be like written law – intellect without appetite – yet safer and more authoritative than the rule of human beings or even the best of them.

The partisan of the laws then presents another criticism of the *pambasileia*. He notes that no one man can oversee many things or, more precisely, "it is not easy for one person to survey many things" (3.16.1287b8). Because the city requires many offices, "there will be a need for a number of persons to be selected as rulers under him" (3.16.1287b9–10), he asks, "what difference is there between having them present right from the beginning and having one person select them in this manner?" (3.16.1287b10–12). Then the argument returns to the contention that the many good are preferable to the simply good (3.16.1287b12), and, the partisan for the laws notes: "Even now there are offices (that of juror, for example) which have authority to judge concerning some matters that the law is unable to determine; for in the case of those it is able to determine, at any rate, no one would dispute that the law would be the best ruler and judge concerning them" (3.16.1287b15–18).

The partisan of the laws admits that the law cannot determine all things. However, he argues, the things that the laws are capable of deciding are commonly agreed to have been done fairly. But the possibility that some things cannot be encompassed by the laws raises the question as to whether "the rule of the law is more choiceworthy than that of the best man" (3.16.1287b20–21).

Aristotle also notes that to "legislate concerning matters of deliberation is impossible" (3.16.1287b22). Aristotle argues that deliberation about happiness or the things conducive to happiness is also impossible (*Rhetoric* 1.15). The laws clearly cannot replace human deliberation (*Politics* 3.11.1282a7–13). Is this the best argument against the laws and for the rule of the best man?

Aristotle has the partisan of the laws address this argument by noting, "it is not necessary for a human being to judge in such matters, but rather that there should be many persons instead of only one" (3.16.1287b23–24). Should not the many, educated by the laws, deliberate over the matters which are of great importance to the city? The partisan for the laws argues that, "[e]very ruler judges finely if he has been educated by the law; and it would perhaps be held to be odd if someone should see better with two eyes, judge better with two ears, and act better with two feet and hands than many persons with many" (3.16.1287b25–29).

The many have many hands, many eyes, and many feet. Their strength is that they have more of the qualities needed for a good judge than any individual. However, the image of a many-handed, many-footed, many-eyed, and many-eared being is that of a monster (*Politics* 3.11.1281a43–b21). Yet is this not also the characterization of the powers of a god, a being that is omnipotent? The evidence for the superiority of greater numbers is that monarchs "create many eyes for themselves, and ears, feet and hands, as well; for those who are friendly to their rule and themselves they make co-rulers" (3.16.1287b29–31).

From the question of best regime and its peak, we see that Aristotle suggests that law being a product of political body that creates it is implicitly inferior to the rule of the best ruler per se. Yet at the same time the claim here that the rule of law allows the less best types to reach a condition by which their collective deliberations when restrained by the rule of law allows them to judge and decide better than the best ones. The rule of law thus allows an inferior type to deliberate more beautifully than the one (or the few) who by their natures and character ought to be simply better judges and deliberators. In this way the defense of the rule of law in the debate between it and the rule of the simply best man at the end of *Politics 3* presents something similar but different from the modern project proposed by Machiavelli and continued by Hobbes, Locke, et al., a means by which the vulgar many are able to outshine the virtuous few (see Garrett 1993; Bookman 1992).

## 3.6 Turn to the Role of Law in Politics 4

From the heights of the philosophic discussion we have in *Politics* 3 we turn to the "nitty gritty" of messy political orders in the middle books of the *Politics*. Even here we see that Aristotle wants to show us that law and the rule of law if it has any import as a political concept, it will do so as a means by which a given political regime, even if a defective and less than perfected form of rule, can escape the extreme consequences of its defective character and lead to a form of moderation that allows what is best in that given political form to lead it to govern the whole better than it would have if it merely did what the ruling part would have wanted without any restraint.

When we turn to *Politics* 4, especially chapters 4–10 we find we have the presentation of various regime types and the different variations that occur within each particular type. We see the different types of democracies (in 4.4), oligarchies (in 4.5 [and in 4.6 a cross comparison between oligarchies and democracy]), aristocracies (in 4.8) and the regime that is called regime (in 4.8 and 4.9)[8] and then finally tyranny (4.10). Yet when one looks what the key factors are of the transition from the penultimate variation within a regime (especially the treatment of democracies in 4.4 and oligarchies in 4.5) to the final variation within a regime we see that it is the absence of the rule of law, where the rulers rule as they want.

Looking at what Aristotle does when he presents the different varieties of democracy and oligarchy in *Politics* 4 and 6 one sees that the law (or the claim of the rule of law being a guide to the regime) only disappears in the final variety of both democracy and oligarchy. In all the forms of democracy and oligarchy earlier presented by Aristotle in *Politics* 4 and 6 the presence of law (meaning that the law or the rule of law is acting within and upon the ruling part of the city) is explicitly noted. It is in the final form of democracy and oligarchy where law is absent or nowhere mentioned and therefore nowhere present that the rule of the ruler is the only law. We note that the forms where law disappears are also said by Aristotle to be similar in character to tyranny (*Politics* 4.4.1292a15–24, 4.4.1292b4–10, 6.4.131926–29). What leads those forms of democracy and oligarchy to get a tyrannical character is the fact that the part which rules the given political community reigns without any restraint, by their desires, their wants, or as the modern put it, by their will. The earlier forms of oligarchy and democracy all had the law – albeit oligarchic or democratic law respectively – acting as a restraint on those who are exercising rule over the community. Hence law moderates by retraining those who rule and thereby moderates the character of their rule and thus prevents such law turning tyrannical.

It seems that the rule of law is the key moderating feature of regimes like democracies and oligarchies, that allows them to reach their higher and more just forms of variation within the regime type. But when law is lost, both democracy and oligarchy go to a variation of the regime type that has more in common to tyranny and other forms of unjust rule than the earlier form of variation within that regime type where the rule of law remains. This is because in both the democratic and oligarchic regime the way of life that each regime in its pure form directs has much in common with tyranny. This was the reason why the older view of the Middle Ages held that the mixed constitution of the so-called 'polity' was the only way that a regime of the many could be a virtuous or good form of rule (see Johnson 1988; Bluhm 1962; Stocker and Langtry 1986; Blythe et al. 1992). Yet given the fact that no political community would ever be composed of parts which wholly shared the common

[8] The regime called regime is translated as polity in most translations of Aristotle's *Politics*. See my argument about so-called Polity in Bates (2003, 102–121). The point I make there is that the confusion of making a regime type out of the name regime points to the idea that the claim that the rule of the many that is called democracy is not really a defective regime as it is presented in both in *Politics* 3 and the first two chapters of *Politics* 4. I suggest that the discussion of so-called polity that happens in 4.7–4.9 is really an overall reflection on the overall nature of the regime itself, or so to speak all regimes per se that have been discussed up to this point. But this argument needs to be developed more than the scope of this paper would permit.

political character, all political communities per se are mixed. This means that the mixed form was not one form among many, but the common condition per se of the political community. Thus when any regime form tries to rule to the extreme of its regime principle it would by necessity have to impose its rule on those parts of which the way of life is either opposed to or in agreement with that of the ruling part and that would lead to the problems that lead to misrule, which is that the rulers rule for the sake of their interest rather than that of the mutual interest of the whole community.

When we look closely at the last two varieties of democracy and oligarchy found in *Politics* 4, we see that the rule of law make the said regimes more just than the stage/variation without it. Thus the rule of law gives a regime the means of moderation that are absent when rulers of any regime per se rule simply by their own authority, or from the dictates of their will simply. This is to say without the restraint that the principle of the rule of law provides to any regime type as a means to escape or avoid the tendency of abuse and misrule that would happen if the given regime's claimant to rule was allowed to rule as it wanted to. The moderating character of law here in *Politics* 4 echos the argument that emerged from the debate/dialogue at the end of *Politics* 3, where we see that the argument of the rule of law on its own is not enough to defeat the rule of the best one, rather it needs to be tied to the rule of the many (or *demos*).

Without the restraint that law is said to give, the many will come to have habits and character so slavish that their ability to deliberate well in a collective manner will be greatly diminished, if not lost totally. Aristotle clearly states that as long as the many are not so slavish they will be better judges than the *aristoi*, the best ones, or even the one best man (Aristotle *Politics* 3.11.1282a13–19; also Bookman 1992). Here the rule of law plays a very important role in moderating elements that although individually greatly inferior in judgment and capability of the best men, collectively they far surpass and exceed even the judgment and capacity of the best man.

So hopefully from our journey here we can see more clearly than before Aristotle's argument about law's role in relation to the regime of a given political community as the way the rule of law emerges as tool of politics by which a statesman can lead his regime towards more just governance. That the rule of law is not to be understood to be a competing form of government or regime but something any regime can adopt or use and by which it can escape the consequences of its own inherently self-destructive character. Thus even though law remains nothing but the given community's understanding of what justice is, the fact that this becomes the norm by which it defines what is just and what is not just, allows all parts of the community to appeal to it and as long as it has authority over all, including those who rule, then there is some level of equity and balance between even those parts of the community who do not fully share in ruling the community but still can benefit from the rule of those who rule over them. Hence law is a form of self-restraining guideline which allows those who lack authority in the given community to restrain those who do rule and hold authority.

Yet although law and the rule of law offer a common point of reference that all parts of the community have in common, Aristotle still recognizes that those who make the law will be that part that has authority and that usually will shape it reflecting its view of what justice is about. Thus law is not something that stands wholly outside of the political, thus 'trumping' it as modern liberal political thought would hold (see Schmitt 1976). Rather Aristotle's presentation of the way law and the rule of law work must be understood as operating within the framework of given political regimes and not as something that transcends them. Thus law is not something META-political, but something like the household that is sub-political, consisting within the political community. Given this fact, one must be very careful of any attempt to treat law as something more than what it is, and at the same time one must not fail to pay attention to the fact that without it (like without the household – another sub-political part) there can be no politics.[9]

## Bibliography

Arendt, Hannah. 1958. *The human condition*. Garden City: Doubleday.
Arnhart, Larry. 1990. Aristotle, chimpanzees and other political animals. *Social Science Information* 29(3): 477–557.
Arnhart, Larry. 1994. The Darwinian biology of Aristotle's political animals. *American Journal of Political Science* 38(2): 464–485.
Aristotle. 1957. *Aristotleis Politica*. Ross, W. D. ed. Oxford: Clarendon.
Bartlett, Robert. 1994a. Aristotle's science of the best regime. *American Political Science Review* 88(1): 143–155.
Bartlett, Robert. 1994b. The realism of classical political science. *American Journal of Political Science* 38(2): 381–402.
Bates Jr., Clifford A. 2003. *Aristotle's best regime*. Baton Rouge: Louisiana State University Press.
Bluhm, William. 1962. The place of 'polity' in Aristotle's theory of the ideal state. *The Journal of Politics* 24(4): 743–753.
Blythe, James M. 1992. *Ideal government and the mixed constitution in the middle ages*. Princeton: Princeton University Press.
Bodeus, Richard. 1991. The law and regime in Aristotle. In *Essays on the Foundations of Aristotle's Political Science*, Carnes Lord and David O'Connor (eds). Berkeley: University of California Press, 234–248.
Bodeus, Richard. 1993. *The political dimensions of Aristotle's Ethics*. Jan Edward Garrett (trans). Albany: State University of New York Press.
Bookman, John T. 1992. The wisdom of the many: An analysis of the arguments of book III and IV of Aristotle's *Politics*. *History of Political Thought* 13(1): 1–12.
Cherry, Kevin M. 2009. The problem of polity: Political participation and Aristotle's best regime. *The Journal of Politics* 71(4): 1406–1421.
Coby, Patrick. 1986. Aristotle's four concepts of politics. *Western Political Science Quarterly* 39(3): 480–503.

[9] The dangers of turning law or the rule of law into a META-political trump is the very point about the nature of liberalism made by critics of liberals such as Carl Schmitt, Antonio Gramsci, Ernesto Laclau, Jacques Derrida, Michel Foucault, Hannah Arendt, Giorgio Agamben, Claude Lefort, Antonio Negri, Paolo Virno, Slavoj Žižek, Alain Badiou, Jacob Taubes, Chantal Mouffe, et al. This is not to say such critics are correct in their stance on the political, but like the 'broken clock' have the ability to be none-the-less correct twice a day.

Coby, Patrick. 1988. Aristotle's three cities and the problem of factions. *Journal of Politics* 50(4): 896–919.
Davis, Michael. 1996. *The politics of philosophy: A commentary on Aristotle's Politics*. Savage: Roman and Littlefield.
de Alvarez. Leo Paul S. 1989. Translation and Introduction. In *Niccolo Machiavelli: The Prince*. Prospect Heights: Waveland Press,
Dreizhnter, Alois. 1970. *Aristoteles' Politica*. Munich: Wilhelm Fink Verlag.
During, Ingemar. 1966. *Aristoteles. Darstellung und Interpretation seines Denkens*. Heidelberg: Winter.
Eidelberg, Paul. 1975. *Discourse on statesmanship*. Urbana: University of Illinois Press.
Ewbank, Micheal B. 2005. Politeia as focal reference in Aristotle's taxonomy of regimes. *The Review of Metaphysics* 53(3): 815–841.
Finley, Moses I. 1985. *Democracy ancient and modern*. Revised edition. New Brunswick: Rutgers University Press.
Forsyth, Murray, and Maurice Keens-Soper. 1988. *A guide to the political classics*. Oxford: Oxford University Press.
Garrett, Jan Edward. 1993. The moral status of 'the many' in Aristotle. *Journal of the History of Philosophy* 31(2): 171–189.
Grant, Sir. Alexander. 1885. *The ethics of Aristotle,* 4th ed., Revised, 2 vols, 1885. London: Longman, Green and Co.
Holmes, Stephen H. 1979. Aristippus in and out of Athen. *American Political Science Review* 73(1): 113–128.
Jaffa, Harry V. 1975. What is politics? An interpretation of Aristotle's *Politics*. In *The conditions of freedom*. Baltimore: The John Hopkins University Press, 9–72.
Johnson, Curtis. 1988. Aristotle's Polity: Mixed or middle constitution? In *History of political thought* 9(2): 189–204.
Kahn, Charles H. 1989. The normative structure of Aristotle's 'politics'. In Patzig 1989.
Keyt, David, and Fred Miller, (eds.). 1991. *A companion to Aristotle's "Politics"*. Oxford: Blackwell.
Kraut, David. 1999. *Aristotle: Politics V and VI*. Oxford: Clarendon.
Kraut, David. 2002. *Aristotle: Political philosophy*. Oxford: Oxford University Press.
Lindsay, Thomas K. 1992a. Aristotle's qualified defense of democracy through 'political mixing'. *Journal of Politics* 54(1): 101–119.
Lindsay, Thomas K. 1992b. Liberty, equality, power: Aristotle's critique of the democratic 'presupposition'. *American Journal of Political Science* 36(3): 743–761.
Lintott, Andrew. 1992. Aristotle and democracy. *Classical Quarterly* 42(1): 114–128.
Lockyer, Andrew. 1988. Aristotle: The *politics*. In *Forsyth and Keens-Soper* 1991, 37–68.
Lord, Carnes (ed.). 1984. *The politics of Aristotle*. Chicago: University of Chicago Press.
Lord, Carnes. 1987. Aristotle. In *History of political philosophy*, 3rd ed. Leo Strauss and Joseph Cropsey, (eds). Chicago: University of Chicago Press, 118–154.
Lord, Carnes, and David O'Connor (eds). 1991. *Essays on the foundations of Aristotelian political science*. Berkeley: University of California Press.
Manent, Pierre. 1994. *An intellectual history of liberalism*. Princeton: Princeton University Press.
Manent, Pierre. 2010. *Les metamorphoses de la cite: Essai sur la dynamique de la Occident*. Paris: Flammarion.
Masters, Roger. 1989. *The nature of politics*. New Haven: Yale University Press.
McCoy, Charles N.R. 1963. *The structures of political thought*. New York: McGraw Hill.
McCoy, Charles N.R. 1989. Democracy and the rule of law. In *On the intelligibility of political philosophy*, James V. Schall and John J. Schrems (ed). Washington, DC: Catholic University of America Press 39–48.
Mulgan, R.G. 1977. *Aristotle's political theory*. Oxford: Claredon Press.
Mulgan, R.G. 1991. Aristotle's analysis of oligarchy and democracy. In *David Keyt and Fred D. Miller* 1991, 307–322.
Nagel, Brendon. 2006. *The household as the foundation of Aristotle's polis*. New York: Cambridge University Press.

Newman, William L. (ed.). 1973. *The politics of Aristotle*, 4 vols. New York: Arno Press.
Nichols, Mary P. 1991. *Citizens and statesmen*. Savage: Roman and Littlefield.
Nussbaum, Marta. 1994. *The therapy of desire: theory and practice in hellenistic ethics*. Princeton: Princeton University Press.
Pagden, Anthony, (ed). 1987. *The languages of political theory in early modern Europe*. Cambridge: Cambridge University Press.
Patzig, Günther (ed). 1989. *IX Symposium Aristotelicum: Studien zur Politik des Aristoteles*. Göttingen: Vandarhoeck and Rupreht.
Popper, Karl. 1945. *The enemies of the open society*, 2 vols. London: Routledge and Kegan Paul.
Preus, Anthony. 1979. Eidos As Norm in Aristotle's Biology. *Nature and System* 1: 79–101.
Quinn, Michael. 1990. Aristotle on justice, equality and the rule of law. *Polis* 9(2): 170–186.
Rhodes, James M. 1991. Right by Nature. *Journal of Politics* 53(2): 318–337.
Robinson, Richard, (ed. and trans.). 1962. *Aristotle's politics, Book III and IV*. Oxford: Clarendon Press.
Rubinstein, Nicolai. 1987. The history of the word *politicus* in early modern Europe. In *Pagden* 1987, 41–56.
Saxonhouse, Arlene. 1992. *The fear of diversity*. Chicago: Chicago University Press.
Schmitt, Carl. 1976. *The concept of the political*. George Schwab (trans). New Brunswick: Rutgers University Press.
Simpson, Peter. 1993. Review of Eckart Schütrumpf's Aristotoles Politik, uberstetzt und erlautern. *American Journal of Philology* 114(2): 320–323.
Simpson, Peter. 1997. *The politics of Aristotle – Translation and introduction*. Chapel Hill: University of North Carolina Press.
Simpson, Peter. 1998. *A philosophical commentary on the politics of Aristotle*. Chapel Hill: University of North Carolina Press.
Stocker, Michael, and Bruce Langtry. 1986. Aristotle and polity. In *Political thinkers*, ed. David Muschamp. New York: St. Martins Press.
Strauss, Leo. 1953. *Natural right and history*. Chicago: University of Chicago Press.
Strauss, Leo. 1978. *The city and man*. Chicago: University of Chicago Press.
Strauss, Leo. 1988. *What is political philosophy: And other studies*. Chicago: University of Chicago Press [Originally, The Free Press, 1959].
Strauss, Leo. 1989. *An introduction to political philosophy: ten essays by Leo Strauss*. Hilail Gildin (ed). Detroit: Wayne State University Press.
Susemihl, Franz, and R. Hicks. 1894. *The politics of Aristotle*. London: Macmillan.
Swanson, Judith A. 1992. *The public and the private in Aristotle's political philosophy*. Ithaca: Cornell University Press.
Swanson, Judith A., and C.David Corbin. 2009. *Aristotle's politics*. London: Continuum Press.
Swazo, Norman. 1991. The authentic *Tele* of politics: a reading of Aristotle. *History of Political Thought* 22(3): 405–420.
Voegelin, Eric. 1978. *Anamnesis*. South Bend: University of Notre Dame Press.
Von Fritz, Kurt. 1954. *Theory of the mixed constitution in antiquity*. New York: Columbia University Press.
Yack, Bernard. 1985. Community and conflict in Aristotle's political philosophy. *The Review of Politics* 47(1): 169–188.
Yack, Bernard. 1993. *The problems of a political animal*. Berkley: University of California Press.
Zuckert, Catherine. 1983. Aristotle on the limits and satisfactions of political life. *Interpretation* 2(2): 185–206.

# Chapter 4
# The Best Form of Government and Civic Friendship in Aristotle's Political Thought: A Discussion Note

**Ki-Won Hong**

## 4.1 Introduction

The place of law in Aristotle's political theory has been discussed by many authors. One of the remarkable studies recently published on the subject would be Frank's (2006) challenging interpretation of constitutionalism and the rule of law in the Greek philosopher. He argues that many scholars' traditional understanding of Aristotle as the source of the argument opposing the rule of law to the personal rule should concede to Aristotle's own understanding of the rule of law as issuing from political power's practical wisdom. Very significant in the political atmosphere where radical distinction of the rule of law from democracy is often put forward on the polarized understanding of the concepts, this new interpretation will not be, though, exempt from some criticism, in that the approach is constructed on the denial of the unprescriptiveness of Aristotle's constant distinction between the rule of law and personal rule (*Pol*. III, 15). I think that what C. A. Bates, Jr. shows in Chap. 3 through his exhaustively detailed analysis of *Politics* III,15–17 can give a useful answer to fix the problem Frank's interpretation raises.

Among many topics Bates is dealing with in his discussion, two questions particularly drew my attention. First, he points out that Swanson (1992) made a "gross misrepresentation" of the text in question by reading "law" as referring to "natural law," rather than to statutory laws or customary law. Secondly, by reconstituting the arguments developed in the text into a form of dialog between a "partisan" of the rule of law and the other "partisan" of the rule of wise king, Bates succeeds not only in clearly showing what the point of Aristotle's argument exactly is, but also in

A previous version of this discussion note was published in the *Korea Comparative Public Law Association Journal* 13/1 (February 2012, 229–245).

K.-W. Hong (✉)
Ph.D. in Law, University of Aix-Marseille, France

L. Huppes-Cluysenaer and N.M.M.S. Coelho (eds.), *Aristotle and The Philosophy of Law: Theory, Practice and Justice*, Ius Gentium: Comparative Perspectives on Law and Justice 23,
DOI 10.1007/978-94-007-6031-8_4, 

restituting the authority of the traditional understanding of Aristotelian distinction between the rule of law and personal rule.

There will not be much to discuss with respect to the first topic, for Bates is totally right, I think, when he shows that the word 'law' (νόμος) was conceptually distinguished from the word 'nature' (φύσις) by Aristotle as well as by other Greek philosophers.[1] And a careful reading of the *Politics* will lead to the understanding that all eight books are in fact a presentation of how to build a good city upon man-made laws (convention), not on natural law.[2] It is not surprising, then, that Aristotle seems to say that convention has little to do with natural law. We even see him frequently affirm that it is conformable to natural order that the inferior people obey the superior.[3] So to speak, natural law as a norm which regulates political institutions doesn't exist in Aristotle's political thought.

My discussion note will mainly concern, therefore, the second topic, i.e. the place of law and its role in Aristotle's political theory.[4] One of the excellent observations of Bates in this respect is that he examines the question with regard to Aristotle's theory of political regimes. Skeptic about democracy, Aristotle seems to indirectly criticize its arguments by putting them in duel with arguments for the rule of law. Bates is still right in showing that law is presented as playing a role to moderate defects of deviant regimes—democracy, for instance, is susceptible to factional conflicts—, and that Aristotle's discussion of the mixed regime purports to present a model of constitution free from these defects. Yet there are some places in which Bates advances an interpretation to which not every reader may be ready to agree. He firmly identifies the argument for the rule of law with the democratic argument, or with the "justification for the rule of the many over the laws." In doing so, he portrays Aristotle as a partisan of democracy. The most difficult problem raising from Bates' explanation is that, by considering the rule of law as an element common to all "decent" regimes, he seems not to pay attention to the particular significance of the mixed regime in Aristotle's political thought.

It is important to note, however, that the *Politics* consistently distinguishes a pure form of kingship or aristocracy from those "under the law" (*Pol.* III, 15–16). For Aristotle, kingship and aristocracy can exist without the rule of law, whereas this latter principle is conceived as an innate constitutional element of mixed regime. I hope this discussion note will contribute to show that the *Politics* presents mixed regime as the best form of government and that such a reading would be a way to fully understand Aristotle's political theory as the continuation of the project he

---

[1] On this topic, see Fritz and Ernst (1950, 38–40).

[2] Bartlett (1994) explains such a position of Aristotle as a denial of the superiority of "divine legislation," in favor of the guidance supplied by "unaided reason."

[3] In many places, Aristotle says that there are those who are "naturally" to rule and those who are "naturally" to be ruled (*Pol.* 1252a30, 1259b32–33, 1284b33–34). Recall his famous argument for "natural slavery" (1254a17–1254bb, 1255b1–15, 1278b33) Cf. Ambler (1985, 173–174, 178).

[4] Yet the topic of nature or natural law in Aristotle will be discussed below when necessary.

declared at the end of his *Nicomachean Ethics*.[5] Concerned for bridging again politics and ethics, the two realms of Aristotle's philosophy, I will also try to explain the principle of mixed regime in terms of civic friendship, a goodness recommended to attain the true end of a city.

## 4.2 The Best Form of Government: Theory and Practice

A constitution, defined as "the organization of a city [or *polis*], in respect of its offices generally, but especially in respect of that particular office which is sovereign in all issues" (*Pol.* 1278b9–11, trans. Barker), has its particular form according as it concerns the common interest or only the personal interest of the ruler(s). Constitutions of the first type are "right" ones, and those of the second "wrong" ones, says Aristotle (1279a19), each of the two subdividing according to the number of the ruler(s) into three forms of government: kingship, aristocracy, and polity[6] for the right constitutions and tyranny, oligarchy, and democracy for the wrong ones[7] (1279a22–1279b10).

Aristotle's cautious way of description of each form of government gives rise to the controversy over the question which constitution among the right ones he thinks to be the best or, at least, looks on with favor. For some authors, Aristotle's commendation of monarchical rule is based on the fact that he thinks it possible that the best political regime is "a monarchy run with a rational efficiency that leaves little or no scope for citizen participation" (Newell 1987, 160). There are, of course, some passages in the *Politics* in which Aristotle suggests such a possibility (III, 15–18; IV, 2, 1289a26). But, as Bluhm (1962, 751–752) has already pointed out, we need not take these passages so seriously, as the kingship ruled by a king of virtue, reminiscent of the Platonic ideal form of government, is in fact rejected by Aristotle because of

---

[5] *EN* X, ix, 1181b19: "we will consider what institutions are preservative and what destructive of states in general, and of the different forms of constitution in particular, and what are the reasons which cause some states to be well governed and others the contrary. For after studying these questions we shall perhaps be in a better position to discern what is *the best constitution absolutely*, and what are the best regulations, laws, and customs for any given form of constitution" (trans. Rackham. Italics by me). It is regrettable that the connection of the *Politics* with the *Nicomachean Ethics* is not much discussed in recent studies on Aristotle, while some early scholars (Loos 1897, 317; Fritz and Ernst 1950, 43, 52) were, with much interest, conscious of its importance.

[6] The word 'polity' in this discussion note refers to the Greek 'πολιτεία,' which is translated in English, generally, as 'constitutional government.' That is the case with Rackham and Barker.

[7] Another passage from the *Nicomachean Ethics* (VIII, x, 1–3), too, established a similar classification of political regimes: kingship, aristocracy, timocracy for normal forms of constitution, and tyranny, oligarchy, democracy for perverted or corrupt ones. This raises a very hard question of interpretation, for, while identifying timocracy with "constitutional government or Republic," the author abruptly inserts a phrase saying: "The best of these constitutions is Kingship, and the worst Timocracy" (trans. Rackham). To give a clear answer to this question is out of the scope of this discussion note.

its lack of reality.[8] A more difficult problem arises when many authors stick to the interpretation that with regard to happiness and moral life—an important purport of the city—, democracy is the "true" *polis*[9] (Lintott 1992, 116–117). We must beware, however, that the deviant character of democracy in Aristotle's classification should not be neglected in favor of the ideologized concept of democracy of modern authors.

By the way, the answer to the question above may be given rather easily than it appears as complex as in the controversy, if we carefully follow Aristotle's rhetorical argumentation on the polity (*Pol.* III, 3–11), in the last passage (1296b2) of which he concludes, although implicitly, that it is the polity that is the best constitution. The polity as one of the right constitutions, different from democracy in that the first pursues the common interest and the second the interest of the multitude, takes the form of a mixture, paradoxically, of two wrong constitutions: oligarchy and democracy (1293b33). As each of the right constitutions in its pure form hardly exists in reality, Aristotle seeks instead how the best form of government could be possible in the practical sense. And, for the critic of Platonic idealism, this possibility of the practically best has a general importance for most communities (Galston 1994, 398, n. 282):

> We have now to consider what is the best constitution and the best way of life for the majority of cities and the majority of mankind (*Pol.* 1295a25, trans. Barker).

Here we need to classify the six constitutions anew into two groups, i.e. unitary regimes and the mixed regime.[10] Kingship, aristocracy, tyranny, oligarchy, and democracy are classified into the first group[11] and the polity into the second. It is well known (Coby 1988, 904) that for Aristotle, it is almost a natural law that only a well-constructed regime is stable and that the first rule of such a good construction is to have a regime suitable to its people, since, as the Greek philosopher articulates with his famous phrases, "the city belongs to the class of things that exist by nature" and "man is by nature a political animal" (1253a2, trans. Barker). It is on this basis that Aristotle continues to explain what sort of constitution is desirable for what sort of civic body (*Pol.* IV, 12–13). Thus, there may be a people whose stage of social development makes them choose kingship for their proper political regime,[12] while it may be aristocracy that is suitable to

---

[8] Coby's (1988, 909–910) thesis of Aristotle as proponent of aristocracy may also be rejected for similar reasons, but to a relatively restricted extent, for it is true that Aristotle thinks that aristocracy is more practicable than kingship.

[9] Cf. Wolff (1997, 106–123): "la politique aristotélicienne est *démocrate*." Aubenque (1993, 255–264) seems to try to find a justification of democracy in Aristotle's theory by stressing that Aristotelian polity resembles much the democracy of today.

[10] This twofold classification is partly borrowed from Coby's (1988, 902–917) three-city schema: unitary regimes, mixed regimes, and ideal regime.

[11] To mark the difference between the unitary and the mixed and, especially, to explicitly distinguish the mixture of oligarchy and democracy from these two regimes, Lindsay (1992b, 757) specifies democracy and oligarchy as unitary regimes by adding a corresponding adjective to the name of regime like "radical" democracy and "unmixed" oligarchy.

[12] For Aristotle (1284b34, trans. Barker), it would also appear to be the "natural course" that all others should "pay a willing obedience to the man of outstanding goodness." But, some passages later (1287b36–), he explains that "there is no society which is meant by its nature for rule of the tyrannical type," the worst form of government.

another people's conditions. But this naturalist relativism doesn't go further so as to take it for granted that kingship and aristocracy will be stable forever or, at least, for a considerably long time.[13] The problem with kingship and aristocracy is that like other things in nature,[14] these constitutions are subject to the natural law of coming-to-be and passing-away. Aristotle holds that the dynamics of every political society is pictured by the ideological confrontation between the standard of excellence and that of equality,[15] whether these ideas may be or not a part of the organizing elements of the concerned political regime. Kingship and aristocracy—constitutions ruled on the assertion of excellence, respectively, of one man or the few—are constantly challenged by the assertion of equality from the multitude and, consequently, liable to move towards an egalitarian regime along with the numerical increase of "people of similar quality," as Aristotle witnesses in his days (1312b38–). For a constitution to be perpetual, it should be therefore more than a unitary regime: it must be a mixture of some constitutions, the pure form of each constitution existing rather in theory than in reality.[16]

Oligarchy, the perversion of aristocracy, is constituted on the standard of excellence in terms of wealth, while democracy, the perversion of the polity, on that of equality (*Pol.* IX, 9). None of the two is a normal constitution. However, by mixing themselves, in other words, by moderating the principle of excellence with that of equality, they can remedy the defects of each other to form a polity, which is for Aristotle the practically best constitution.[17] Thus, the combination of two principles puts the polity on the very "mean" or "just middle"[18] between the regime of excess and that of deficiency. And in Aristotle's political theory correlated with his ethics in this respect, it follows that the polity is naturally thought to be ruled by the middle class, who represents in effect the synthesis of aristocratic quality and democratic quantity.[19] Aristotle says (1295b30, trans. Barker):

> A city aims at being, as far as possible, composed of equals and peers, which is the condition of those in the middle, more than any group. It follows that this kind of city is bound to have the best constitution since it is composed of the elements which, on our view, naturally go to make up a city.

To be short, "the best form of association is," concludes Aristotle, one where power is vested in the middle class "without which the regime turns either into an

---

[13] And it would be important to add that, as Frank (2006, 45) rightly points out, the legal aspect of Aristotle's relativism doesn't justify either a positivist reading of his theory on the conformity of laws with the end of constitution. For such a misreading, see Fritz and Ernst (1950, 53, 56).

[14] The treaty in which Aristotle treats this question from the viewpoint of natural science is *De generatione et corruptione*.

[15] Coby (1988, 910). Cf. Lindsay (1992a, 114) terms this confrontation 'distinction between virtue and vice.'

[16] In this respect, Jaeger's (1962, 263–275) and Stocks' (1927) analysis of the composition of the *Politics* is still valid in that mutual references between books 2 and 3 and books 7 and 8 are explained by their theoretical approach, which is distinctive from the empirical approach of books 4–6. For criticism of Jaeger's view, see Lord (1981, 469–470).

[17] Examples of such a remedy: *Pol.* VI, 6–7; IV, 9, 1294b1–40.

[18] For the concept of the mean, see *EN* 1106b36–1107a2.

[19] 1296b17: "Quality and quantity both go to the making of every city."

"extreme democracy," an "unmixed oligarchy," or a "tyranny"" (1295b34–1296a6, trans. Barker). The middle type of constitution as the best form of government is for him no other than the mixed regime of oligarchy and democracy.

## 4.3 Civic Friendship: Ethics of Mixture for the Best Form of Government

It is needless to say that when Aristotle reiterates that "man is a political animal," he means by that more than that men organize a society to have shelter. He reminds that "the end of the city is not mere life; it is, rather, a good quality of life"[20] (1280a30, trans. Barker). The perfection of a city and the men who are living therein can only be expected when the city sets itself to continuously promote human goodness (ἀρετή), and the argument shows in addition that the society Aristotle presupposes is not one of a small number of virtuous people, but of a mass of "ordinary people," who will be able to approach the level of the virtuous through education (1295a25–33). Composed of diverse classes with a majority of ordinary people rather than homogeneous people of virtue, the city needs an ethical principle regulating its members' mind and attitudes to bring forth among them a harmonized order of civil life. The political principle of the mixed regime we have just examined above is thus complemented by another principle of mixture, which Aristotle articulates in terms of "friendship" (φιλία) in general (*EN* VIII–IX)[21] and "civic friendship" (πολιτικὴ φιλία) in particular [22] (*EE* 1242a–1243b).

Regarding the way people treat each other rightly or wrongly, friendship is "either identical or very close" to justice (*EE* 1234b30; *EN* VIII, ix). And, just as the city must maintain justice by determining for its people what is just (*Pol.* 1253a35), so the city should see it as a "special task" to promote friendship (*EE* 1234b23), since this virtue is one of the greatest goods to make possible a better quality of human life.[23]

In terms of motivations, there are three kinds of friendship. One kind is based on virtue, another on utility, and a third on pleasure (*EE* 1236a10–33; *EN* VIII, iii), each of the three being subdivided into the friendship based on superiority and that on equality (*EE* 1238b15–1239a, 1242b1–5; *EN* VIII, vii). And another classification

[20] See also 1278b15–1278b29 and 1280b6.

[21] We take 'friendship' to translate 'φιλία' as is the conventional usage in many English translations such as Aristotle's (1934, 2011). It must be noted, however, as Stewart (1892, II, 262) has already reminded us, that "Aristotle's φιλία is a wider term than *Friendship*."

[22] The question of friendship and its relation to justice and state in Aristotle was fully studied by Stern-Gillet (1995, 147–169). Instead of summarizing what she has done, the second section of this discussion note will be limited to show that civic friendship is presented in the Greek philosopher as the ethical principle for, and of, the mixed constitution.

[23] *EN* VIII, i, 1155a20, trans. Rackham: "it is one of the most indispensable requirements of life."

of friendships is made according to the nature of the parties of friendship: those between kinsfolk, comrades, partners, and, lastly, civic friendship.[24] It is the latter that is presented, in Aristotle's ethics, as the goodness by and for which relationships between private citizens are established in the city. Civic friendship, based on utility on the one hand, explains why people are brought together to form a city and, based on equality on the other, it concretizes the political principle of the polity by way of a rotational participation in the offices (1242a–1242b29). Equality being thus stressed, civic friendship may look like a goodness suitable for democracy, as equality constitutes, in fact, the principle of democracy in Aristotle's political theory (*Pol.* 1280a10), and equal share in office and honors is held to be a just right of people (1280a10, 1289a9–32).

However, it must not be forgotten that equality and participation in public affairs through rotation are also effective in the mixed regime of oligarchy and democracy.[25] To the contrary of its apparent reference to democratic principles, Aristotle's conception of civic friendship is related to what the late Winthrop (1978b, 162) has once defined as Aristotle's "non- or antidemocratic recognition of men." In other words, his appreciation of civic friendship as a goodness which can make a better quality of life presupposes—regardless of his explicit appraisal of the superiority of the virtuous—a heterogeneous composition of society, which necessitates seeking and imposing a civic ethics as well as designing the best form of government with a view to let all the people live in harmony and durable peace. Kingship and aristocracy, too, are right constitutions. But they are liable to conflicts and changes either due to internal causes[26] or to "failure to combine different elements properly" (trans. Barker).[27]

Aristotle the natural scientist recognizes that the things that are "not susceptible to change at all [...] may well by nature be the best things of all," but Aristotle the human scientist doesn't forget to distinguish these things from those "within the scope of human action" (*EE* 1217a31–36, trans. Kenny). In the *Politics*, kingship is often presented as a form of government which has its origin in, and is conformable to, natural order,[28] while aristocracy, in the course of linear evolution of political regimes, is closer to this origin than other constitutions (1286b8–21), as it is, just

[24] *EN* IX, iv gives another fivefold classification of friendships according to feelings.

[25] 1295b20; 1298b5–10 in the deliberative power, 1300a35 in the executive, and 1301a18 in the judicial.

[26] That's the case of kingship. *Pol.* V, 10–11: "Such causes may take two forms. One is dissension among the members of the royal household: the other consists in attempting to govern in a tyrannical fashion by claiming a larger prerogative without any legal restrictions" (1312b40, trans. Barker).

[27] That's the case of aristocracy *Pol.* V, and VII.

[28] *EE* 1241b28; *Pol.* 1252b20, 1284b32–34: "The only alternative left—and this would also appear to be the natural course—is for all others to pay a willing obedience to the man of outstanding goodness. Such men will accordingly be the permanent kings in their cities" (trans. Barker). In his two excellent studies on Aristotle's political theory, Lindsay (1992a, 114–116; 1992b, 750) includes obedience to the few, too, in this "natural course" alternative of Aristotle. But it must be noted that, as we have just seen from the citation, Aristotle's "natural course" refers more to kingship than to aristocracy.

like kingship, still "based on merit" (1310b31–35, trans. Barker). A natural history of political regimes, if we may be allowed to term Aristotelian depiction of the evolution as such, is no other than a history of institutional deterioration from the pure to the impure, which pairs in effect with the growth of democratic elements in the city. This doesn't mean, however, that Aristotle is disdainful of common people or democratic values like equality. On the contrary, he is mostly hopeful about their contribution to realize a better quality of life, which is not possible in a pure democracy, but can be realized in a mixed regime. For:

> [w]hen they all meet together, the people display a good enough gift of perception, and combined with the better class they are of service to the city.[29]

In the two *Ethics*, this principle of mixture is articulated in terms of friendship. The perfect kind of friendship is, of course, friendship based on virtue, says Aristotle (*EN* VIII, iii, 6). But the problem is that this sublime kind of friendship is rarely found in real civil life. Furthermore, in the real political world, friendship between the superior and the inferior does not always work so well as ideologically recommended. It is in this respect that Aristotle (*EN* VIII, x, 3) explains that deviance of kingship and aristocracy into, respectively, tyranny and oligarchy results from the failure of the ruler(s) to govern in a manner appropriate to the friendship, respectively required in each political regime. Fortunately, an effective cure or prevention of this political instability can be provided by encouraging civic friendship, i.e. the ethical goodness of the mixed regime, because friendship based both on utility and equality can serve as an antidote against political and social factions: the friendship of utility by making steps toward civic justice (*EE* 1242a12) and the friendship of equality by creating a change toward the rule of law (1242b31–35).

For Aristotle, the problem of faction will neither be solved by pursuing Plato's remedy to get rid of the probable causes of faction by putting all the members of the city in family relationship,[30] nor by perpetuating kingship or aristocracy, in which the textile of political and social relationship would be woven on the basis of friendship of virtue. A more effective alternative is therefore proposed by him through a bridge between his ethics and politics. That's civic friendship, namely the ethical support of the polity.

## 4.4 Conclusion

As Bates rightly points out, Aristotle's political institutions are constructed on, and moderated by, human laws. Nature conditions our faculties, of course, and we are constraint to live within the limits which nature imposes. But by building a city on

[29] 1281b35–37: "... (just as impure food, when it is mixed with pure, makes the whole concoction more nutritious than a small amount of the pure would be)" (trans. Barker).

[30] *Pol.* II, 1–6. Cf. Coby (1988, 915–916).

law and justice, which consequently possesses a better quality of life, we can overcome our doomed destiny (Winthrop 1978a, 1208):

> Man, when perfected, is the best of animals; but if he be isolated from law and justice he is the worst of all (*Pol.* 1253a30–35, trans. Barker).

The alternative proposed by Aristotle for this end is to design the best form of government that is generally practicable for most people as well as for most cities (*Pol.* IV, 11). And this is no other than a mixture of oligarchy and democracy. When the principles of democracy are mixed with, and thus moderated by, those of oligarchy, the polity becomes a city in which participation of more people in public affairs is assured (Lindsay 1992a, 106–108). Given that, the polity is also considered as the most stable form of government, while other forms are very susceptible to decline and corruption due to their factional tendency.

Civic friendship plays an important role of the goodness which regulates the people of the polity and leads their way of life to approach the civic justice. Ordinary people, who will compose the dominant middle class in the polity, can thus have an elevated spirit with virtue through education of such a civic ethics. For Aristotle, to construct the best form of government is, in effect, to provide practical means to have most people live a life of better quality.

## Bibliography


Ambler, W.H. 1985. Aristotle's understanding of the naturalness of the city. *The Review of Politics* 47(2): 163–185.

Aristotle. 1934. *(EN) Nicomachean ethics.* Revised edition. Trans. H. Rackham. Cambridge: Harvard University Press.

Aristotle. 1998. *(Pol.) Politics.* Trans. E. Barker, revised with an introduction and notes by R. F. Stalley. Oxford: Oxford University Press.

Aristotle. 2002. *(AC) The Athenian constitution.* Trans. with introduction and notes P. J. Rhodes. London: Penguin Books.

Aristotle. 2011. *(EE) The Eudemian ethics.* A new trans. A. Kenny. Oxford: Oxford University Press.

Aubenque, P. 1993. (sous la direction de), *Aristote, Politique: Études sur la* Politique *d'Aristote.* Paris: Presses Universitaires de France.

Bartlett, R.C. 1994. Aristotle's science of the best regime. *The American Political Science Review* 88(1): 143–155.

Bates Jr., C.A. 2003. *Aristotle's "best regime": Kingship, democracy, and the rule of Law.* Baton Rouge: Louisiana State University Press.

Bluhm, W.T. 1962. The place of the "polity" in Aristotle's theory of the idea state. *The Journal of Politics* 24(4): 743–753.

Coby, P. 1986. Aristotle's four conceptions of politics. *The Western Political Quarterly* 39(3): 480–503.

Coby, P. 1988. Aristotle's three cities and the problem of faction. *The Journal of Politics* 50(4): 896–919.

Frank, J. 2006. Aristotle on constitutionalism and the rule of law. *Theoretical Inquiries in Law* 8(37): 37–50.

Fritz, K. von, and Ernst, K. 1950. Introduction. In Aristotle, *Constitution of Athens and Related Texts.* New York: Hafner Publishing, 3–66.

Galston, M. 1994. Taking Aristotle seriously: Republican-oriented legal theory and the moral foundation of deliberative democracy. *California Law Review* 82(2): 329–399.
Jaeger, W. 1962. *Aristotle: Fundamentals of the history of his development.* 2nd ed., Trans. with the author's corrections and additions R. Robinson. London: Oxford University Press.
Lindsay, T.K. 1992a. Aristotle's qualified defense of democracy through "political mixing". *The Journal of Politics* 54(1): 101–119.
Lindsay, T.K. 1992b. Liberty, equality, power: Aristotle's critique of the democratic "presupposition". *American Journal of Political Science* 36(3): 743–761.
Lintott, A. 1992. Aristotle and democracy. *Classical Quarterly* 42(1): 114–128.
Loos, I. 1897. The political philosophy of Aristotle. *The Annals of the American Academy of Political and Social Science* 10(1): 313–333.
Lord, C. 1981. The character and composition of Aristotle's *Politics*. *Political Theory* 9: 459–478.
Lord, C. 1982. *Education and culture in the political thought of Aristotle*. Ithaca: Cornell University Press.
Mara, G.M. 1987. The role of philosophy in Aristotle's political science. *Polity* 19(3): 375–401.
Mathie, W. 1987. Political and distributive justice in the political science of Aristotle. *The Review of Politics* 49(1): 59–84.
Murphy, J.B. 2002. Nature, custom, and reason as the explanatory and practical principles of Aristotelian political science. *The Review of Politics* 64(3): 469–495.
Nederman, C.J. 1994. The puzzle of the political animal: Nature and artifice in Aristotle's political theory. *The Review of Politics* 56(2): 283–304.
Newell, W.R. 1987. Superlative virtue: The problem of monarchy in Aristotle's 'Politics'. *The Western Political Quarterly* 40(1): 159–178.
Stern-Gillet, S. 1995. *Aristotle's philosophy of friendship*. Albany: State University of New York Press.
Stewart, J. A. 1892. *Notes on the* Nicomachean ethics *of Aristotle*, 2 vols. Oxford: The Clarendon Press.
Stocks, J.L. 1927. The composition of Aristotle's *Politics*. *The Classical Quarterly* 21(3–4): 177–187.
Swanson, J.A. 1992. *The public and the private in Aristotle's political philosophy*. Ithaca: Cornell University Press.
Winthrop, D. 1978a. Aristotle and theories of justice. *The American Political Science Review* 72(4): 1201–1216.
Winthrop, D. 1978b. Aristotle on participatory democracy. *Polity* 11(2): 151–171.
Wolff, F. 1997. *Aristote et la politique*. Paris: Presses Universitaires de France.

# Chapter 5
# Controversy and Practical Reason in Aristotle

**Nuno M.M.S. Coelho**

## 5.1 Introduction: The Antilogical Context

It is extremely important to understand an author based on his own texts. This may be obvious, but it is especially difficult when we talk about Aristotle.[1] Because reading involves contextualising, no work by Aristotle can be separated from its context. Again, we must pay attention to texts – both Aristotle's and those of his contemporaries – to avoid simply imposing established prejudices on Aristotle.

Our starting point is the following question: what is the context from which Aristotle speaks? This question may clarify other questions that are important to our goals: what is the goal of *Nicomachean Ethics*? Why was it written? These questions can help us to examine another decisive question from the hermeneutic point of view: what is its argumentative structure?

Scholars usually state that Aristotle portrays an ideologically uniform society from which he comprehends human ethical tasks. Both communitarism and its opponents seem to assume this interpretation, which is rooted in Hegel's view of Greek culture.[2] This is a key concept in the self-description of Modernity as plurality, diversity

---

[1] Aristotle has been read and explained from many points of view, and sometimes it is difficult to distinguish what is really Aristotelian from the things we know about "Aristotle." Aristotle may fit both a Thomistic and a Heideggerian point of view, to cite two major philosophers who have used him as the foundation of their own philosophies. It is impossible to read texts except with our own eyes. We must keep this hermeneutic advice in mind: if we cannot abandon our prejudices, we must attempt to criticize them, or we will not be able to learn anything from texts.

[2] According to paragraph 150 (and its Note) in *Principles of Philosophy of Right*, virtue "contains nothing more than conformity to the duties of the sphere to which the individual belongs" (integrity) "Virtues are ethical reality applied to the particular." From this, "what a man ought to do, or what duties he should fulfill in order to be virtuous, is in an ethical community not hard to say. He has

N.M.M.S. Coelho (✉)
Faculdade de Direito de Ribeirão Preto, University of São Paulo,
Avenida Bandeirantes 3900, Ribeirão Preto, São Paulo 14040-906, Brazil
e-mail: nunocoelho@usp.br

L. Huppes-Cluysenaer and N.M.M.S. Coelho (eds.), *Aristotle and The Philosophy of Law: Theory, Practice and Justice*, Ius Gentium: Comparative Perspectives on Law and Justice 23,
DOI 10.1007/978-94-007-6031-8_5, 

and tolerance (and of Post-Modernism as fragmentation and difference), which is differentiated from ethics of the Ancient and Middle Ages and their cultural and axiological uniformity.

However, this is not what actually arises from the texts of the fifth and fourth centuries. Athens in those times was the site of open discussion about everything. With the emergence of new ways of thinking, all beliefs and traditional concepts were systematically challenged. A few generations before Aristotle, the so-called pre-Socratics contradicted the universe's mythical foundations and proposed new principles to explain the behaviour of the cosmos. *Logoi,* resulting from the observation of the *physis,* replaced the narrative basis of life and world. Discussion began as early as the sixth century because there was little agreement about the *archai* proposed by the new scientists. Is it water, as Tales claims, or fire, as Heraclitus claims? The philosophical point of view is polemical from its beginning.[3] The thinkers of this first generation attacked traditional perspectives (see *Xenophon*, DK 21 B 11; *Heraclitus*, DK 22 B 5 or DK 22 B 42, according to which Homer deserves to be whipped) and attacked each other (*Heraclitus*, DK 21 B 81, calls Pythagoras "the king of charlatans"). The ruin of the traditional grounds of life[4] gives place to an enormous discussion.[5]

The *Sophists* (in Plato and Aristotle's terminology) immediately deepened the doubts and the discussion. They did not abandoned physical problems (as the *physiologoi* did not ignore social ones), but their speeches were shocking because they focused on political, moral and legal issues. They questioned the existence of gods,[6] the superiority of

---

to do nothing except what is presented, expressed and recognized in his established relations." In a similar interpretation of the ethical experience among Greeks, Günther states that acting well presupposes the correct interpretation of predetermined contextual forms of life and established practices. He concludes this from his reading of Gadamer, who reads *phronêsis* as something that does not sketch a new *ethos* but rather clarifies and specifies given normative contents (2004, p. 267). From the perspective of MacIntyre (1988, p. 141), in Greek ethics, the "standards of rational action directed to the good and the best" can only be expressed in types of activity wherein the goods are hierarchically ordered and the individuals occupy well-defined roles. "No practical rationality outside the *polis* is the Aristotelian counterpart to *extra ecclesiam nulla salus*."

[3] It is interesting to note that many times, the pre-Socratics identified the *archai* of the universe according to polemical principles. The same occurred in the rising medical science (see Vlastos, 1947).

[4] Jan Patočka, a very interesting thinker in the phenomenological tradition, explains that philosophical culture is marked by questionability: "History begins in Patočka's view, where that 'distance' is established, where that 'liberation' takes place, and where man starts to explicitly pose questions which were unnecessary in myth." To these questions, new and never before posed, there are no ready answers. 'The problematic character not of this or that but of the whole as such, as well as of the life that is rigorously integrated into it.' (HE 25) emerges. The discovery of this new, all-inclusive problematicity is seen by Patočka as a shock, which fundamentally changes the way of life which people up until that time lived, changing their world and man himself" (Chvatík 2004).

[5] This polemical scenario offers the starting point for the investigations of Aristotle in *Metaphysics*.

[6] Xenophon had already clearly stated that men created the gods according to their own image (see DK 21 B 16), but the fifth century would go further, to investigate why men created the gods. According to Democritus, they were led by fear of natural events; to Gorgias, it was gratitude (see Guthrie 1969, 1971).

Greeks,[7] and even the justice of law.[8] As specialists now warn, there was never a school to unify the sophists. Instead, they multisonorously fought each other[9] in public demonstrations of wisdom, sagacity and eloquence (see Casertano 2004).

Protagoras, possibly the most representative intellectual of Classical Greece, presented the formulas "man is the measure of all things"[10] and "on every issue, there are two arguments opposed to each other".[11] Some educational texts, such as the *Tetralogies* of Antiphon and the *Dissoi Logoi*, show that the ability to identify *antilogoi* was a key element of success in Athenian public life. Indeed, the *antilegein* was the principle that organised disputes in the political arena and in legal courts (in processes on which honour, citizenship and life often depended[12]). Democracy, the increase of the right of citizenship and the enlargement of powers of popular courts made this issue increasingly present in the daily life of Athenians.[13]

Socrates, Plato and Aristotle attempted to address the relativism of their times, endeavouring to overcome the disseminated antilogical perspective. Their attitudes, however, were influenced by Greek taste for controversy. Socrates was an obstinate critic, exercising his *elenchos* against every conception presented by the tradition or by his contemporaries, the sophists. Similarly, Plato insistently attacked not only the ideas of other philosophers (to which he many times dedicated his dialogues) but also traditional poetry (Homer) and theatre.

Plato and Aristotle were often feral in their treatment of the sophists. Even among themselves, however, there was polemic and controversy; it is enough to recall the famous passage in which Aristotle states (*EN* I.6.1096 a), against "the authors of the Theory of Ideas", that it is "desirable, and indeed it would seem to be obligatory,

---

[7] See Antiphon, DK 44 B.

[8] This was already an important problem to the Pre-Socratics and certainly was the main theme in fifth-century debates. See, for example, the Protagorean "great speech" (Plato, *Protagoras*, 320d), the *Tetralogies* of Antiphon, and all the investigations of Socrates. It is also the key problem in tragedy (recall *Antigone's* Sophocles). It is not by chance that this is the central subject of the Platonic reaction.

[9] Most of the hard work of grasping the significance of the lasting texts of sophistry is concerned with rebuilding the debates in which they were engaged. See, for example, the work of Untersteiner (1954) and Gagarin (2002).

[10] As we read in Plato (*Theaetetus*, trans. Harold N. Fowler 152a) "but one which Protagoras also used to give. Only, he has said the same thing in a different way. For he says somewhere that man is 'the measure of all things, of the existence of the things that are and the non-existence of the things that are not.'" Diogenes Laertius (*Lives of Eminent Philosophers* IX.51. trans. R.D. Hicks): "Furthermore he [Protagoras] began a work thus: 'Man is the measure of all things, of things that are that they are, and of things that are not that they are not.'" Also see Sextus Empiricus (*Against the mathematicians* VII.60).

[11] Diogenes Laertius, (*Lives of Eminent Philosophers* IX.51. trans. R.D. Hicks): "Protagoras was the first to maintain that there are two sides to every question, opposed to each other, and he even argued in this fashion, being the first to do so."

[12] About the role of litigation in Athenian life, see Christ 1998.

[13] Also theater (comedy and especially tragedy) exercises *antilegein*.

especially for a philosopher, to sacrifice even one's closest personal ties in defence of the truth."[14]

We must agree that this is not a picture of a homogeneous world.

## 5.2 *Antilogiae* and the Philosophical Method in Socrates, Plato and Aristotle

It is not clear what Protagoras really meant by his human-measure theory and the affirmation of *antilogiae*, but it represents much more than an isolated position. The way his statements were read by Plato and Aristotle tells us about its meaning at the time, when relativism seemed a common perspective on truth, justice and the gods.

Part of the reaction against that position was developed in the courts through legal processes against philosophers, such as the ones that led to the death of Socrates[15] and probably to the escape of Protagoras.[16] In the intellectual field, this position challenged a strong criticism that drove important discoveries. It is no exaggeration to suggest that the Socratic discovery of the concept,[17] the Platonic second navigation,[18] and the Aristotelian invention of Logic (based on his insistence on the principle of non-contradiction[19]) were strongly propelled by their need to fight the antilogical (and relativist) perspective.

The names of these three major thinkers of Classical Greece are often heard as indicating the same philosophical perspective about truth.[20] From a larger perspective, Socrates, Plato and Aristotle took on a common task: facing the legacy of relativism disseminated by the sophists, which mastered the political arena and the judicial courts in Athens.

However, the incentive they received from their enemies was not only negative; their methods were strongly determined by the atmosphere of controversy in which they lived. Socrates methodically asked about everything, and many times, he honestly could find no answer to his own questions, accepting *aporia* as a result of conversation. All of his investigations were developed through oral, face-to-face debates. But his strategy was grounded on an assumption totally contrary to

---

[14] Aristotle also criticizes the ideas of Socrates: *EN* VI.13.1144 b 15–20.

[15] As we can suppose from *The Clouds* (Aristophanes) and from the Platonic *Apology to Socrates*, it was not easy for Socrates' contemporaries to distinguish him from the other thinkers of the fifth century. We must remember that the platonic construction of the sophists was performed *pari passu* to the reinvention of Socrates as the paradigm of philosophical thought. See Marques (2000).

[16] See Dodds (2004)

[17] See Jaeger (1986) and Reale (1990).

[18] On the invention of ontology by the postulation of ideas, see Lima Vaz (1999).

[19] Let us assume the formulation of the law of non-contradiction used by Schiappa (2003, p. 192) in his analysis of this process: "It is a property of being itself that no being can both have and not have a given characteristic at one and the same time." Schiappa borrows from MacIntyre (1967).

[20] See Lima Vaz (1999).

Protagorean *antilogiae*. Confutation, the first step in his method, imposes the test of self-consistency on any opinion to be considered. False knowledge, which must be destroyed, is revealed as such through the perception that the consequences and starting points of the *logos* are not compatible.

The influence of the atmosphere of dialogue and controversy on the Platonic method is also evident. He also called *dialectics* the philosophical effort towards truth and developed all of his scientific texts as dialogues. Any contradiction can (must) be finally overcome by reason on the horizon of science, the true knowledge.

In Aristotle, the dialogical and antilogical perspective is even more important. It was not *tout court* denied as an expression of error and falsehood, but it was assumed, in principle, to structure part of reality and the sciences.[21] This brings us to the Aristotelian distinction between practical and theoretical sciences. In the practical sciences – politics, economics, law, ethics and rhetoric – we do not deal with a universe of axiomatic and undisputed principles. In this horizon (in which things have principles that vary), we must instead deal with conflicting possibilities and viewpoints. The principles (*archai*) of reasoning are problematic. We do not simply think from *axiomata*, but from *endoxa*, whose essential trait is to be a *logos* questioned by other *logoi*, by other *endoxa*. It is not initially obvious which of the conflicting *logoi* is correct. Which is the *orthos logos*? This question requires investigation, and the inquirer must be prepared to perform a much more complex reasoning to discover the truth.[22]

Two new sciences are proposed to address the intellectual tasks in this field. Topics,[23] which help one to find, identify and select *endoxa*, and *Dialectics*, which help one to think from them.[24]

Of course, Aristotle is neither Protagoras nor Antiphon, who may have accepted (or strongly stated) the antilogical structure of truth, which signified (for the sophists, as shown by Plato and Aristotle) the destruction of any truth. In neither scientific

[21] On *phainómena* as the starting point of Aristotelian methods and the rescue of the Protagorean and tragic anthropocentrism, see Nussbaum (1986). On Dialetics as general method in Aristotelean works, see Irwin (1988).

[22] This clarifies the meaning of *proairesis*, a central concept in ethical theory: "Deliberation seems to be implied by the very term *proaireton*, which denotes something *chosen before* other things" (*EN* III.2.1112 a, trans. Rackham).

[23] The first words of the book expose the antilogical context in which Topics is exercised: "Our treatise proposes to find a line of inquiry whereby we shall be able to reason from opinions that are generally accepted about every problem propounded to us, and also shall ourselves, *when standing up to an argument*, avoid saying anything that will obstruct us. First, then, we must say what reasoning is, and what its varieties are, in order to grasp dialectical reasoning: for this is the object of our search in the treatise before us" (*Top.* I.1 trans. Pickard).

[24] Dialectics has no meaning outside the horizon of controversy: "For it is not every proposition nor yet every problem that is to be set down as dialectical: for no one in his senses would make a proposition of what no one holds, nor yet make a problem of what is obvious to everybody or to most people: for the latter admits of no doubt, while to the former no one would assent" (*Top.* I.9.104 a 5–10). On the goals of Dialectics, see *Top.* (X and XI), which defines the 'dialectical proposition' and the 'dialectical problem'. It clarifies that this method aims at situations of divergent opinions, which give room for conflicting syllogisms. It also clarifies the kind of commitment of Dialectics to truth.

nor practical sciences could Aristotle accept that any statement can be truly made and refuted, that something is simultaneously true and false, or that everything is false.[25] Although he assumed that practical fields are intrinsically problematic, he fought relativism and endeavoured to find an adequate conception of truth. Aristotle's dialectic is a method of discovering truth in the context of controversial points of view, which always begins from them.[26] The dialectical development of conflicting arguments must show their weakness or strength and must help to find truth.[27] The way Aristotle does this often recalls Socrates and his verification of the coherence of thought. It is necessary to know, among conflicting *logoi* (among conflicting *endoxa*), which is better, and Dialectics help us to determine this with ratiocinations and questions: do the consequences of the *logos* deny its departure point? Does this *logos* depend on unacceptable principles or lead to unacceptable consequences? If so, it is false.[28]

The openness of Aristotle's system to the horizon of things whose existence (*arche*) depends on the human being – and the kind of truth that fits opinions, decisions and actions – allows the Philosopher to give a place to Rhetoric, the "counterpart of

---

[25] His *Sophistic Refutations* is dedicated to unmasking the strategies to make the infringement of the law of non-contradiction possible or to unmask apparent contradictions of a true statement. These strategies are closely connected to the Aristotelian concept of sophistry: "The art of the sophist is the semblance of wisdom without the reality, and the sophist is one who makes money from an apparent but unreal wisdom. (…)" (*SE*. 1.165a 20–25, trans. Pickard).

[26] His Topics aims at the same task. Topical correctness is the first step towards identifying sophisms and fallacies. The correct identification of *endoxa*, for example, allows for the identification of reasoning as merely polemic, which does not deserve to be taken seriously: "Reasoning is 'contentious' if it starts from opinions that seem to be generally accepted, but are not really such, or again if it merely seems to reason from opinions that are or seem to be generally accepted. For not every opinion that seems to be generally accepted actually is generally accepted. For in none of the opinions which we call generally accepted is the illusion entirely on the surface, as happens in the case of the principles of contentious arguments; for the nature of the fallacy in this is obvious immediately, and as a rule even to persons with little power of comprehension" (*Top*.I.1.100b 20–25, trans. Pickard).

[27] The commitment to truth does not mean ignorance of the specificities of truth in different horizons (scientific and dialectical): "For purposes of philosophy we must treat of these things according to their truth, but for dialectic only with an eye to general opinion" (*Top*. XIV. 105b 30–35). Dialectics aims at the truth that is possible on the horizon of things whose principles vary; this is different from stating that anything can be proclaimed or refused on this horizon. The consciousness of the specificity of truth in the different horizons of thought is related to the consciousness of the lack of exactness in ethical science.

[28] See, for example, *Top*. V, on the "sophistic method." He finishes this paragraph: "Any one who has made any statement whatever has in a certain sense made several statements, inasmuch as each statement has a number of necessary consequences: e.g. the man who said 'X is a man' has also said that it is an animal and that it is animate and a biped and capable of acquiring reason and knowledge, so that by the demolition of any single one of these consequences, of whatever kind, the original statement is demolished as well. But you should beware here too of making a change to a more difficult subject: for sometimes the consequence, and sometimes the original thesis, is the easier to demolish." Otherwise, the mistake may be revealed by the simple elucidation of the meaning of the words (see *Top*. XVIII.108a 20–35).

Dialectic" (*Rhet.* I.1, trans. J. H. Freese). Its horizon is marked by controversy, which is useful to people engaged in criticising or upholding arguments, defending or accusing. (*Rhet.* I.1) In its own way, Rhetoric is also committed to truth in practical contexts; it is the "faculty of discovering the possible means of persuasion in reference to any subject whatever". It is different from Dialectics because Rhetoric is concerned with persuasion and concrete circumstances. It is the exercising of *logos* in the context of acting, in which a decision is required among possibilities. It is a speech about how *logos* can be effective in its participation in dialogical (or, more than this, antilogical) situations, in which there are always other conflicting *logoi*.

## 5.3 Dialectic as Methodological Principle in *Nicomachean Ethics*

Politics, Ethics, Rhetoric and Dialectics share the same horizon.[29] Aristotle returns in Book VI of *Nicomachean Ethics*[30] to the division between thought dedicated to things whose principles are immutable (*epistêmê* and *sophia*) and thought concerned with things whose principles vary (*technê* and *phronêsis*) to frame the reasoning about ethical problems in the final context, similar to Dialectics (and Rhetoric). It is no *epistêmê*, wherein reasoning must only derive linear consequences from given and undisputed premises. Similar to Dialectics and Rhetoric, thought in Ethics is concerned with deciding and deliberating. Deliberating is needed because there is no previously given answer. Or, better, because there are many possible answers, and one must be chosen. Deliberative rhetoric helps people to be persuasive in discussions about what to do in the future. So does Ethics, which is concerned with deliberating and its main question: what should I do in this situation? As a dialectical science, it attempts to identify which opinion is better (that is, the most reasonable and plausible) in the context of conflicting *logoi*.

The link between Ethics and Dialetics is developed in different works. Many of the central problems of Ethics are discussed in dialectical and rhetorical texts, revealing the involvement of central ethical concepts in debates and antilogical confrontations in the language community. Happiness, good, convenience, forms of government, justice, pleasure, emotions, friendship, fear, shame, virtue, vice, character,

[29] "Rhetoric is as it were an offshoot of Dialectic and of the science of Ethics, which may be reasonably called Politics. That is why Rhetoric assumes the character of Politics, and those who claim to possess it, partly from ignorance, partly from boastfulness, and partly from other human weaknesses, do the same. For, as we said at the outset, Rhetoric is a sort of division or likeness of Dialectic, since neither of them is a science that deals with the nature of any definite subject, but they are merely faculties of furnishing arguments" (*Rhet.* I.2.1356a).

[30] 1139a: "Let us now similarly divide the rational part, and let it be assumed that there are two rational faculties, one whereby we contemplate those things whose first principles are invariable, and one whereby we contemplate those things which admit of variation (…)."

luck and fate are discussed in Rhetoric[31] from the perspective of persuasion and dissuasion in intersubjective situations.

Also the method in *Nicomachean Ethics* is clearly dialectical. This clarifies the task of reason in Ethics, whose situations are controversial.

First, Aristotle works topically to establish a map of considerable and clashing opinions (*endoxa*). He collects and interacts the conceptions of the majority, of the wise or the elderly, of the specialists, of the poets or dramatists, and of the other Philosophers. Eventually, he adds his own thesis.[32] These *endoxa* serve as starting points for further investigation. The development of each *logos* and the comparison among them must face *aporiai* and show their weaknesses or strengths. The unfolding of the argument may show that this opinion, beyond the first impression, is not really acceptable because it conflicts with another *logos* that one cannot consider excluding.[33]

The beginning of *Nicomachean Ethics* describes the plurality of conceptions of life and human good in Athens (*EN* I.4 and 5). Aristotle acknowledges that everyone agrees that happiness is the final human goal. But when he asks what "happiness" means, there is great dissension. The passage presents the scenario of alternative beliefs that works as a starting point and leitmotiv of the whole work. This dissension makes Ethics possible and necessary. If the meaning, end or principle of life were necessary (in the sense of the epistemic sciences), there would be no possibility or necessity of choosing. One must deliberate exactly because there are disputing conceptions of life to challenge decisions.[34]

Some seek money, others pleasure, friends, health, power or beauty. People dispute About life fulfillment, and Aristotle participates in this discussion. The Book, he

---

[31] *Rhet.* Book II is entirely dedicated to this subject. *Top.* also focuses on many of these themes, which offer material on the many types of reasoning discussed there. Both works, *Rhet.* and *Topics*, are concerned with preparing the reader to make dialectical arguments in deliberation based on a confident use of ethical and political *endoxa*.

[32] A "thesis", in a dialectical sense: "A 'thesis' is a supposition of some eminent philosopher that conflicts with the general opinion" (*Top.* I.11.104b 19–20, trans. Pickard).

[33] Let us take an example from *EN.* I.5.1095b 15 (trans. Rackham): from "the generality of men and the most vulgar", Aristotle gets the *logos* that "identif[ies] the Good with pleasure"; they "accordingly are content with the Life of Enjoyment." But the unfolding of this *logos* shows that living that way is living as a slave, and this challenges the Greek common opinion according to which a good life is a free life. This is enough to dismiss the first opinion as false.

[34] If there were not more than one point of view on the matter, the inquiry would be outside the boundaries of dialectical research: "Not every problem, nor every thesis, should be examined, but only one which might puzzle one of those who need argument, not punishment or perception. For people who are puzzled to know whether one ought to honor the gods and love one's parents or not need punishment, while those who are puzzled to know whether snow is white or not need perception. The subjects should not border too closely upon the sphere of demonstration, nor yet be too far removed from it: for the former cases admit of no doubt, while the latter involve difficulties too great for the art of the trainer" (*Top.* I.11.05. trans. Pickard). In the argumentation of *Nicomachean Ethics*, a good example of opinion that is not included in the dialectical game is *eudaimonia* as the end of human existence. This is not discussed, even by Aristotle. The dialectical investigation is required when the controversy begins.

says, is necessary not only to know about human good but to make people better. It is critique and counsel about living. His is the participant perspective; his attitude is not that of the moralist (who preaches) but of the dialectical inquirer. He faces moral issues as open for discussion. The starting points of thinking, in this sense, are not apodictically established. Despite incontrovertible concepts, we must think from conflicting opinions – that is, from *endoxa*, reasonable arguments supported by the majority or by the elder or the wiser, poiting to different conclusions. Dialectic deals with opinions that deserve consideration, as it is not yet decided which one must be discarded. This is exactly what reasoning must discover.

## 5.4 A Short Incidental Remark (or Warning)

The dialectical method exercised in *Nicomachean Ethics* incorporates different *logoi* using different strategies. Thus, Aristotle collects the *endoxa* he finds among his contemporaries to compose the dialectical scenario of the issue on which he wants to focus. We must not ignore the fact that conflicting arguments (*antilogoi*) are frequently the departure point of argumentation in *Nicomachean Ethics*, and their development structures the work.

This helps us to avoid serious mistakes. It would be wrong to assume these opinions as Aristotelian beliefs or scientific truths when they are presented only to be attacked or, at least, to be questioned or verified.

The notion of natural justice is the most important example. This concept, which is an important *topos* in Greek culture, is brought to the dialectical inquiry about justice in Book V and is exalted by the natural law tradition as the very perspective of Aristotle. Thus, it can invoke his authority in favour of natural law causes. However, this is hardly compatible with the argumentation we find in *Rhet.*, in which both written and unwritten law are rendered problematic by the antilogical structure of dialectical and rhetorical contexts. In *Rhet.* II.15. they are considered among "inartificial proofs, for these properly belong to forensic oratory. These proofs are five in number: laws, witnesses, contracts, torture, oaths." His concern, when he speaks of the laws, is the "use [that] should be made of them when exhorting or dissuading, accusing or defending." There is no definitive position in *Rhet.* about the superiority of written or unwritten law. Aristotle counsels the speaker in a legal process to invoke "the natural law" if his opponent grounds his claim on a written law, and the opposite if the opponent grounds his claim on natural law.[35] It would be absurd to quote these statements to show Aristotle as a jusnaturalist or a positivist. It is possible that the references to natural law in *Nicomachean Ethics* are

[35] "For it is evident that, if the written law is counter to our case, we must have recourse to the general law and equity, as more in accordance with justice (…). But if the written law favors our case, we must say that the oath of the dicast "to decide to the best of his judgement" does not justify him in deciding contrary to the law, but is only intended to relieve him from the charge of perjury, if he is ignorant of the meaning of the law (…)" (*Rhet.* II.15.1375a–b, trans. Freese).

not Aristotelian positions but rather *endoxa* to be problematized.[36] I do not want to address this point in this study, but if it is true, the claim that Aristotle is representative of jusnaturalism is weakened.

This also occurs in *Nicomachean Ethics*, where Aristotle constantly presents existing opinions (*endoxa*) as argumentative possibilities. This is very clear in some parts of the work, but not in others. We must pay attention to phrases such as "people usually state that…" or to references to poets, politicians or philosophers, but it is not always easy to recognise these statements. It is important to be aware of the dialectical structure of the book and to not naively assume that everything we find in the work is Aristotle's point of view.

## 5.5 Community, Language and Controversy in Key Aristotelian Concepts: *Anthropos, Polis, Eudaimonia*

I propose reading *Nicomachean Ethics* as follows: (1) it is a book written in an antilogical culture (in which many people believe that anything can be stated and contradicted). Instead of a homogeneous world, Aristotle describes and writes to a controversial and polemic society in which people do not share the same opinions about human good but rather dispute it. The plurality of opinions is the starting point of his ethical theory; and (2) because of this, Aristotle applies the dialectical method to address ethical problems because he assumes its dialogical (or antilogical) structure.

Let us now consider some key concepts of Aristotle's Philosophy to see how they presuppose the language community and its controversial structure.

As we have seen, the conception of *eudaimonia*, as presented in *Nicomachean Ethics,*[37] is utterly involved in controversy, and this polemic is the starting point of ethical investigation and practical challenges. From another perspective, we can realise the polemical essence of *eudaimonia* in the context of the agonistic Greek culture, wherein the search for success was seen as participation in a competition. This comes from Homeric times (remember Achilles and Agamemnon risking the Achaeans for the ownership of a girl; or the dispute for Achilles' shield)[38] and

[36] See Vega (2010).

[37] In *Rhet.* I.5. Aristotle does not support one of the conceptions of *eudaimonia* of his time. They are presented as parts of *eudaimonia*, and they can all be useful to the speechmaker in persuading or dissuading. *Eudaimonia* is "good birth, plenty of friends, good friends, wealth, good children, plenty of children, a happy old age, also such bodily excellences as health, beauty, strength, large stature, athletic powers, together with fame, honor, good luck, and virtue." It is also self-sufficiency, pleasure, safety, nobility, wealthy, reputation, honor, etc. We must also notice that in this section, *eudaimonia*, the central concept in Ethics, is declared the central concept in Rhetoric, "for all who exhort or dissuade discuss happiness and the things which conduce or are detrimental to it."

[38] "Most poetry and prose from at least as early as Homer through the classical period was created for or performed in a competitive environment, often in a formal contest. The Sophists entered fully into this competitive spirit, challenging earlier authors and striving to outdo their contemporary rivals" (Gagarin 2002, p. 19).

increased in the context of democracy, in which life fulfilment depended on success in political and judicial battles.[39]

Let us briefly recall two other concepts.

The famous definition of *anthropos* as the animal that has the *logos* (*zôon logon echôn*) puts *antilogiae* at the heart of the Aristotelian concept of human being. Man is not defined as a gregarious animal. This is not enough; there is something more than living together to specify him: he is capable of speech, which is not "mere voice" but the capacity to distinguish right and wrong, good and bad.[40]

Notice that the conception of language employed by Aristotle directly refers to antilogical use; human language and rationality (*logos*) are oriented towards controversy. The singularity of the species lies in its capacity to use language in this sense. Having and exercising *logos* means discerning (and discussing and disputing) between opposites, the profitable and the useless, the just and the unjust. This changes human beings, from a simply gregarious animal, into a political animal, unlike other ones, such as the bee, the ant or the wasp.[41]

This "controversial principle" we find in the Aristotelian concept of the human being has a strong influence on his Politics. It defines a political community, the nature and the ends of written laws, and the purpose of political theory. The *polis* is a community characterised by difference; it is a place where different people coexist.[42] The *polis* is not a community of bees but of beings capable of speech and of meaning. These two aspects show how untruthful it is to present community, in Greek Philosophy, as

---

[39] "Athenian courts, rather than providing a forum for the resolution of disputes and avoidance of further conflict, instead furnish an arena which seek out to pursue and intensify antagonisms" (Cohen, 1995, p. 8–9). "Athenian law was less concerned to secure 'justice' in a narrowly legal sense than to negotiate balance of social and political power between the two litigants" (Gagarin 2002, p. 145).

[40] *Pol.* 1253a: "And why man is a political animal in a greater measure than any bee or any gregarious animal is clear. For nature, as we declare, does nothing without purpose; and man alone of the animals possesses speech. The mere voice, it is true, can indicate pain and pleasure, and therefore is possessed by the other animals as well (for their nature has been developed so far as to have sensations of what is painful and pleasant and to indicate those sensations to one another, but speech is designed to indicate the advantageous and the harmful, and therefore also the right and the wrong; for it is the special property of man in distinction from the other animals that he alone has perception of good and bad and right and wrong and the other moral qualities, and it is partnership in these things that makes a household and a city-state" (trans. Jowett).

[41] In the first paragraph of *Rhet.* Aristotle stresses the controversial nature of human beings to suggest the universality of Rhetoric: "All men in a manner have a share of both [of Rhetoric and of Dialectics]; for all, up to a certain point, endeavor to criticize or uphold an argument, to defend themselves or to accuse."

[42] Politics II.1.1261a: "Yet it is clear that if the process of unification advances beyond a certain point, the city will not be a city at all for a state essentially consists of a multitude of persons, and if its unification is carried beyond a certain point, city will be reduced to family and family to individual, for we should pronounce the family to be a more complete unity than the city, and the single person than the family; so that even if any lawgiver were able to unify the state, he must not do so, for he will destroy it in the process. And not only does a city consist of a multitude of human beings, it consists of human beings differing in kind. A collection of persons all alike does not constitute a state."

something uniform (as Modern and some Contemporary Philosophy does). In contrast, the experience of division and secession was the main trait of political experience in the fifth and fourth centuries. Thus, the preservation of the community became the main concern of Aristotelian political theory.[43] Law and the tribunals are understood according to this atmosphere of division and with the aim of attaining *koinônia*.

Another Aristotelian concept that is strongly marked by the antilogical structure of language and reasoning (*logos*) is *orthos logos*. Let us devote a specific section to this topic, which will help us to deepen the analysis of practical reason from the perspective of its integration in the language community in the special (antilogical) sense we understand it here.

## 5.6 *Phronêsis* and Its Inquiry: Which *Logos* Is the *Orthos Logos?*

The *orthos logos* is a key concept in Ethics. It is associated with the very ethical question: how should one act? Ethical questions relate to the good in action. Aware of the limits of Ethics (again: it is not *epistêmê*, and it neither aspires nor admits its precision[44]), Aristotle the ethicist cannot simply say what is right. This answer always hangs on the situation, and Ethics, as a practical science, can only offer general indications.[45] Therefore, practical science can only show the target; acting well is acting according to the mean (*mesotês*). That is, the moral task is measuring desire; acting well is desiring neither too much nor too little from security, honour, pleasure, or money in the context of a given situation. The task of practical reason is to find the *mesotês* under the circumstances.

Many people find this theory unhelpful because it abandons the actor to decide for himself. Aristotle simply admits this: "This bare statement however, although true, is not at all enlightening."[46] This point is important; Aristotle does not ask of Ethics more than it can offer. Nothing (neither the treatise on Ethics nor even the law) can take the place of the person in the context of action; she must realise the circumstances, think and find an answer that fits the situation.

The entire moral problem is transposed to this task. This makes Book VI absolutely central, and thus *orthos logos*; it is dedicated to the rational discovery of the right thing to do.

Book VI is dedicated to clarifying the notion of *orthos logos*. Thus, it begins: "We have already said that it is right to choose the mean and to avoid excess and

---

[43] See Bates (2008).

[44] *EN* I.1.1094a 11–13.

[45] An important example of the limits of ethical theory is the discussion of the limits of responsibility, which cannot be determined without considering the particular circumstances of action. *EN* V.6.1135a 20–31.

[46] *EN* 1138b 20.

deficiency, and that the mean is prescribed by the right principle. Let us now analyse the latter notion."

To do so, Aristotle dialectically presents a series of related concepts, *endoxa,* and one of these candidates is the right kind of reasoning to address practical situations. The issue in this Book is the kind of reasoning that can identify the *orthos logos* in the context of acting. There are some possibilities (*endoxa* supported by different people) that must be seriously considered: is it a kind of scientific knowledge? None less than Socrates and Plato had stated so. Is it a kind of sensual perception? Is it technical expertise? Is it deliberation?

His answer is well known. *Orthos logos* is the result of deliberation, which leads to decision and prepares for action. It is strictly linked to *phronêsis*, a central concept that influenced Roman *Jurisprudentia* and that would give the emerging science of law its essential features. This is a very important concept in the Western history of philosophy.

*Phronêsis*, one of the virtues of the rational dimension of the soul, is excellence in discovering the right principle of action. It is the intelligence that makes one able to find the *orthos logos* in concrete practical situations. This allows us to say that the *orthos logos* rests at the heart of Ethics. At the same time, it brings to the concept of *orthos logos* the difficulties in understanding how practical reason works, especially knowing which criterion can be used to determine which decision is right.

*Phronêsis* is a complex concept because it is narrowly connected to other intellectual virtues, which it seems sometimes to comprise (such as *gnomê*, *nous*, *synesis* and *deinotês*) or with which it maintains important relations and homologies (such as *epiteme*, *sophia*, and *technê)*. In contrast, *phronêsis* is strongly connected to the non-rational dimension of the soul, especially to desire (*orexis*), sensation (*aisthêsis*) and passion (*pathos*).

Both *technê* and *phronêsis* are excellencies in thinking things whose principles are mutable because their principles lie in the human being.[47] The principle of the house (its project, its draft) lies in the architect, and the principle of action (decision) lies in the person who is acting. But although technical thought leads to something exterior to itself (the work, the creation), *phronêsis* does not; nothing exterior to the person who deliberates results from deliberation.

In contrast, *epistêmê* and *sophia* deal with objects whose principles are immutable. They consider the parts of the universe that are not determined by human activity, such as the movement of stars or the march of the seasons.

The openness to human influence is the main difference between the horizon of theoretical sciences and the horizon of technique and praxis. Here, human activity is responsible for things being the way they are. These paths of thinking do not simply watch their objects (as in theoretical thinking); they build their objects. This is the field of freedom and responsibility. Human beings' decisions are the cause of things, and the task of thought is to find out this principle.

---

[47] *Met.* VI.1.1025 b: "The first principle of things that are practical is free-will in the agent (...)."

Things, in practical horizon, can be done in different ways. There are different possibilities of acting, and *this* unfolds the dialectical field. More than one possibility of acting seem good, and different people advocate different conduct as righteous.

Even in the context of the soul of a single person, there is a debate about the right thing to do. During the process of deliberation, different possibilities are proposed as principles of action. Deliberating is the process of considering which one is the best. Among different *logoi*, which one is correct? Which is the *orthos logos?*

How does one decide among different *logoi*, among conflicting proposals of the mean?

As we can see, the problem remains open. Some people believe that in the *orthos logos* can be found the criterion of the goodness of acting. However, among different possibilities, one should choose the one that corresponds to the *orthos logos*. The task is finding this.

## 5.7 Orthos Logos *in a Controversial Context*

We can see that *orthos logos* is not a criterion to be used to deliberate but is the result of practical reasoning. Thus, it is what practical reason must discover. It is the outcome of good deliberation and the principle of virtuous behaviour.

There are two passages to corroborate this reading, although they are translated by some scholars to obscure *orthos logos* as the result of practical research.

In *EN* II.6.1107a, the mean (which practical reason must find and affirm) is given by practical wisdom: "Virtue then is a settled disposition of the mind determining the choice of actions and emotions, consisting essentially in the observance of the mean relative to us, this being determined by principle, that is, *as the prudent man would determine it*."[48]

In *EN* VI.13.1144b 20–25, it is also stressed that *orthos logos* is the result of *phronêsis*: "A proof of this is that everyone, even at the present day, in defining Virtue, after saying what disposition it is and specifying the things with which it is concerned, adds that it is a disposition determined by the right principle; and *the right principle is the principle determined by Prudence* (…). It appears therefore that everybody in some sense, divines that Virtue is a disposition of this nature, namely regulated by Prudence. This formula however requires a slight

[48] Different translations change everything here. See, for example, the translation by W.D. Ross: "Virtue, then, is a state of character concerned with choice, lying in a mean, i.e. the mean relative to us, this being determined by a rational principle, *and by that principle by which the man of practical wisdom would determine it*." At first glance, the last words could suggest that *orthos logos* preexists the exercise of practical reason, working as a model or a source of truth to be used by the *phronimos*. In this sense, the expression would be exempt from any antilogical or even dialogical coloration.

modification. Virtue is not merely a disposition conforming to right principle, but one cooperating with right principle; *and Prudence is right principle in matters of conduct*" (trans. Rackham).

Now we will consider the expression *orthos logos*.

Many of the different meanings of the word *logos* can fit Aristotelian speech. In the expression *orthos logos*, *logos* can mean (*Lidell-Scott Greek Lexicon*) *process of thought, discussion, debate, deliberation, thinking, reasoning, reflection, dialogue*, or *debate*, pointing to the entire process of discovering the truth in practical situations. In a different sense, it can mean *measure, reckoning, relation, correspondence, proportion*, or *ratio*, pointing to a kind of comparison of magnitudes that is very close to *phronêsis*, as research on the mean, which prepares decision. *Logos* can also mean *ground, purpose, reason, statement of a theory, argument, proposition, idea, expression, phrase, thesis, hypothesis, provisional ground, explanation*, or a *speech* delivered in court or assembly. In this sense, *logos* fits the proposition that reason offers to the irrational dimension of the soul (desire, which can embrace or refuse it) in the deliberation process. In this sense, *logos* may also point to an arena of clashing arguments.

In another group of meanings, *logos* can mean *consideration, value, esteem, counsel, rule, principle, law, word of command*, or *behest*, pointing to the normative nature of the counsel given by reason to desire in the process of a decision.

Yet, the expression *orthos logos* seems to indicate something else. Why not simply *logos*, but *orthos logos*? The expression was not invented by Aristotle; according to his own words, it is a common expression.[49] Indeed, the expression is a legacy from sophistry in its ethical and rhetorical discussion.[50] Kerferd recalls the use of the expression in the dialogue between Protagoras and Pericles (DK 80 A 10) and in a fragment of Antiphon (DK 87 B 44 A). This earlier use of the expression by the sophists should warn us about its antilogical compromises, if it does not work as a vaccine against its jusnaturalistic interpretation.

In the expression *orthos logos*, the adjective *orthos* possibly indicates that this is one *logos* among others. The adjective is necessary to distinguish one *logos* (the one that hits the target, finding the mean and proposing it to the desiring dimension of the soul) from others in a conflict of possibilities of acting, of counsels, of *logoi*. It directly refers to the context of *logoi* that compete to become a decision. The person who is acting has different choices; which one of them is the right one?

---

[49] *EN* VI.13.1144b 20–25, trans. Rackham: "A proof of this is that *everyone, even at the present day*, in defining Virtue, after saying what disposition it is and specifying the things with which it is concerned, adds that it is a disposition determined by the right principle; and the right principle is the principle determined by Prudence." Also see *EN* II.2.1103b 30–35: "Now the formula 'to act in conformity with right principle' is common ground, and may be assumed as the basis of our discussion."

[50] "But persuasion consists in making one view appear preferable in at least some respect. One way was to class the preferred argument as *orthos* – 'upright', 'straight', 'right' or *orthoteros* 'straighter', 'more correct and so on', and it is clear that the concept of *orthotes* and of an *orthos logos* was important" (Kerferd 1981, p. 102). See also Schiappa (2003).

## 5.8 The Ground of Practical Truth and the Community of Practice and Language: The *Phronimos*

Thought, in the context of practice, is dialectical. Under the circumstances of a specific situation, different principles appear; which of them is the righteous principle? Which of them indicates the mean of desire (*mesotês*)? Of course, nothing we have said thus far answers this question, and the problem remains open. The practical question remains: where is practical truth? How should one act *hic et nunc*?

There is an answer in *Nicomachean Ethics*. It is interesting to note that this answer sends the reader to the arena of discussion of the community of language. Deciding well, Aristotle states, is acting as a virtuous and wise person would act. The *phronimos* is the criterion of correctness in deliberation. As we saw in *EN* II.6.1107, the *orthos logos* is the *logos* that would be found by the reasonable and good man.

In *EN* II.4.1105a 5–10, there is a similar argument, which affirms that the just man is the criterion to determine if an action is just: "Thus although actions are entitled just and temperate when they are such acts as just and temperate men would do, the agent is just and temperate not when he does these acts merely, but when he does them in the way in which just and temperate men do them."

This establishes an invincible circularity. The *phronimos* is defined as a person who deliberates well; he has *phronêsis*, the good habit of finding the *mesotês* in practical situations, and so he is expert at finding and following the *orthos logos* during his life. At the same time, good deliberation is defined as the reasoning that a *phronimos* would make. In this sense, the *orthos logos* is the *logos* that would be affirmed by the *phronimos* in a particular context.

It is difficult to deny this circularity in the argumentation of *Nicomachean Ethics*. However, there is something to learn from this because it indicates important things about practical reason. In particular, I would like to direct attention to the fact that this circularity allows Aristotle to avoid an escape valve from the communitarian and linguistic context in which the practical investigation develops. Thus, nothing could come to the rescue of the person who is acting or could take his place in his own and untransferable task of deciding – nothing like an Idea, or Nature, or any definitive help from *epistêmê* or *sophia*.

Practical reason is enclosed in the linguistic context. The reference to the *phronimos* as the criterion of practical truth in the concrete process of deliberation signifies the calling of language-practice community to participate in individual investigations about the right thing to do. The *phronimos* is the person who is praised by the community for his good decisions,[51] and the community's values, opinions and examples are reintroduced into the debate that will lead to ethical decisions.

There are many interpretations of the meaning of the *phronimos* as the criterion of the good. Some criticise Aristotelian ethics as a conservative and even authoritarian theory of practical truth. Truth is anything that the empowered old and rich male says. This interpretation, however, ignores the fact that being *phronimos*, for

[51] The examples of *phronimoi* always indicate someone to whom the Greeks reserved great glory for his participation in the construction of the community. The paradigm of the *phronimos* is Pericles, the most important Athenian statesman (see *EN* VI.5.1140 b 7–11).

Aristotle, does not mean simply being able to repeat past decisions given, but rather to find, *hic et nunc,* the correct answer from a conscientious awareness of the circumstances of each particular situation. It is not just following whatever the law states; the law can be defective. It may even be necessary to disobey the law and to find the *orthos logos* by oneself.

The reference to the *phronimos* as the ground of practical truth does not signify that the right decision simply repeats established values and norms or established practices. We are invited to think as the *phronimos* **would** think, which is completely different. We do not have space to discuss this subject here, but a meditation about *epieikeia.* (*EN* V.10) can be helpful to clarify that the right thing to do, for Aristotle, is not the repetition of established standards, legal or consuetudinary. Law may be defective because each particular situation is new, and it is impossible for the legislator to predict everything. The same is true about decisions made by the historical *phonimoi* in the past. The *orthos logos* is the adequate answer to the **present** situation, which is always new (or, at least, possibly new).

It is indeed a good counsel to observe what wise people have done before. Certainly, some help may come from the past (from the experience of the *polis*) and from the law. But this may not be enough. So the question remains open: how can one know if it is necessary to follow the law or the example of the past, or whether it is a situation that requires equity?

Thus, the reference to the *phronimos* does not give the actor an answer about the right thing to do. The question remains open, and it must remain open. The *orthos logos,* the principle of action, as the appropriate response to a situation, must be found by the person who acts. Any clarification from ethical science can only provide general indications. The same is true for law and the example of wise and virtuous people: they establish a frame to comprehend ethical tasks, but they do not simply provide an answer. The *orthos logos* is always to be found with the help of these people (including the law and the paradigms), but the answer is not already there.

Of course, this conclusion is a truism from a dialectical and ethical perspective. If the answer preexisted practical research, deliberation would not be necessary. The dialectical effort is renewed in each practical situation.

This makes the dialectical debate about *eudaimonia* endless. Of course, a good solution *hic et nunc* will contribute to the discovery of practical truth in future situations. If one is successful in facing practical tasks, becoming recognised as *phronimos* and virtuous (*eudaimôn*), future generations will honour and regard him as a *criterion* for acting. However, the future generations of the language-practice community will always have, as we do now, the task of being *phronimoi* themselves, the task of finding the *orthos logos* in their own contexts.

The reference to the *phronimos* is not an answer, but a challenge. As Aristotle states, "The best are the ones who can find the answer by themselves"[52] (*EN*. I.4.1096b 10).

[52] Other scholars think the *phronimos* is a philosopher. The problem here is the ultimate assimilation between *phronêsis* and *sophia/epistêmê*. Why would Aristotle be so concerned about distinguishing them? Considering the *phronimos* a kind of scientific expert removes *phronêsis* from the field of Dialectic and moves it to the field of apodictical reasoning (demonstration). However, as Aristotle frequently states, this is unacceptable. Ethics does not belong to the same horizon as Geometry but rather belongs to Politics, Rhetoric, and Dialectic.

## 5.9 The Search for Practical Truth in the Paradigm of Dialogue: Soul as *Agora*

I have stated that Aristotelian Ethics presupposes a community of language marked by controversy and struggle. In this horizon (which frames Dialectics and Rhetoric), ethical tasks are faced, and practical reason is mobilised.

I would like to suggest another perspective from which practical thought seems to be especially linked to a concept of language community in the controversial sense. For this last topic, let us give attention to the process of deliberation as it is developed in the soul of the individual to determine how Aristotle employs the language community's decisions and processes to clarify the way each singular moral decision is built.

The analogy between the city and the soul is common in Greek political philosophy.[53] In my opinion, when Aristotle makes this comparison, he implies and mobilises a conception of community distinguished by *antilogiae* and controversy, and this presupposition is relevant to his description of practical reasoning.

Aristotle presents the different elements that participate in the investigation of practical truth.[54] A decision is a deliberated desire, or a desired deliberation.[55] According to this formula, a decision is a sort of amalgam of desire and thought[56] that requires the participation of two different dimensions of the soul: the sensitive-appetitive (in which sensibility, passion, desire, and movement are grounded) and the rational part. A chosen action always results from the contribution of these two dimensions of the soul (rationality and desire).[57]

The right decision is the embracing of desire and the principle proposed by correct deliberation.[58] In a way, this relationship is problematic because desire and

[53] Recall Plato (*Republic*. IV).

[54] This is not the result of an isolated power, function or activity of the soul. He dissents from Socrates, to whom all virtues are aspects of an only virtue (wisdom): "Socrates' line of enquiry was right in one way though wrong in another; he was mistaken in thinking that all the virtues are forms of Prudence, but right in saying that they cannot exist without Prudence (…) he said that they are all of them forms of knowledge" (*EN* VI.2.1139b).

[55] *EN* VI.2.1139b 5: "Choice may be called either thought related to desire or desire related to thought; and man, as an originator of action, is a union of desire and intellect."

[56] *EN* VI.2.1139b: "Now the cause of action (the efficient, not the final cause) is choice, and the cause of choice is desire and reasoning directed to some end. Hence choice necessarily involves both intellect or thought and a certain disposition of character [for doing well and the reverse in the sphere of action necessarily involve thought and character]. Thought by itself however moves nothing, but only thought directed to an end, and dealing with action."

[57] This complex conception of the soul turns the human challenge of self-construction into a double task: *eudaimonia* requires being both prudent (*phronimos*) and ethically virtuous. It is necessary to develop both dimensions of the soul. Actually, it is impossible to be virtuous without being *phronimos* or vice versa. *EN* VI. 12,1114a: "It is clear that we cannot be prudent without being good;" *EN* VI.12.1114b: "True Virtue cannot exist without Prudence" because character and practical intelligence are developed in the same situations. Both are required each time one must define one's behavior in each intersubjective situation.

[58] *EN*. VI.2.1139a, 20–25: "Desire must pursue the same things as principle affirms."

thought do not necessarily agree. Instead of coming to terms, they may dissent, exposing the soul to tension.

All difficulties of practical life arise here. There is always a risk of failing to find an adequate answer to a concrete situation. In this case, a lack of (practical) intelligence causes a poor decision. But there is another possibility of failure: even if reason finds the mean (*mesotês*), acting well still depends on its embrace by desire.

The description of the relation between these two different dimensions of the soul suggests that they establish a dialogue in which the rational dimension faces the task of persuading the desiring dimension to wish what is correct. This relationship works according to a political or deliberative model and not to an imposing one.[59] This means that the most that reason can do is to offer counsel to desire.

This counsel and its acceptance by desire occur in an excited context. Deciding is always related to passion, pleasure and pain. *Pathos* is an inevitable ingredient of ethical situations. Unlike epistemic investigations, deliberation is the search for truth under the pressure of time[60] and the influence of passions in the context of an existential situation.[61] In the process of deliberation, the soul seems to work as the *agora*.[62]

Our last argument considers how practical deliberation in the soul alludes to the structure of deliberative processes.

---

[59] According to Aristotle, the effort for *eudaimonia* is an effort to put the soul in order. Acting well is acting according to the right soul's hierarchy. The order of the soul is clear from a political paradigm and obviously reflects the aristocratic preferences of the Philosopher. The rational dimension must rule over the desiring dimension, but their relation is not thought of as a tyranny. The best that reason can do to impose its *logos* is to offer its advice to desire. Desire does not necessarily follow the indications of reason; it must be persuaded. If reason fails to convince desire, man fails in acting well. Triumph in the task of doing the right thing in practical situations depends on success in the rhetorical task of convincing oneself.

[60] The search for practical truth always occurs under the pressure of time. People do not have all of the time they wish to decide because a decision, to be good, must attend to the request of opportunity (*kairos*). This temporal dimension integrates practical truth (if the decision is delayed, it is *ipso facto* wrong).

[61] The search for practical truth is always in a specific existential context. It is necessary to consider the many aspects of the problem to determine the right answer. The *phronimos* is sensitive to relevant circumstances and is able to distinguish how to act from a clever perception of the people involved (he does not deal with children as he does with an old man), the time and place, the available means *etc*. This is connected to the limits of exactness (*akribes*) of Ethics, which cannot establish outside of the situation (in general) what is right or wrong. Rhetoric is considered less exact and strictly connected to a concrete situation. In *EN*, Aristotle states how ridiculous it is to require from a rector the same exactness we desire from a mathematician. The problem is the disregard of circumstances. In *Rhet.* Aristotle emphasizes the circumstantiality of argumentation. Persuasion always occurs in a concrete context. The choice of the arguments one should use is completely influenced by the context of argumentation; it depends on who the speaker is and who the hearer is, for example.

[62] Let us recall Plato's *Apology to Socrates* to quickly understand how feelings and passions appear in the *Agora*. Defenders can implore for commiseration, judges can become furious. Passions are an unavoidable ingredient in any rhetorical effort, such as in the context of acting. The search for the principle of acting requires one to deal with one's own passions. In fact, dealing with *pathos* is the starting point and the inevitable task of Ethics.

The conclusions of practical thought are not indifferent to the means available to the person who thinks and who must act. At the same time, practical conclusions are not indifferent to their consequences. This shows that thinking in the horizon of praxis is a multifaceted process that endures over time. Deliberating before acting requires time because it considers the situation in its particular circumstance (acting well is desiring the *mesotês* related to the context) from different perspectives (it is necessary, for example, to think about its feasibility – do we have the means to accomplish it? – and about its consequences). This submission of the problem and its possible answers to different perspectives, which draw upon practical reasoning, characterises rhetoric as well. In fact, this is the main trait of rhetoric, as Aristotle understands it.

Thinking about the right thing to do is not a linear ratiocination. Notwithstanding the use of syllogism in practical reason, it does not exhaust everything involved in deliberation. A brief consideration of what happens when we deliberate shows that deciding is more complex. It involves a progression through coming and returning to allow the problem to be considered from various perspectives. An important part of this is thinking about the action we would like to select from the perspective of the means we have to enact this possibility. In the search for the principle of acting, people must always consider their own capacity to accomplish their decisions. This is an unavoidable requisite of truth in the practical horizon; Aristotelian Ethics is committed to the efficacy of acting. Acting well is not only wishing well but also the ability to put it into practice. The same structure can be found in rhetorical argumentation. It is marked by the analysis of the problem from diverse perspectives, which include the analysis of its consequences and its viability.

In the same way, practical thought is shown as a non-linear ratiocination. It comprehends a comparison between possibilities from the point of view of the consequences. Good deliberation depends on understanding the costs of different possible decisions. Aristotelian Ethics, contrary to Kantian Ethics, is sensible to the consequences of acting and requires that possibilities be balanced from this perspective, as politicians (and rhetoricians) do.

## 5.10 Conclusions: On the Contextualism of Aristotle

The way the community is conceived and mobilised in Aristotelian Ethics deserves reconsideration of contemporary Philosophy.

Of course, there is no Ethics without the presupposition of a community within which people exercise senses, passion, desire, memory and reasoning. A practical decision is a historical challenge that finds in history its possibility and its grounds.

In Aristotle, however, there is no community conceived as a homogeneous group. Instead, life in common, in his time, was subject to vivid disputes. His was a world marked by doubts and contradictions, rivalry and controversy, in both the political and the intellectual arenas. This situation drove Aristotle to understand the exercise of practical reason in a context of difference and dialogue.

The community Aristotle mobilises is a language community marked by the contrast of opinions, a horizon on which the question of the right thing to do does not dispose of given and non-problematic answers. On the contrary, it always challenges the capacity to deliberate, to find truth, in each particular situation, and considers different (and clashing) possibilities. It is not simply the repetition of established practices or the application of a supposed uniform and consolidated *ethos* but rather participation in the intense and endless debate about the meaning of human life.

"What a man ought to do, or what duties he should fulfil in order to be virtuous", in the horizon of Aristotelian Ethics, is anything but an easy task.

## Bibliography

Antiphon the sophist. 2002. *The fragments*. Trans. Gerard J. Pendrick. Cambridge: Cambridge University Press.
Antiphon. 1965. *Discours suivis des fragments d'Antiphon le Sophiste*. Trans. Louis Gernet. Paris: Société d'Edition "Les Belles Lettres".
Antiphon. 1997. *The speeches*. Trans. Michael Gagarin. Cambridge: Cambridge University Press.
*Antiphonte Tetralogie*. 1969. Trans. Fernanda Decleva Caizzi. Milan-Varese: Istituto Editoriale Cisalpino.
Aristotle. 2012. *(SE) On sophistical refutations*. [Internet classics archive]. http://Classics.Mit.Edu/Aristotle/Sophist_Refut.Html
Aristotle. 1926a. *(Rhet.) Rhet*. Trans. J.H. Freese. Cambridge/London: Harvard University Press/William Heinemann.
Aristotle. 1926b. *(Top.) Topic*. Trans. W.A. Pickard. Cambridge/London. Harvard University Press/William Heinemann Ltd.
Aristotle. 1933/1989. *(Met.) Metaphysics*. Trans. Hugh Tredennick. Cambridge/London: Harvard University Press/William Heinemann.
Aristotle. 1934. *(En) Nicomachean ethics*. Trans. H. Rackham. Cambridge/London: Harvard University Press/William Heinemann.
Aristotle 1984. *(Pol.)* "Politics" *The complete works of Aristotle*, vol. 2. Trans. Jowett, Benjamin. Ed. Jonathan Barnes, 1987–2129. Princeton: Princeton University Press.
Bates, Clifford. 2008. The fundamental political problem of legislators addressed in Politics 2. *Revista da Faculdade de Direito de Conselheiro Lafaiete*, 23–36.
Casertano, Giovanni. 2004. *Sofista*. Napoli: Guida.
Christ, Matthew R. 1998. *The litigious Athenian*. Baltimore: The Johns Hopkins University Press.
Chvatík, Ivan. 2004. The heretical conception of the European legacy. *Essays of Jan Patočka*, 55–72. Lisbon: Phainomenon.
Cohen, David. 1995. *Law, violence and community in classical Athens*. Cambridge: Cambridge University Press.
de Lima Vaz, Henrique C. 1999. *Escritos de filosofia IV – Introdução à ética filosófica 1*. São Paulo: Edições Loyola.
Dodds, E.R. 2004. *The Greeks and the irrational*. Berkeley/Los Angeles/Oxford: University of California Press.
Diels, Hermann, and Walther Kranz. 1989. *Fragmente der Vorsokratiker*. Hidessheim: Weidmann.
Gagarin, Michael. 2002. *Antiphon the Athenian, oratory, law, and justice in the age of the sophists*. Austin: University of Texas.
Guthrie, W.K.C. 1969/1971. *The sophists*. London: Cambridge University Press.
Guthrie, W.K.C. 1971. *The sophists*. New York: Cambridge University.

Irwin, Terence. 1988. *Aristotle's first principles*. Oxford: Clarendon Press.
Jaeger, Werner. 1986. *Paideia. The ideals of Greek culture*. Trans Gilbert Highet. New York: Oxford University Press.
Kerferd, G.B. 1981. *The sophistic movement*. Cambridge: Cambridge University Press.
Laertius, Diogenes. 1925. *Lives of eminent philosophers*. Trans R. D. Hicks. Cambridge: Harvard University Press, Loeb Classical Library.
Laertius, Diogenes, and R.D. Hicks. 1972. *Lives of eminent philosophers*. Cambridge: Harvard University Press.
Liddell, Henry George, and Robert Scott. 1940. *A Greek-English lexicon*. Oxford: Clarendon.
MacIntyre, Alasdair. 1967. Ontology. *The encyclopedia of philosophy, Vol. 5*, 542–543. New York: Macmillan.
MacIntyre, Alasdair. 1988. *Whose justice? Which rationality?* Notre Dame: University of Notre Dame Press.
Marques, Marcelo Pimenta. 2000. O sofista: uma invenção platônica? *Kriterion* 102: 66–88.
Nussbaum, Martha C. 1986. *The fragility of goodness: Luck and ethics in Greek tragedy and philosophy*. New York: Cambridge University Press.
Plato. 1921. Theaetetus. *Plato in Twelve Volumes, Vol. 12*. Trans. Harold N. Fowler. Cambridge, MA/London:Harvard University Press/William Heinemann Ltd.
Reale, Giovanni. 1990. *A history of ancient philosophy II*. Trans John R. Catan. New York: State University of New York Press.
Schiappa, Edward. 2003. *Protagoras and logos: A study in Greek philosophy and Rhet*. Columbia: South Carolina Press.
The first sophists. 2000. Trans. Robin Waterfield. New York: Oxford University Press.
The great sophists. 2003. Trans. John Dillon and Tania Gergel. Harmondsworth: Penguin Books.
Untersteiner, Mario. 2008. *I Sofisti*. Milano: Bruno Mondadori.
Vega, Jesús. 2010. Aristotle's concept of law: Beyond positivism and natural law. *Journal of Ancient Philosophy*, IV(2).
Vlastos, Gregory. 1947. *Classical Philology* 42(3):156–178

# Chapter 6
# Aristotelian Ethics and Aristotelian Rhetoric

**Marcel Becker**

## 6.1 *Introduction*: *Nicomachean Ethics* and *Art of Rhetoric* as Part of Practical Philosophy

In the court of law people are accused of crimes, and subsequently they argue their case. In indictments as well as in pleas, people use a wide variety of means to convince the judge and the jury. During such court proceedings, rhetorical skills are and were very important, both in the twenty-first century and in Aristotle's age. Aristotle mentions the court as one of the three places where rhetoric is by preference put into practice. However, in the twenty-first century, as in Aristotle's time, rhetoric is not undisputed. Plato disqualified it as a perverted skill, and in contemporary speech rhetoric is often synonymous with using (cheap) tricks and is seen as a way of speaking in which important arguments are obfuscated.

In our search for an appropriate assessment of the place of rhetoric in courts, we see that the history of philosophy offers a variety of descriptions of what rhetoric is as well as a variety of notions of what rhetoric should be. In such discussions, the relation between rhetoric and ethics turns out to be of vital importance. Is rhetoric a morally neutral skill that can be used equally for good and bad things? Or does rhetoric in itself have an inclination to promote morally good or bad actions? The answers people give to these questions touch the very foundations of the nature of good and evil and of the relation between language and reality. That rhetoric holds such a fundamental place in the western canon is demonstrated by the discussions between Socrates and the sophists, which are widely recognized as the starting point of philosophy.

When Aristotle wrote *Nicomachean Ethics* and *Art of Rhetoric*, he was working out of the opposition between Plato and the sophists. He took up a nuanced position:

---

M. Becker (✉)
Department of Philosophy, Faculty FFTR,
Radboud University Nijmegen, Nijmegen, The Netherlands
e-mail: M.Becker@ftr.ru.nl

L. Huppes-Cluysenaer and N.M.M.S. Coelho (eds.), *Aristotle and The Philosophy of Law: Theory, Practice and Justice*, Ius Gentium: Comparative Perspectives on Law and Justice 23,
DOI 10.1007/978-94-007-6031-8_6, 

when certain conditions are fulfilled rhetoric is a necessary and useful skill in public life. Nevertheless his position has aroused many questions about the normative orientation of rhetoric and the relation between ethics and rhetoric. We will ask two questions:

- Does *Art of Rhetoric* bear in it an orientation towards the good life, as described in the *Ethics*?
- Is there something rhetorical in the way Aristotle presents his arguments in the *Ethics*?

This paper is driven by the conviction that an answer to these questions requires that we do justice to the Aristotelian framework of practical philosophy, to which also *Politics* belongs. A problem is that in the rehabilitation of Aristotelian practical philosophy the unity of his framework is disguised by the fact that the *Nicomachean Ethics* has by far received the most interest, and is often treated as a separate work. This, however, cannot be justified. In the opening chapters of the *Ethics* Aristotle announces that he is starting a new discipline, political science (*EN* 1094a30 hê politikè), that has as its object *eudaimonia*. His first step is an investigation into the 'ethical' part of political science: a treatise on virtue. It is important for the statesman to have knowledge about the human soul and virtue (*EN* 1102a7 and 1102a19).

At the end of *Nicomachean Ethics* Aristotle announces the second part of his program, the 'political' part in the narrow sense of the word. The transition from *Ethics* to *Politics* is completed when Aristotle describes, in the opening chapters of the *Politics*, that a human being is a political animal by nature. Aristotle also explicitly states that *Art of Rhetoric* is embedded in his project of practical philosophy: Rhetoric is the 'paraphues' ('loot'/ 'offshoot/appendage') of political science (*Rhet.* 1356a20–21). However, there is no transitional chapter from *Ethics* or *Politics* to *Art of Rhetoric*. The place of the book in the system is not clear.

Unfortunately, *Art of Rhetoric* has received minor attention among philosophers compared to *Ethics* and *Politics*. The book has been discussed in journals like *Rhetoric Society Quarterly* and *Rhetorica* more than in journals of (practical) philosophy. In those discussions the rhetorical skills Aristotle presents are treated in isolation. Therefore there are few contributions that compare the different ways in which Aristotle treats key moral concepts in *Ethics* and *Politics*. And rarely is Aristotle's treatment of the political and judicial debate in *Art of Rhetoric* compared to his description of institutional arrangements in the *Politics*.

Our first step will be to discuss Aristotle's concept of rhetoric and the ways in which he relates it to his *Ethics*. Answering this question inevitably requires that we explain how Aristotle develops a middle position between Plato and the sophists concerning the normative orientation of rhetoric and its instrumental use. Subsequently a discussion of rhetoric as part of practical philosophy will enable us to lay bare an orientation towards the good life in *Art of Rhetoric*. In the second part of the paper we will answer the question about the rhetorical calibre of ethics. Core characteristics of ethics as practical philosophy will bring us to the conclusion that *Ethics* has a persuasive character and therefore is intrinsically connected to *Art of Rhetoric*. This connection sheds a better light on the contribution of rhetoric to the process of truth-finding as it is practised in courts.

## 6.2 Rehabilitation of Rhetoric

Aristotelian practical philosophy must be understood against a historical background. In Athens in the fifth century BC, the landscape of political discourse had thoroughly changed. Originally the *polis* had been characterized by a tribal culture in which blood relationships were primary. After the reforms of Solon and Cleisthenes membership was a matter not primarily of family relationship but of living on the area of Athens. Rapid economic developments were responsible for social mobility, and the pluralism of an open, egalitarian 'polis' came into being. In this new climate social relations became more politicized: conflicts were solved within the framework of political institutions, by debate and discussion. So politics and rhetoric were conceptualized in close connection (Miller 1993, 215–216).

The sophists took up an extreme position. According to them politics can be reduced to rhetoric. The outcome of deliberation in the public sphere is determined by the ability of the speakers to present their arguments in a persuasive way to the audience. Rhetoric, being the faculty that enables a person to argue on both sides of a question, can be used by anyone to serve his own interests. The good orator is able to delude and 'hoodwink' listeners. The philosophical consequences of this position are made explicit in the debate between the sophists and Socrates. When the outcome of a discussion is completely dependent on a persuasive presentation of arguments, the danger of the victimization of truth is just around the corner. Plato is horrified by the sophists' use of rhetoric. He introduces a sharp distinction between rhetoric and politics and denies any merit to rhetorical skills that are not embedded in the broader project of philosophy, which is the quest for goodness, beauty and truth.

As with many lines of thought in Aristotle's philosophy, his treatment of rhetoric must be understood against the background of Plato's struggle with the sophists. Against the sophists he states that rhetoric does not dominate politics but that it is subordinated to it. In opposition to Plato, who compares rhetoric with flattery and cooking (Gorgias 465d7–e1),[1] Aristotle starts a project of 'rehabilitation of rhetoric' as a relatively independent art.

The first step in Aristotle's project of giving rhetoric its proper place consists in a limitation of the art of rhetoric to three domains: deliberative, forensic and epideictic (*R* 1158b5): in courts, political discussion and public praise. Aristotle warns anyone who is tempted to widen the range of application of rhetoric not to do so, 'more than its legitimate subjects of inquiry have already been assigned to it' (*Rhet.* 1359b6). Each domain is characterized by modes of persuasion that imply specific rules, fixed habits and standards, and rhetoric has to take these into account.

However, the limitation to these domains does not alter the fact that Aristotle's elaboration of rhetorical skills allows for their instrumental use. For instance, Aristotle presents his knowledge of moral behaviour of people in such a way that this knowledge can be used for both good and bad goals. He suggests in *Art of*

[1] Indeed Plato did give an account of 'real' rhetoric, or rhetoric based on knowledge, but in this account rhetoric is just a part of philosophy. The real orator would be a philosopher-king.

*Rhetoric* that this knowledge be used in a good way, but he leaves possibilities for misuse open. He even presents several rhetorical arguments that rely on moral ideas he himself rejects (Irwin 1996). In *Art of Rhetoric* knowledge of moral behaviour seems to be of interest only because the audience considers it important, and because the orator must take its valuation into account in his attempts at persuasion.

Take for instance Aristotle's treatment of character ('ethos') as one of the three modes of persuasion. The orator as described by Aristotle does not intend to convey an objective impression of his character to the audience; he 'only' wants to make himself seem trustworthy ('acsiopistis') to his listeners. The orator needn't have the discussed qualities in reality; he only has to *appear* to have them (*Rhet.* 1404b18–26; see also Sprute 1994, 124). As opinions about what it means to be virtuous or trustworthy often diverge, the presentation of character has to be adapted to the specific audience. Another example of instrumental use is given when Aristotle discusses friendship. He opens the short treatise on friendship with a definition almost similar to the one that is given in the *Ethics:* 'wishing for anyone the things which we believe to be good, for his sake but not for our own' (*Rhet.* 1381a1–3). But he ends the treatise with suggestions about the instrumental use of friendship: 'It is evident, then, from what we have just said, that it is possible to prove that men are enemies or friends, or to make them such if they are not' (*Rhet.* 1382a17–19). Friendship has become nothing more than the suggestion that someone has a friendly attitude towards someone else.

The possibilities for instrumental use suggest that we have to differentiate between a morally neutral theory of rhetoric on one hand and the application of the theory on the other. Any use of rhetoric is immediately open to suspicion. To test whether things really are that simple we must study rhetoric as part of practical philosophy.

### 6.2.1 Rhetoric as the 'Antistrophos' of Dialectic and as an Offshoot of Politics

The suggestion that rhetoric is a skill that can be used instrumentally comes under suspicion when we have a close look at the characteristics of rhetoric. There is a difference between rhetoric and other 'technai' that can be used instrumentally, like strategy and economy. The military and economic domains are clearly defined and fenced off domains of human action, and knowledge of these domains is 'used' by a statesman. But the 'use' of rhetoric cannot be like that. Rhetoric as it is by nature, is not confined to particular modes of action that can be used for arbitrary goals the statesman has in mind. It is a faculty of 'logos' at work in the public sphere, and 'logos' is a core-characteristic of the truth-finding human being. This at least brings us to the presumption that 'rhetoric' cannot be used instrumentally and has a direct relation with formulation of the goals. We will investigate this presumption by discussing the relation of rhetoric to dialectic and politics.

In the opening phrase of *Art of Rhetoric* Aristotle describes rhetoric as a counterpart ('antistrophos') of dialectic. His words are clearly directed against Plato, who says that rhetoric is the 'antistrophos' of cookery. The notion 'counterpart' suggests that it is 'not an exact copy, but making a kind of pair with it', like for instance the antistrophê to the strophê in a choral code.[2] The 'antistrophê' is the part of the song that the chorus sings when they return to the place in the song where they had left when first singing the 'strophê'. Dialectic and rhetoric are two different faculties. Dialectic is a science ('epistêmê') through which practical truth is discovered (Top. 101-2-4). It is the method of all reasoning. But the dialectic line of reasoning cannot succeed in conveying truth to an average audience. People are not always convinced by truth or scientific arguments, let alone by complicated lines of thought. Rhetoric is the ability of discovering the means to make those truths as persuasive as possible to an average audience.

The orator does not expect an audience to extensively go through the arguments on each side (*Rhet.* 1357a2–5), but this does not mean that there is a red line between rhetoric and dialectic. 'It belongs to the same capacity (*dynamis*) to see the truth and what is like the truth' (*Rhet.* 1355a14–18). There is continuity between *endoxa*, to which the people usually are directed, and the truth. This is because rhetoric does not aim at a kind of persuasion that is opposed to truth. It is aimed at convictions that answer the implicit question about the facts of the matter (*Rhet.* 1355a15, 1355a31; Engberg-Pedersen 1996, 125).

The art of convincing bears in it an orientation towards truth. This is often undervalued, as rhetoric usually is translated as *persuasion*. This translation focuses exclusively on the part of rhetoric that aims to cause the audience to have some belief or other, irrespective of whether or not that belief is true. However, the translation of 'peitho' as conviction suggests more than a notion about certain facts (*Rhet.* 1355b15–16) and makes a clear connection between conviction and truth. This connection makes it more understandable that Aristotle gives rhetoric a place in the judicial process, where truth finding is one of the major aims.

The connection between dialectic and rhetoric is confirmed when Aristotle says that too many writers about rhetoric neglect the importance of enthymemata. Enthymemata refer to our truth seeking faculties (*Rhet.* 1355a15–20). Being the 'body of proof' (*Rhet.* 1354a13 'sôma tês pisteôs') they are the heart of rhetoric[3] (*Rhet.* 1354a15). Rhetoric is essentially an art of reasoning that consists of proofs being conveyed through the enthymemata.

The relationship between rhetoric and practical truth-finding is also confirmed when Aristotle discusses deliberation as one of the three domains of application. This domain clearly contains a political dimension. Aristotle calls the pursuit of deliberative rhetoric as 'nobler and more worthy of a statesman' (*Rhet.* 1354b24–25, Schütrumpf 1994) than the other two forms. This is not surprising. The background principles of statesmanship are discussed by political science. And rhetoric is a

[2] Commentary of the translator in: Aristotle (1926).

[3] 'Rhetoric is useful, because the true and the just are naturally superior to their opposites' (1255a18).

'paraphues' ('loot'/ 'offshoot/appendage') of political science. (*Rhet.* 1356a20–21) This description means that rhetoric is nourished by political science.

The orator who tries to persuade audiences of ordinary people uses knowledge that is generated within 'politikê', although he does not have full knowledge of background principles. It makes the arbitrary use of rhetoric problematic. If rhetoric is to carry out its function properly, it must do justice to its origin. In cases where the rhetorical practices are alien to the orientation towards truth that characterizes 'politikê', the skill suffocates its roots. This suggestion is confirmed when Aristotle says that what is naturally true and what is naturally better is always easier to prove and more convincing. (*Rhet.* 1355a38) A speaker who is sincerely concerned with the matter itself will almost automatically be oriented toward what is true and better. Thus he would be closer to the perspective of a thorough investigation than a normally accomplished orator will be. The underlying idea is clear: people have a tendency to accept true assertions (as laid bare in political science), and Aristotle urges the orator to concentrate on these. Again we see the close relation between rhetoric and truth finding.

The narrow relation between 'politikê' and rhetoric deeply influences Aristotle's view on the role of emotions in *Art of Rhetoric*. Aristotle was confronted with Plato's rejection of emotions on one hand and the sophists' embracing of radical emotions on the other. In this debate he takes a nuanced stance, leaning on the moral psychology developed in *Ethics*. At first glance his view on emotions is puzzling. In the opening chapter of *Art of Rhetoric* he attacks the (sophist) writers of rhetorical handbooks for dealing with emotional appeals. The orator should not be tempted to give in to selfish purposes or invoke the uncontrollable emotions of the people. Yet he proceeds to discuss emotional appeals fairly extensively in book II. This apparent inconsistency can be explained when we look at Aristotle's elaboration on the relation between practical rationality and emotions in *Ethics*. It is through emotion that people are impelled to action. But their role should be limited: emotions must be cultivated and modelled.[4]

The same moral psychology can be distinguished in *Art of Rhetoric*. In his discussion on the role of enthymemata, Brunschwig goes so far as to suggest that what dialectic is to the private and conversational use of language, rhetoric is to the public use of language (Brunschwig 1996). This can be questioned seriously. Of course, the enthymeme, the logical argument, has priority. The argument is the body of the proof. However, logical arguments alone do not suffice to energize the will of the people. In speeches for a larger audience emotions play a more important role and appeals to emotions are necessary. But their use must be embedded in an orderly exposition of arguments. Aristotle says that the style will have due proportion ('to prepon') 'if it is emotional and in character and corresponds to the underlying subject matter ('pragmata') (*Rhet.* 1408a10–11). When the appeal to emotions fits with the content of the speech, 'ethos' and 'pathos' can play their role. The emotional appeal is not simply an extra-rational enchantment. Rather, it is guided by the argument.

[4] Elaborated in book III of *EN*.

'Pathos' has its proper place 'under (the guidance of)' rationality, 'hupo tou logos' (*Rhet.* 1356a14).

Parallel to this, how the speaker represents his character must be done 'through the argument'[5] ('dia tou logou'). As Aristotle argues, '[f]or justice should consist in fighting the case with the facts alone, so that everything else that is beside demonstration is superfluous' (*Rhet.* 1404a5–6). Aristotle reproaches contemporary orators who make use of material 'outside the subject matter' (*ta ecso tou pragmatos*). They manipulate an audience's prejudices, fears and other emotions.[6] Unfortunately the audience often is susceptible to sham arguments, particularly when they have a bad ('mochtêria') character. Such uses of rhetoric would lead to regrettable situations.

In a particular sense Aristotle's treatment of the emotions explicitly confirms that truth-finding is part of rhetoric. In cases where there is room for doubt, confidence in the speaker is important, since we are inclined to rely on those we have confidence in (*Rhet.* 1356a6–8); *ethos* is the most effective means of persuasion (*Rhet.* 1356a13). Because the person with a good character knows what he is talking about, he tells us frankly what he himself thinks and gives us the best possible advice (*Rhet.* 1378a6–19). Engberg Peterson concludes: "The hearers want to know. That's why they believe in a speaker whom they take to be a good man." People do not want to be cheated; they want the truth. We see yet again that in the rhetorical situation the audience is fundamentally trying to find out the truth (Engberg-Petersen 1996, 136–138).

Because of their importance and their power, Aristotle treats emotions carefully. Since emotions are the driving factors behind vice as well as virtue, he considers it a challenge to prevent harmful appeals to the emotions. The orator has knowledge of emotions but his appeal to emotions should not arouse vegetative appetites, nor harmful emotions. Each of these gives the animal passions sway over human reason. With the help of the moral psychology presented in *Ethics* we better understand Aristotle's efforts to diminish the danger rhetoric can have. The result is a concept of rhetoric in which the oversimplified emotional appeal to the audience and the use of harmful tactics is reduced.

### *6.2.2 Criteria Implicit in Rhetoric*

The idea that the art of rhetoric succeeds better when it does justice to the human orientation towards truth and human good is confirmed when we see that for Aristotle exercising rhetorical skill has a value in itself. Each of the three modes of rhetoric speech (deliberative, forensic and epideictic) takes place in a domain that is essential for the development of the human being: the political community ('the common interest'), judicial processes ('the just') and those situations in which we

[5] Cooper (1994, 197): 'and not merely by what he says'.

[6] Engberg-Petersen (1996) hints at several places where this 'austere conception' of rhetoric can be found.

praise the merit people have for the community ('the noble'). In each of these modes the use of language and rhetoric is related to the goals to be reached. To gain clarity about this it is important to know more about the relation between the goals and the way they are reached. For this it is necessary to look at Aristotle's theory of action in *Ethics*.

A key element to Aristotle's theory of action is his formula that 'phronêsis' is related to the 'ta pros to telos'. The latter expression is sometimes translated as 'means to a goal'. But in speaking about human action Aristotle denies the existence of an external means-goal distinction. 'Eudaimonia' consists of doing and living well ('eu zên kai eu prattein'). The orientation towards the good, as couched in the 'hexis', is vague and needs to be made concrete in action. The discussion led to the conclusion that the formula can better be translated as 'means conducive to a goal', 'instances' or 'constituents' of the goal. This line of thought is confirmed when Aristotle says that in order to reach a goal, we should use the 'kallista' (best; tidy) means (*EN* 1112b16–17). This idea can also be applied to *Rhetoric*. Language, as a communication tool, is not morally neutral, but a venue through which the good is realized. In each of the three domains good speech is that kind of speech that does the most justice to the good(s) that is(are) at stake. Each of the three kinds of rhetoric is an instance of practical rationality functioning well (Garver 1994, 55–60).

It might be refuted that Aristotle's theory of action (and the formula 'pros to telos') is presented in the chapter about 'phronêsis'. Rhetoric is a 'technê', and Aristotle clearly distinguishes 'technê' from 'phronêsis'. However, this distinction should not be drawn too sharply. Every art has two ends. On one end, the art should realize external goals (in rhetoric it is 'persuading'; in medicine it is 'curing the patient'). On the other end it should comply with the 'internal standards' that guide the activity, the so-called 'guiding ends'.

The art of rhetoric, an activity of human 'logos', is 'arguing'. Thus, artful persuasion, done according to the standards, is 'energeia', i.e. an exercise of human capacities that realises its form, and therefore is good in itself (Garver 1994, chapter I). That this is the case with argumentation can be seen by the fact that when a listener accepts the tenability of an argument, the same thing – the same form – exists in both speaker and audience. It is furthermore of note that this movement of forms is the final result of a development, a growth process, in which proving the argument is done *throughout* the discourse. Its proof is 'all over the blackboard'.

Aristotle contrasts this process with the way the sophist uses emotional appeals. An emotional appeal is not 'energeia' but 'kinêsis'. It exists fully in the mind of the audience when the movement is over. The movement itself, however, is not a growth process; it is directed towards a result that suddenly 'is there'. "When I am angry I am no longer being angered, or being induced to become angry" (Garver 1994, 37). In other words, in arousing emotional appeals like anger, the process is aimed at a goal that is only present in the end. In this way, understanding rhetoric as a passing of forms may do more justice to the Aristotelian theory of action.

The guiding end of the 'technê' is realised by acting according to the internal standards. Of course, acting according to these standards does not guarantee that the action is successful. By following the internal goal, the external goal ('the given end') will probably be reached, but this is not sure. The relation between internal and external goal is contingent. In the same way that a good doctor is not always able to cure the patient, and that bad doctors (i.e. doctors that do not comply very well to the accepted standards) sometimes 'luckily' cure patients, the techniques of internal goals aren't always the best ways to reach the external goal. The role and importance of chance ('tuchê') can never be excluded. 'Art' does not drive out chance completely. But this possibility should not distract the orator from the criteria that guide his art. His first responsibility is to use his skill the way it is intended.

This does not mean that Aristotle pleads for a rigid format for rhetorical skills. The use of rhetoric must be tailored to the situation. In the perfect community there exists a large susceptibility for arguments, and debate and discussion are of high quality. But rhetoric is able to make truth sound convincing to an average audience, or to an audience with defective moral knowledge. In the less than perfect communities persuasion requires skills that are remote from the kind of argument that characterizes the discipline 'politikê'. Aristotle does not deny this possibility, as we can learn in *Politics*. In this book he makes clear that good political action is always related to circumstances. In less than perfect conditions it is allowed to do what is necessary. Sometimes forms of rhetoric that are strongly based on pathos and ethos are needed in order to secure victory for the just cause. "Though it might not quite correspond to Aristotle's ideal of rhetoric (...), by the mere force of circumstances rhetoric has to deal with and recommend these ways of persuasion" (Sprute 1994, 126).

Our conclusion must be that Aristotle's limitation of rhetorical activity to three domains, his description of rhetoric as an offshoot from politics, his view on emotions and his elaboration of rhetoric as 'technê' all imply that the art of rhetoric is directly related to ethics. Being a 'faculty of speech' rhetoric is an intrinsic good, like health, wealth, bodily strength and the like. Understanding rhetoric in this way has important implications, since rhetoric is used as a tool by human beings in truth-finding. But the described properties of the 'technê' of rhetoric do not remove the possibility that a morally bad orator will use his skill for morally problematic issues (as is the case with health and bodily strength). Aristotle's conviction that rhetoric possesses intrinsic goodness should be distinguished from the suggestion that 'to be accomplished an orator need to be a morally good person', or 'to be a good orator one has to have full ethical-political knowledge'. Rather, Aristotle appeals to the reader not to use rhetoric in a morally bad way (*Rhet.* 1355a31–1355b10). Those who use rhetoric for bad things don't do justice to rhetoric as a part of practical philosophy.

Our consideration of rhetoric as part of practical philosophy has shed light on the ethical dimension of rhetoric activity. The next step is to ask the question about the relation between ethics and rhetoric the other way around: does *Ethics* contain rhetorical elements? Can the way Aristotle presents his arguments in Ethics be described as rhetoric? A proper answer to these questions requires that we look closer at fundamental characteristics of ethics as part of Aristotelian practical philosophy.

## 6.3 From Rhetoric to Ethics

At first glance it seems a not too difficult task to discuss the rhetorical calibre of Aristotle's *Ethics*. We have to describe the notions and standards Aristotle develops in *Art of Rhetoric*, and subsequently see whether they are used in the *Ethics*. But at a closer look, there seem to be good reasons to not even try to do what we intend to do. Speaking persuasively in public and writing an ethical-political treatise are activities of a completely different kind. Because of this difference it is not obvious that the techniques and skills Aristotle presents in *Art of Rhetoric* are applied in the *Ethics*. Moreover, as we have seen in the opening chapter of *Art of Rhetoric,* Aristotle limits the use of rhetoric to three domains. What allows us to enlarge the application-reach of this art?

Again the broader background of practical philosophy contains the answer. As we have seen Aristotle mentions rhetorical deliberation about political affairs as one of the three domains. As part of practical philosophy ethics is directly connected to this kind of deliberation. We are justified to ask whether on a deeper level the connection between the books implies that *Nicomachean Ethics* has a rhetorical calibre. Is ethics as Aristotle describes it an activity with a persuasive dimension?

The object of practical philosophy is to discuss what gives meaning to human life and human action. Action takes place in the domain of contingency, the domain of the variable things, the things 'that are not necessarily as they are' ('ta endechomena'). In this domain man has freedom to influence matters; things are 'within our power' ('eph' hêmin' *EN* 1112a30, 1112b3, 1112b27 1112b33). But the way we interfere in the course of things cannot be predicted, nor written down in precise rules; every situation is different. Things are unstable, changeable and incalculable (Klein 1988). This line of thought is elaborated in the *Ethics* but it is also visible in *Art of Rhetoric*. Rhetoric is not about 'what is necessary', but about 'things which may (...) be other than they are' (*Rhet.* 1357a22 'peri tôn endechomenôn'; *Rhet.* 1357A 25–1357B10). Rhetoric has to do with probabilities[7] that are affected by our action[8] (*Rhet.* 1356b28–1357a7).

Since Aristotle develops methods of the particular sciences according to the special needs of the subject-matter of each, his ideas about the contingency of human action imply guidelines about the method to be followed in practical philosophy. His most well-known idea concerns the exactness of the guidelines. Practical philosophy cannot and ought not provide us with fixed standards of what is considered to be 'morally right'. Instead, it provides rough guiding principles. 'We must therefore be content if (...) we succeed in presenting a broad outline of the truth; when our subject and our premises are merely generalities, it is enough

[7] 1355a17 'pros ta endoxa'; cfr. 1356b33 'endoxon', cfr. 1357a14., 1359a30: 'things which may possibly happen or not'-endechetai'.

[8] 1359b1: our examination is limited to finding out whether such things are possible or impossible for us to perform.

if we arrive at generally valid conclusions'[9] (*EN* 1094b19–21). A closely related characteristic of practical philosophy is that it does not uncover a universally valid truth. There is no 'point zero' from which actions can be judged. Also in this respect practical philosophy differs from scientific knowledge, which aims to formulate universally valid laws.

These characteristics of practical philosophy bring ethics closer to rhetoric, and by looking at the truth finding process in the *Ethics*, the way persuasion operates will be better understood. As is said, in the beginning of the *Nicomachean Ethics* Aristotle denies the possibility of presenting fixed universal rules for good conduct. The real and deeper meaning of a statement cannot be given on objective grounds; it is not because of an eternally valid argument that people are convinced. As there is no point zero, the first step to gaining ethical knowledge is to determine prevalent notions of basic concepts as they prevail in a particular community.[10] Aristotle develops a critical dialogue with common sense opinions in his community. He compares the force of various understandings of specific ethical concepts and he explores basic relationships between a plethora of ethical concepts with the aim of establishing a coherent whole. Through this critical discussion the concrete convictions of people are brought to a higher level.[11]

Note that there is a parallel between this methodology and the way Aristotle describes the good orator in *The Art of Rhetoric*. The orator starts by taking seriously the opinions of the audience, and has to adjust to the people in order to place them in the proper frame of mind (*Rhet.* 1377b23; Triadafilopoulos 1999, 749–750). His statements are connected to the beliefs of the audience. At the same time the orator is willing to change the opinions of the audience: the conclusions need not to reflect the audience's original opinion. As Aristotle says in both *Ethics* and *Art of Rhetoric* popular opinion often contains a portion of truth, but it is rarely fully correct.

The fact that Aristotle develops his thoughts in interaction with the community becomes particularly clear in his use of 'endoxa'-principles as a starting point. 'Endoxa'-principles are reputable opinions. They are neither linguistically necessary (as an analytical philosophical approach would suggest), nor in themselves necessary (as a Kantian inspired philosopher would suggest), nor biological necessities. Their force is derived from the fact that they mirror generally accepted understandings. But the appeal to *endoxa* is not a simplistic common sense approach. As Klein convincingly shows, they are a starting point for a critical and nuanced discussion. For instance in the opening chapter of the *Ethics* Aristotle subtly plays with the combination of two kinds of *endoxa* principles: substantial premises about the

[9] As Jaeger (1957) elaborates, the difference is clearly visible in the use Plato and Aristotle make of examples from medicine.

[10] Jaeger (1957) states that Aristotle's examples from medicine are meant to support this claim. According to Aristotle theoretical knowledge described in books is only good when it is supplemented by experience. Ultimately the doctor is educated by a doctor.

[11] I've elaborated this interpretation of Aristotle's methodology in Becker (2004).

content of good life and regulative principles about the structure of happiness as the highest goal of human action (that it is self-sufficient, proper to the human being, and regulating all our actions). This combination guides his inquiry and results in deeper knowledge (Klein 1988).

Aristotle demands that the conclusions of his discussion are accepted as guidelines for the good life. As we have seen, his arguments are about human action that takes place in the domain of variable things, where man influences reality. The ultimate goal is not to lay out a theoretically correct theory of ethics; it is to make people good. The practical philosopher refers to an audience and he asks it to be actively involved. And here rhetoric comes in. An ethical treatise is directed towards the disposition of the audience. It directly or indirectly intends to incite someone to action. Because of this desire to induce action, in a treatise on practical philosophy the contact between the 'speaker'/philosopher on one hand and the 'audience' /pupils on the other is crucial (McCabe 1994, 156). The developments in opinions and judgements are dependent on how arguments are presented to people. To state it in rhetorical terms, an enthymeme is more than a 'simple logical inference' (Cooper 1994, 667), it is necessarily about what interests people and about what affects them. Note that enthymeme is derived from 'thymos', which traditionally indicates the seat of both feeling and thought.

The similarities in methodology are confirmed by the similar sources that Aristotle uses. In *Art of Rhetoric* he rarely refers to famous rhetoricians. He does not really give us the feeling that he listened much to orators.[12] Instead, he usually refers to epideictic statements and borrows his examples from vastly different sources: poetry, written prose, drama, comedy, proverbs, and anecdotes of every kind. Attention is given most to common sense opinions, poets like Homer, and playwriters. The similarity is also confirmed when, in *Ethics*, Aristotle emphasizes how important it is for good behaviour that other people give good examples. This emphasis is essentially related to the contingent character of truth: truth is hidden in the actions of concrete persons. The importance of giving the good example can be understood better when we have in mind the ideas Aristotle presents in *Art of Rhetoric* about *êthos* and *pathos* as persuasive elements. The character of the agent and his emotional influence contribute to the perception and appreciation of the values he shows.

## 6.4 Conclusion

Aristotle mentions the court as one of the three stages where rhetoric is used. This raised questions about the degree to which rhetoric is susceptible to misuse, and about its relation to truth-finding. With these questions in mind we discussed core

[12] Trevett even speaks about 'a curious neglect of genuine political and forensic oratory'. Quoted in Graff (2001).

elements of Aristotle's project for the rehabilitation of rhetoric: the relation rhetoric has to dialectics and politics, the criteria implicit in rhetoric as an art and the role of emotions in rhetoric. It turned out that rhetoric is directly connected to practical philosophy.

The activity of an orator must be understood and judged from an ethical-political background. Rhetoric is not morally neutral and indeed has an intrinsic connection to truth-finding. Moreover there turned out to be a direct connection between discovering and communicating moral truth on one hand and rhetoric on the other. In dealing with cases concerning the opposition between good/bad and justice/injustice, it is impossible to decide from a neutral standpoint what is good, bad, just or unjust. Practical truth has a contingent character, and arguments necessarily have a persuasive dimension and a rhetorical character.

This fact, however, does not remove the possibility that rhetoric is used for morally bad things. Rather, it is a challenge to do justice to rhetoric's proper character, for only when this is done is the exercise of rhetoric constitutive of good life. Due to its special relationship with the 'logos' and its use in the public sphere, rhetoric turns out to be intrinsically linked to citizenship. The judicial court is one of the venues where rhetoric's importance can be seen.

## Bibliography


Aristotle. 1926. *(Rhet.) The art of rhetoric*. Trans. H. Freese. London: Loeb Classical Library

Aristotle. 1928. *(Top.) Topics*. Trans. G. Mure. Oxford: Oxford University Press.

Aristotle. 1934. *(Pol.) Politics*. Trans. H. Rackham. London: Loeb Classical Library.

Aristotle. 1936. *(EN) Nicomachean ethics*. Trans. H. Rackham. London: Loeb Classical Library

Becker, M. 2004. Virtue ethics, applied ethics and rationality twenty-three years after *after virtue*. *South African Journal of Philosophy* 23(2): 25–42.

Brunschwig, J. 1996. Aristotle's rhetoric as a 'counterpart' to dialectic. In *Essays on Aristotle's rhetoric*, ed. A. Oxenberg-Rorty, 142–173. London: Berkeley.

Cooper, J. 1994. Ethico-political theory in Aristotle's rhetoric. In *Aristotle's rhetoric, philosophical essays*, ed. D. Furley and A. Nehamans, 193–210. Princeton: Princeton University Press.

Engberg-Pedersen, T. 1996. Is there an ethical dimension to Aristotelian rhetoric? In *Essays on Aristotle's rhetoric*, ed. A. Oxenberg-Rorty, 116–141. London: Berkeley.

Garver, E. 1994. *Aristotle's rhetoric, an art of character*. Chicago: University of Chicago Press.

Graff, R. 2001. Reading and the 'written style' in Aristotle's rhetoric. *Rhetoric Society Quarterly* 31(4): 19–44.

Halliwell, S. 1994. Popular morality, philosophical ethics and the rhetoric. In *Aristotle's rhetoric, philosophical essays*, ed. D. Furley and A. Nehamans, 211–230. Princeton: Princeton University Press.

Irwin, T. 1996. Ethics in the *rhetoric* and in the *ethics*. In *Essays on Aristotle's rhetoric*, ed. A. Oxenberg-Rorty, 142–173. London: Berkeley.

Jaeger, W. 1957. Aristotle's use of medicine and Aristotle's use of medicine as a method in his ethics. *Journal of Hellenic Studies* 77(1): 54–61.

Klein, S. 1988. An analysis and defense of Aristotle's method in NE I and X. *Ancient Philosophy* 8(1): 63–72.

McCabe, M. 1994. Arguments in context: Aristotle's defense of rhetoric. In *Aristotle's rhetoric, philosophical essays*, ed. D. Furley and A. Nehamans, 129–165. Princeton: Princeton University Press.

Miller, C. 1993. The *polis* as rhetorical community. *Rhetorica* 11(3): 211–240.

Plato. 1925. *Gorgias*. Trans. W. Lamb. London: Loeb Classical Library.

Schütrumpf, E. 1994. Some observations on the introduction to Aristotle's rhetoric. In *Aristotle's rhetoric, philosophical essays*, ed. D. Furley and A. Nehamans, 99–116. Princeton: Princeton University Press.
Sprute, J. 1994. Aristotle and the legitimacy of rhetoric. In *Aristotle's rhetoric, philosophical essays*, ed. D. Furley and A. Nehamans, 117–128. Princeton: Princeton University Press.
Triadafilopoulos, T. 1999. Politics, speech, and the art of persuasion: Toward an Aristotelian conception of the public sphere. *Journal of Politics* 61(3): 741–757.

# Chapter 7
# Is There Any Theory of Value in Aristotle's Ethics?

**António de Castro Caeiro**

Anachronism is always present when we address a question to an author. That doesn't mean we shouldn't try to removing it out of sight. Raising questions concerning what could be an Aristotelian theory of value (Werttheorie), doesn't necessarily mean that Aristotle is a precursor of the "Wertphilosophie".[1] This would not only be a sheer anachronism, it would be completely wrong to think of Aristotle as developing a theory of value from a practical or technical standpoint (Gordon 1964; Caston 1998; Chapura 2009).

This would mean that, after having established a cognitive content of φύσις, Aristotle would have shifted to other horizons as he did indeed. So according to the traditional schema we would have: λόγος vs. ἄλογον, meaning that Philosophy unfolds it's investigation towards several horizons of beings depending upon the specific access we need to get there. "Logic" approaches both: θεωρία and πρᾶξις, i.e. the διάνοια but also the ἠθικαὶ ἀρεταί (Halloran 1982; Heath 1988). On the other hand, what is sheer ἄλογον belongs to the sphere of the πάθη or to what is present only for a little while: the αἴσθησις. ΕΠΙΣΤΗΜΗ, ΗΘΟΣ, ΠΑΘΟΣ as the disciplines of philosophy (Coby 1986; Collins 2004; Cua 2003; Dickie 1923) are not only a stoic heritage, they come from the classical philosophy of Plato and Aristotle and are still present in later developments in the history of philosophy as for instance, Kant.

Therefore, when Nietzsche, Kierkegaard, Rickert, Scheler, Husserl and Heidegger address to the question of value they all have this cartography in their minds, even

[1] During a research semester spent in Oriel College (Oxford), Professor David Charles raised some questions concerning what could be an Aristotelian approach to the question of value (Werttheorie). He doesn't think Aristotle is a precursor of the "Wertphilosophie". This would not only be a sheer anachronism, it would be completely wrong to think of Aristotle as developing a theory of value from a practical or technical standpoint.

A. de Castro Caeiro (✉)
Philosophy Department of the Faculty of Social Sciences and Humanities,
New University of Lisbon, FCSH/Universidade Nova de Lisboa, Lisbon, Portugal
e-mail: acaeiro@mac.com

L. Huppes-Cluysenaer and N.M.M.S. Coelho (eds.), *Aristotle and The Philosophy of Law: Theory, Practice and Justice*, Ius Gentium: Comparative Perspectives on Law and Justice 23,
DOI 10.1007/978-94-007-6031-8_7, 

if only to do away with it. What a theory of value sketches is several strata where beings are structurally modified. The first and primordial stratum would have been merely a sense-content, the material components of the physical world, filtered by the αἰσθήσεις. From there on the θεωρία would project a world content where φύσις would come up to be understood through an epistemological point of view. So the chain: αἴσθησις, μνῆμαι, ἐμπειρία, τέχνη, ἐπιστήμη in Metaphysics A. In addition to this primordial stratum, we would have a second-order one, for instance a technical one. A Tree becomes Timber, getting thereby a value. As timber a tree can be transformed by the carpenter into a table or a chair. So the technical value is a second-order content bestowed upon the timber. It is not difficult to understand that if our agency where constituted by sheer reactions, we would act without foreseeing or forecasting what would be or what will be the end result we would get into. So agency as proactive action is worthwhile because it tries not to underestimate the meaning of what we are through the meaning of our actions.

There would be also practical values from another order. Human situations define practical existential resources we may or may not have, to oppose fear and behave courageously or to resist temptation and get a grip on ourselves, staying in control of the situation.

> We assume therefore that excellence (ἀρετὴ) is the practical quality (πρακτική [ἕξις]) of acting in the best way in relation to pleasures and pains (περὶ ἡδονὰς καὶ λύπας), and that perversion (κακία) is the opposite. (…) There are three things that are the motives of choice (αἵρεσις) and three that are the motives of avoidance (φυγαί); namely, the glory (τὸ καλόν), the useful (τὸ συμφέρον), and the pleasure (τὸ ἡδύ), and their opposites, the disgrace (τὸ αἰσχρόν), the damage (τὸ βλαβερόν), and the painful (τὸ λυπερόν). Now in respect of all these the good man is likely to go right (ὁ ἀγαθὸς κατορθωτικός) and the bad to go wrong (ὁ δὲ κακὸς ἁμαρτητικός) (*EN*, 1104b27–34).

On the practical horizon we can get a more hierarchical organization of the goals we are up to. Let us call them teleological values for the sake of the argument. The ὄρεξις, intentional project, we live by, is motivated positively by goals we choose to pursue: the glory (τὸ καλόν), the useful (τὸ συμφέρον), and the pleasure (τὸ ἡδύ). On the contrary, we try to avoid the opposite ones: the disgrace (τὸ αἰσχρόν), the damage (τὸ βλαβερόν), and the painful (τὸ λυπερόν). The πρᾶξις is for Aristotle defined at its basic level by pleasure and pain and the way we relate to them, even if at a most basic level. We pursue pleasure (ἡδονή) and avoid or try to escape pain (λύπη).

We thus understand not only cognitively, or theoretically, what pleasure and pain are all about, as if we didn't take any action in order to chase or avoid them. Pleasure and pain trigger respectively promises and menaces. The practical horizon Aristotle determines is future-constituted. It is from the future that pleasure opens up our expectations. It is also from the future that danger springs, leaving us afraid of what might happen.

Aristotle considers different objects of pursuit and avoidance (Fortenbaugh 1964; Garver 1982). At the basic level we have τὸ ἡδύ and τὸ λυπερόν. Sweetness and bitterness describe the existential practical world we live in. But as humans we have an understanding of what is of advantage and what can harm us: the useful (τὸ συμφέρον) and the damaging (τὸ βλαβερόν). Those goals are respectively chosen or avoided. But they are at an absolutely different level from the one where we experience pleasure and pain as such. What is useful needs not be necessarily pleasant. At least not at first sight.

On the other hand, it is perfectly possible that we derive pleasure from what is damaging. The most radical and extreme objects of choice are on the practical horizon: the glory and the disgrace: τὸ καλόν and τὸ αἰσχρόν. Both these objects constitute the most radical way of pursuit and the most important thing we need to avoid. We can also think that they have nothing to do with what is useful or damaging. We can understand, however, that they can be a source of pleasure and pain. Isn't glory sweet and defeat bitter?

Our access to these object values is not theoretical or at least not cognitive only. They give rise to κινήσεις. We move towards pleasure, it is useful, it brings us glory. We move away from pain, from what is damaging, from what brings us disgrace and defeat. These actions or movements are emotional or felt with affection. They are πάθη. They let us understand the way it feels like. On the other hand, they are embedded in different layers. We are to interpret where pleasure, advantages, and glory will lead us. We need to interpret where pleasure leaves us, what is damaging for us, what disgraces us and is shameful to us. The question is: what do we get when we go after pleasure, the useful and the glory? What do we really escape when we avoid pain, the damaging and disgrace? (Kreager 2008; Mara 2000; Mulgan 1990).

This is Aristotle's way of understanding what we may call "value". We gain access to what is a promise of a "better future", when we have shaped a better version of ourselves. On the contrary, we know where we are heading to, when we anticipate the menaces and danger of what is painful, damaging and disgraceful. This assessment is "our" assessment of the situations and circumstances we are in. Or perhaps it is rather an assessment ἐπ' ἐμοί, that depends solely on my life and on the way life lets me understand what I must do or how I must be, in order to become what I am supposed to be. To be the way I am.

Our assessment of ourselves can underestimate or overestimate our worth. The situation is that of evaluating the relation between merits or awards and the efforts made. There are: with no effort, every gain; with all the effort, no gain; more effort, less gain; less effort, more gain. The grounding situation is that which we are in: of saying that we or others didn't deserve what happened to them, that they really deserved it, either good or bad, pitying them or being happy for them. Life has us in constant assessment of punishments and rewards, what does it take to get what one gets, whether deserved or not. All values have a price, an evaluation, an assessment, a scale, a trade, a giving and a receiving, the complex relation between giver and receiver, the effects on the receiver and on the giver, the way the receiver assesses what has been done for him, gladly receiving what he receives but hating that he needed the giver, etc. Aristotle's discussion on how worthy or unworthy is someone of what he or she is receiving, and the respective self-evaluation – if it is just or unjust, deserved or undeserved –, begins with the identification of the μεγαλόψυχος, as the type of person that is not only worthy of what he receives, but knows it (Turner 1970; Ward and Aristotle 2001; Waters 1969; Woodson 1970).

So here is the ranking:

1. The μεγαλόψυχος: "Now a person is thought to be great-souled if he claims much and deserves much; (δοκεῖ δὴ μεγαλόψυχος εἶναι ὁ μεγάλων αὑτὸν ἀξιῶν ἄξιος ὤν, 1123b1–2)".
2. The ἠλίθιος: "He who claims much without deserving it is foolish (ὁ γὰρ μὴ κατ' ἀξίαν αὐτὸ ποιῶν ἠλίθιος, b2–3)".
3. The σώφρων: "He who deserves little and claims little is modest or temperate, but not great-souled, since to be great-souled requires greatness (ὁ γὰρ μικρῶν ἄξιος καὶ τούτων ἀξιῶν ἑαυτὸν σώφρων, μεγαλόψυχος δ' οὔ· ἐν μεγέθει γὰρ ἡ μεγαλοψυχία, b4–6)".
4. The χαῦνος (empty or frivolous) "He that claims much but does not deserve much is vain (ὁ δὲ μεγάλων ἑαυτὸν ἀξιῶν ἀνάξιος ὢν χαῦνος, b8–9)"
5. The μικρόψυχος (pusillanimous): "He that claims less than he deserves is small-souled, whether his deserts are great or only moderate, or even though he deserves little, if he claims still less. The most small-souled of all would seem to be the man who claims less than he deserves when his deserts are great (δ' ἐλαττόνων ἢ ἄξιος μικρόψυχος, ἐάν τε μεγάλων ἐάν τε μετρίων, ἐάν τε καὶ μικρῶν ἄξιος ὢν ἔτι ἐλαττόνων αὑτὸν ἀξιοῖ. 9–11)".

Now the paradigm is the way of being of the μεγαλόψυχος. He is aware of *what* he deserves because he knows he is worthy of it, by the way he has been. His self-awareness is absolutely grounded. He deems he deserves not only great wards but the absolute greatest of all.

> Though therefore in regard to the greatness of his claim the great-souled man is an extreme, by reason of its rightness he stands at the mean point, for he claims what he deserves; while the vain and the small-souled err by excess and defect respectively. ἔστι δὴ ὁ μεγαλόψυχος τῷ μὲν μεγέθει ἄκρος, τῷ δὲ ὡς δεῖ μέσος· τοῦ γὰρ κατ' ἀξίαν αὑτὸν ἀξιοῖ· οἱ δ' ὑπερβάλλουσι καὶ ἐλλείπουσιν. εἰ δὴ μεγάλων ἑαυτὸν ἀξιοῖ ἄξιος ὤν, καὶ μάλιστα τῶν μεγίστων, περὶ ἓν μάλιστ' ἂν εἴη. (b13–17)

What qualifies the ἀξία? We've defined the ἀξία degrees quantitatively. Can we define it qualitatively? Aristotle says *expressis verbis*: ἡ δ' ἀξία λέγεται πρὸς τὰ ἐκτὸς ἀγαθά. I guess we can broaden τὰ ἐκτὸς ἀγαθά to the bodily and also spiritual capacities. It is true, Aristotle says that τὰ περὶ τὰ σώματα and τὰ περὶ τὴν ψυχήν are higher levels than τὰ ἐκτὸς ἀγαθά. The reason why I consider them here is because our relationship with them is extrinsic: it is an evaluation of what one gets whether by deserving it or not (exterior goods, good looks and intelligence, etc.).

There's a further specification of ἀξία besides the ranking and quantity of τὰ ἐκτὸς ἀγαθά, *lato sensu*. τὰ ἐκτὸς ἀγαθά are interpreted in a scale or hierarchy of honours. The ἀξία does not concern only the contents one possesses, whether deserving them or not, receiving them as benefits for what they have done or conferring them to people whose achievements and actions made them worthy of them. As we will see, it is the way we are, in order to understand the "possession", the "having", "the conferring benefits" or "getting" them, that is here at stake. There is for sure one objective value for someone's inner character. There is a direct proportion between the character one displays and the praise one gets for it. But this proportion is asymmetrical. The features of character, one's own intrinsic disposition, exists as "psychic content", a way of being. On the contrary, external goods, possessions, benefits one has received are at the end of the day "things".

We are indeed always up to something. This means we are trying to get better conditions. But what is the real meaning of getting what one wants? Does this mean we want to become wealthier, more powerful, get a new and better house, a brand new car, holidays in fancy places, be wonderful people, be admired, be successful, etc., etc.? Are we better people when we get what we want? Or are we worse people when we cannot get what we were up to? Are these questions meaningful questions? Maybe we can get what we want, maybe not. Getting, having, acquiring, possessing are ways of relating to objects we can have at our disposal. What those objects are doesn't really matter, as long as the way we relate to them is one of possession or ownership, be it things, real estate, cars, even people. Aristotle has nothing against having things. Not having them can be as damaging as having them. It all depends on the form.

So, for him, to have or to not have is just a manner of relating to the world, broadly speaking: external goods, the body we have, the social skills we developed, our psychological qualities, etc., etc.. Those "goods" could have been available to us all along. We can acquire them as life goes on. But there is an absolute difference between what we ARE and what we HAVE. What we have and have not may affect the way we are, contributing to a more comfortable life or to a difficult one. The inner character or intrinsic feature of our life and soul cannot be possessed or acquired by money or by whatever means we have to buy them.

In this sense, what we are up to is a betterment, or an existential improvement, so to speak, which depends solely on what we do. What we do lies formally in the way we are. It is this better version of ourselves that we are pursuing. This different, better, version we can turn into, is acquired or possessed by a change. We abandon our old way of being and turn to another, better way of being. We are different versions of ourselves getting along with life. There are without a doubt things we are proud of and things that we have done only to bitterly regret them.

We are always reformulating ourselves, re-sketching old versions, getting new ones, because we have a picture of a "nec plus ultra" version of what we can be. This superlative picture of one's own image, or what we can achieve, is always shadowing our future to come, the way we will turn into: τὸ ἄριστον. Only with this *nec plus ultra* picture of our most radical and extreme possibility in view can we understand why we are not yet or already the persons we were to be. We can feel either that we are getting there or that we are forever doomed to not be what we were supposed to be.

If this picture is correct, we can now move on to interpret the figure of the "great-souled man" in Aristotle, as the most radical and extreme human possibility. It is from the standpoint of what Aristotle calls μεγαλοψυχία that we can sketch other possibilities of being in relation to what they are, and being the way they are, what do they deserve. In this sense, we are to understand not what we want and deserve because of what we have done, but what we want to turn into, what kind of change we need to perform in order to perceive what we really are, and what we deserve even if we don't get it. How is it that if we happen to get wealth, power, possessions, a long life, a healthy body, a nice family, everything we want to have, we will still understand that these are only goods and things that don't cope with our inner, intrinsic feature of character. No value, not even the most precious one, rewards the intrinsic features of character that mould the way we are. No external good, not even the biggest one, honour, bestowed upon the great-souled man by the most excellent men pays off, or is reward enough to what he is, to the way he is.

> 'Worthy' is a term of relation: it denotes having a claim to goods external to oneself. Now the greatest external good we should assume to be the thing which we offer as a tribute to the gods, and which is most coveted by men of high station, and is the prize awarded for the noblest deeds; and such a thing is honor, for honor is clearly the greatest of external goods. Therefore the great-souled man is he who has the right disposition in relation to honors and disgraces. And even without argument it is evident that honor is the object with which the great-souled are concerned, since it is honor above all else which great men claim and deserve. (ἡ δ' ἀξία λέγεται πρὸς τὰ ἐκτὸς ἀγαθά· μέγιστον δὲ τοῦτ' ἂν θείημεν ὃ τοῖς θεοῖς ἀπονέμομεν, καὶ οὗ μάλιστ' ἐφίενται οἱ ἐν ἀξιώματι, καὶ τὸ ἐπὶ τοῖς καλλίστοις ἆθλον· τοιοῦτον δ' ἡ τιμή· μέγιστον γὰρ δὴ τοῦτο τῶν ἐκτὸςἀγαθῶν· περὶ τιμὰς δὴ καὶ ἀτιμίας ὁ μεγαλόψυχός ἐστιν ὡς δεῖ. καὶ ἄνευ δὲ λόγου φαίνονται οἱ μεγαλόψυχοι περὶ τιμὴν εἶναι· τιμῆς γὰρ μάλιστα [οἱ μεγάλοι] ἀξιοῦσιν ἑαυτούς, κατ' ἀξίαν δέ. 1123b17–22.)

The ἀξία[2] is the τίμή in the sense that exterior goods were somehow the face value of honour. Honour is the biggest, the best, of the external goods (τοιοῦτον δ' ἡ τιμή· μέγιστον γὰρ δὴ τοῦτο τῶν ἐκτὸς ἀγαθῶν). On the other hand, the greatest honour of all is what we bestow to the gods: μέγιστον δὲ τοῦτ' ἂν θείημεν ὃ τοῖς θεοῖς ἀπονέμομεν. This is what is coveted by men of high ranking, and is the

[2] In LSJ ἄξιος, ία, ιον (ος, ον Nonn.D.8.314), for Αγ-τιος is a counterbalancing, cf. ἄγω v1: hence prop. weighing as much, of like value, worth as much as, c. gen. It is a concept relation. One can identify an act of giving and a giver either known or unknown and a receiver that knows who is the dispensator or not, is either grateful or ungrateful. Besides this relation between giver and receiver, there's the thing given or received, whatever that may be: πάντων Ζεὺς ἄξιον ἦμαρ ἔδωκεν Il.15.719. (2) If one isolates the relation between the receiver and what he gets or not, the receiver interprets his or her situation as a result or consequence of deserved or not deserved return. So whatever the situations are, bad or good, we take them already in a connection with something we've done or have not done, even if we ask: what have we done to deserve this? in bad situations. Whatever happens to us is interpreted as deserved or not, as an award or a penalty either as a material, moral, spiritual gratification or falling short of it for what we've done, how we've behaved. Life itself as a whole can be worth living or not. (3) We can determine a positive ranking: πολλοῦ ἄ. worth much, X.An.4.1.28, Pl.Smp.185b, etc.; πλείονος ἄ. Id.Phdr. 235b, etc.; πλείστου ἄ. Th.2.65, Pl.Grg.464d, etc.; παντός, τοῦ παντὸς ἄ., E.Fr.275, Pl.Sph.216c; παντὸς ἄ., c. inf., Ar.Av.797; λόγου ἄ., = ἀξιόλογος, Hdt.1.133, Th.1.73, etc.; σπουδῆς, μνήμης ἄ., Plu.2.35a,172e (4) We can determine a negative ranking: οὐδενὸς ἄ. Thgn.456; ἢ παντὸς ἢ τὸ παράπαν οὐδενός Pl.Phlb.64d; ὀλίγου Id.Grg.497b, etc.; σμικροῦ Id.R.504dd.Lg.692c; μείονος, ἐλάττονος ἄ., X. Vect.4.50, Cyr.2.2.14; πολλαπλασίου τιμήματος ἄ. κτήσεις Arist.Pol.1306b12; also εἰς ὀγδοήκοντα μνᾶς ἄ. worth up to a sum of . ., D.27.10. (5) It is who I am, how I've been doing, what and how I have dealt with life: others, things, myself, that determines my relation with this constitutive way of thinking myself worthy of awards or penalties.

prize awarded for the most noble deeds: καὶ οὗ μάλιστ' ἐφίενται οἱ ἐν ἀξιώματι, καὶ τὸ ἐπὶ τοῖς καλλίστοις ἆθλον·

Aristotle takes κατ' ἀξία to be τὰ ἐκτὰ ἀγαθά as prizes of τιμή. The most radical τιμή is the tribute we pay to the gods. This is what men of ranking strive to obtain. Τιμή is the ἆθλον for the most noble and glorious deeds. We can still unpack this consequence: even this μέγιστον ἀγαθόν, τὸ ἐπὶ τοῖς καλλίστοις ἆθλον, is exterior, for it is conveyed by others, having thus the formal way to relate to honors. We can bestow honors and pay tribute to others, but having the honor bestowed to us by others, and by the ones that are better than ourselves, this is the extreme shape of honor.

Therefore we have a revised ranking:

1. "The small-souled man falls short both as judged by his own deserts and in a comparison with the claim of the great-souled man (ὁ δὲ μικρόψυχος ἐλλείπει καὶ πρὸς ἑαυτὸν καὶ πρὸς τὸ τοῦ μεγαλοψύχου ἀξίωμα. 1123b24–25)"
2. "The vain man on the other hand exceeds as judged by his own standard, but does not however exceed the great-souled man. (ὁ δὲ χαῦνος πρὸς ἑαυτὸν μὲν ὑπερβάλλει, οὐ μὴν τόν γε μεγαλόψυχον. 23b25–26)".
3. "And inasmuch as the great-souled man deserves most, he must be the best of men; for the better a man he is, the more he deserves, and he that is best deserves most. Therefore, the truly great-souled man must be a good man. Indeed greatness in each of the virtues would seem to go with greatness of soul. (ὁ δὲ μεγαλόψυχος, εἴπερ τῶν μεγίστων ἄξιος, ἄριστος ἂν εἴη · μείζονος γὰρ ἀεὶ ὁ βελτίων ἄξιος, καὶ μεγίστων ὁ ἄριστος. τὸν ὡς ἀληθῶς ἄρα μεγαλόψυχον δεῖ ἀγαθὸν εἶναι. καὶ δόξειεν<ἂν>εἶναι μεγαλοψύχου τὸ ἐν ἑκάστῃ ἀρετῇ μέγα. 23b27–30)"[3]

What we have been covering in particular is not only the analysis of possible configurations of character. There is no doubt that Aristotle suggests two ultimate possibilities of relating with the multiple situations that we endure in life, in all their variety of angles: the excess or defect, which eliminate any possibility of behaving appropriately, and the disposition of the environment. But the environment is as little geometric as excess and shortage are solid boundaries.

Environment is this faraway where boundaries are found well preserved. Excess, the ever so massive destruction of putting out all the stops, beyond the point of no return. Defect is the atrophy that can cause an unprecedented implosion of the self.

He who thinks of himself as the greatest, worthy of everything and to whom everything is owed, he has a relationship of excess with himself. From the height of his own opinion of himself, he is unstoppable. He will never be content. Nothing that he gets will ever be enough. But the distance that keeps him away from contentment, abundance and fulfillment will always be frustrating, disappointing, destructive.

[3] Arist. EN 23b27-30

He who thinks of himself as deserving nothing will never be content with anything either. Everything that he receives is perceived as undue and undeserved. The relationship that the μικρόψυχος has with himself is no less difficult than that of the χαῦνος. They both have a hard time coping with themselves. One due to hypertrophy. The other due to atrophy: both have always undermined any chances of relationship with themselves. Neither of them will ever enjoy and utilize the possibilities that are truly available to them. The μικρόψυχοι are not ignoble (κακοί), they are simply mistaken about themselves, they fail (ἡμαρτημένοι δέ). The μικρόψυχος is worthy and deserving of ἀγαθά, but deprives himself of such possibility (ἑαυτὸν ἀποστερεῖ ὧν ἄξιός ἐστι, 125ᵃ20). The misery in which he finds himself is the result of not valuing himself, not thinking of himself as worthy of anything, lack of self-esteem, as it is fashionable to say nowadays, or consistently bringing himself down: (κακὸν ἔχειν τι ἐκ τοῦ μὴ ἀξιοῦν ἑαυτὸν τῶν ἀγαθῶν, 125a21). The μὴ ἀξιοῦν ἑαυτόν is a ἀγνοεῖν δ' ἑαυτόν (125ᵃ22). The inability to open themselves up to their own selves leaves them with the impossibility of an assessment of oneself, of who they really are. In a radical way, we could say that they are not what they have the potential to be. They are in a broken relationship with themselves.

The pusillanimous figures it out. No question about it. And it's because he figures it out that he feels bothered with anything that happens to him, as if he is ashamed of having whatever he might get, even nothing, because he can't truly think of himself as worthy of anything, and therefore everything is undeserved and he should have nothing.

As far as the vain and the frivolous, those are stupid and clearly they can't see themselves: actually, without possessing any worth, they work hard on acquiring honours, but only to be humiliated afterwards, because they are not up to them (οἱ δὲ χαῦνοι ἠλίθιοι καὶ ἑαυτοὺς ἀγνοοῦντες, καὶ ταῦτ' ἐπιφανῶς· οὐ γὰρ ἄξιοι ὄντες τοῖς ἐντίμοις ἐπιχειροῦσιν, εἶτα ἐξελέγχονται, 125a27–29). They also can't see themselves: ἑαυτοὺς ἀγνοοῦντες.

The frivolous and the pusillanimous paradoxically do not fit in the world. The world of the frivolous is minuscule and the world of the pusillanimous is excessively large. Both of them find themselves in a situation of relationship with the self based on a distorted openness or on a shutting down towards themselves. In regards to the understanding of themselves and of the project of what is out there to be obtained in life, in the taking of what is meant for them, they fail by defect or by excess.

In an ensuing excerpt, Aristotle compares the relationship that the frivolous and the pusillanimous have with their life goals and corresponding projects. The pusillanimous could possibly have the intent of obtaining everything of value, in case they are ambitious enough for that: 1125a22–23: ὠρέγετο γὰρ ἂν ὧν ἄξιος ἦν, ἀγαθῶν γε ὄντων. It just so happens that they are too hesitant, ὀκνηροί. Each of them will pursue those actions which are configured within a system of value, importance and significance. But it is possible that they will divert from the noble actions as well as from the true occupations they have to engage in, in order to achieve them, because they find themselves incapable of performing them, even when their goals are merely within the category of external goods. (ἕκαστοι γὰρ ἐφίενται τῶν κατ' ἀξίαν, ἀφίστανται δὲ καὶ τῶν πράξεων τῶν καλῶν καὶ τῶν ἐπιτηδευμάτων ὡς ἀνάξιοι ὄντες, ὁμοίως δὲ καὶ τῶν ἐκτὸς ἀγαθῶν.)

Only apparently is the μεγαλόψυχος afflicted with "delusions of grandeur". This fits more closely the description of the frivolous. The magnanimous makes things big due to the way he sees them and the way he acts, how he behaves towards everything because of the way he perceives himself. He understands that he is not bothered by anything, he doesn't compare himself, doesn't envy. Not in that sense, at least, of what can be possessed. The magnanimous knows that we can't possess anything in this world. His existential move is played in a different tournament. His life is as radical and extreme as it is possible, because his commitment is absolutely with his ambition of honour.

We need to understand well the equation: ambition of honour, because easily he will let us hear about a project of life which is petty, afterwall. The ambition of honour is the pursuit of himself. The constitutive intent of each human being of conversion into oneself. The project of oneself which generates a tension of non-indifference towards oneself. The conversion of each one into oneself depends on the understanding of the possibility of obtaining what really matters. Such an obtaining that is not the acquiring of possessions or riches or power, but rather the ownership of oneself in which each one turns their possibility into a reality. It's for that reason that we can understand in advance that any honour obtained is asymetric and out of proportion to the "worth" of each human being. A honour, even when granted by those that are better than the one that is receiving it, is always "a thing". And not even the entire world is proporcional to life. And even less likely it is for those to whom being alive is the escape from their own belittling, the correction of ambition, and the sublimation of their own possibilities.

We had already read in 1124b8–9: that the magnanimous, when he needs to face danger, he doesn't hold back because he doesn't believe that there is any merit in staying alive at all cost (ὅταν κινδυνεύῃ, ἀφειδὴς τοῦ βίου ὡς οὐκ ἄξιον ὂν πάντως ζῆν). What is the existential picture that outlines such na extreme possibility?

We read in the beginning of *Nicomachean Ethics* that "it is rightly said' that everything without exception is set in motion towards a good: καλῶς ἀπεφήναντο τἀγαθόν, οὗ πάντ' ἐφίεται." (94a2–3). The first line of the Metaphysics reads "every human being without exception is constitutively (φύσει) exists in a exertion to seeing, τοῦ εἰδέναι ὀρέγονται" (MF 980$^{a}$1). Each one of us is in a ἐφίεσθαι, in a ὀρέγεσθαι, in a thrust, a tension of strain. Each one of us has already been exceeded, since always, by a superlative and absolute version of ourselves. This super absolute and superlative of myself may very well be a mirage. It is however always an indelible presence throughout our entire life. It is there that we have always and each time found ourselves. Since always pushing us out of a past and pulling us into the future of self. Or, on the other hand, the past is always there awaiting the time of death, as the future has since always been our "have been". It is the interpretation and the unscrambling of this simple relation to which we are bound by nature, of completely fulfilling ourselves, that alows the existential configuration outlining of the μεγαλόψυχος presented by Aristotle.

A configuration of the μεγαλόψυχος is marked by a tension towards objectives, purposes and goals, κατ' ἀξίαν. Aristotle still calls this objective of the magnanimous: τιμή (ἡ μὲν οὖν μεγαλοψυχία περὶ τιμὴν ἐστι μεγάλην, 125$^{a}$34). What honor is this that is questioned here? We can't follow here *pari passu* Aristotle's argument. It points to a paralellism between a μεγαλοψυχία and a μεγαλοπρέπεια (magnificence) an excellence which heightens generosity (125b1). To receive and grant honour is similar to receive and grant benefits, to pay a price and receive rewards.

> Just like there is a disposition of the correct measure, an excess and a defect in regards to giving and receiving money (ἐν λήψει καὶ δόσει χρημάτων), the same happens with the ambition of honor (ἐν τιμῆς ὀρέξει). In other words, there is a point in which we are aware that we receive or grant too many or too little honours. There is also a point in which we are aware of the origin of the appropriate possibility of relationship with honour (τὸ μᾶλλον ἢ δεῖ καὶ ἧττον, καὶ τὸ ὅθεν δεῖ καὶ ὡς δεῖ) 1125b5.

Aristotle opposes the one who has an incontrollable ambition of honour, the φιλότιμος, to the one that is in a constant destruction of such possibility: ἀφιλότιμος. The double possibility of obsessive behaviour (φιλία) with honour, in wanting it or avoiding it, might have positive aspects, says Aristotle. It is important to see what we reject and what we pursue. The question will always boil down to knowing what τιμή truly means?

While honour is measured by the value of what is precious as far as possession and power, obsession and its respective behaviour by excess and by defect, are outside the scope that comprises lucidity, ψυχή. We read in *Eudimian Ethics* that the designation "μεγαλόψυχος" points to a certain dimension of human lucidity and of power (ὥσπερ ἐν μεγέθει τινὶ ψυχῆς καὶ δυνάμεως. EE 1132$^{a}$27). It is the most powerful of the dispositions, κρατίστη ἕξις, and sweet, ἡδεῖα (a34). It is in greatness (ἐν μεγέθει) that μεγαλοψυχία resides, 1123b5. The elevation of the soul springing from itself or from one of its possibilities is not a tautology. In the dimension where lucidity or human life heightens itself, the disposition is one of pleasure, of an overwhelming feeling hard to explain. The sweetness that it comprises is not derived from any external content, neither from the body itself, neither from whatever else. Is the pleasure one feels from the enjoyment of radical activation that brings each one of us closer to ourselves. Greatness comes from the soul. Exaltation comes from the soul. The soul is absolutely asymmetrical to whatever it is that exists in the world.

*Ethics* is not written for the magnanimous.

Right from the beginning of *Nicomachean Ethics*, we can guess the enigma of the *Epigram of Delos*, 99$^{a}$26–7 based on the configuration of the magnanimous ways, radical and extreme. Everything that exists in absolute conformity with what will be: – this is the most splendorous thing that exists; there is nothing better than enjoying good health, but there is nothing more sweet than finding what we have loved since ever: – κάλλιστον τὸ δικαιότατον, λῷστον δ' ὑγιαίνειν· ἥδιστον δὲ πέφυχ' οὗ τις ἐρᾷ τὸ τυχεῖν.[4]

---

[4] This is an interpretation grounded on the book of Mário Jorge de Carvalho 2009.

## Bibliography

Bywater, I. 1894. *Aristotelis ethica Nicomachea*. Oxford: Clarendon Press.
Ross, W.D. 1924. *Aristotle's metaphysics*, 2 vols. Oxford: Clarendon Press.
Susemihl, F. 1884. *Aristotelis ethica Eudemia*. Leipzig: Teubner.

## Others

Carvalho, Mário Jorge. 2009. Mário Jorge de Carvalho: Die Aristophanesrede in Platons Symposium. Würzburg: Die Verfassung des Selbst, Königshausen & Neumann.
Caston, V. 1998. Aristotle and the problem of intentionality. *Philosophy and Phenomenological Research* 58(2): 249–298.
Chapura, M. 2009. Scale, causality, complexity and emergence: Rethinking scale's ontological significance. *Transactions of the Institute of British Geographers* 34(4): 462–474.
Coby, P. 1986. Aristotle's four conceptions of politics. *The Western Political Quarterly* 39(3): 480–503.
Collins, S.D. 2004. Moral virtue and the limits of the political community in Aristotle's "Nicomachean ethics". *American Journal of Political Science* 48(1): 47–61.
Cua, A.S. 2003. The ethical significance of shame: Insights of Aristotle and Xunzi. *Philosophy East and West* 53(2): 147–202.
Dickie, W.M. 1923. Anticipations in Aristotle of the four experimental methods. *Philosophical Review* 32(4): 401–409.
Fortenbaugh, W.W. 1964. Aristotle's conception of moral virtue and its perceptive role. *Transactions and Proceedings of the American Philological Association* 95: 77–87.
Garver, E. 1982. The meaning of ΘΡΑΣΟΣ in Aristotle's ethics. *Classical Philology* 77(3): 228–233.
Gordon, B.J. 1964. Aristotle and the development of value theory. *Quarterly Journal of Economics* 78(1): 115–128.
Halloran, S.M. 1982. Aristotle's concept of ethos, or if not his somebody else's. *Rhetoric Review* 1(1): 58–63.
Heath, M. 1988. Aristotle's poetics. *The Classical Review* 38(2): 231–233.
Kreager, P. 2008. Aristotle and open population thinking. *Population and Development Review* 34(4): 599–629.
Mara, G.M. 2000. The logos of the wise in the politeia of the many: Recent books on Aristotle's political philosophy. *Political Theory* 28(6): 835–860.
Mulgan, R. 1990. Aristotle and the value of political participation. *Political Theory* 18(2): 195–215.
Turner, M. 1970. The heroic ideal in Sidney's revised arcadia. *Studies in English Literature* 10(1): 63–82.
Ward, L., and Aristotle. 2001. Nobility and necessity: The problem of courage in Aristotle's "Nicomachean ethics". *The American Political Science Review* 95(1): 71–83.
Waters, D.D. 1969. Prince Arthur as Christian magnanimity in book one of the Faerie Queene. *Studies in English Literature* 9(1): 53–62.
Woodson, T. 1970. Thoreau on poverty and magnanimity. *PMLA* 85(1): 21–34.

# Chapter 8
# Intellectual Excellences of the Judge

**Tommi Ralli**

## 8.1 Introduction

In her confirmation hearings for the United States Supreme Court, Sonia Sotomayor responded to Republican critics who questioned President Obama's first nomination. Singling out speeches where Judge Sotomayor had highlighted her gender, ethnicity, and compassion, even once remarking in a law-school address that a "wise Latina woman" might reach better conclusions in some cases than a white male would, her critics voiced concerns over judging with one's heart. Senator Sessions said, "Call it empathy, call it prejudice, or call it sympathy, but whatever it is, it's not law. In truth it's more akin to politics. And politics has no place in the courtroom."[1] The empathy the critics cited, she replied, helped her grasp a case and did not override the law. "My personal and professional experiences help me to listen and understand," she told the Senate Judiciary Committee, "with the law always commanding the result in every case" (Baker and Lewis 2009). The judge's philosophy had to, and did in this case, come down to "fidelity to the law."

The generalised terms used within that debate, such as "law" and "politics," offer limited access to those specific points pertaining to listening and understanding. There is, to be sure, wisdom in the old story about the jurors of the French Revolutionary Tribunal, whom Anatole France divided into two groups, the unemotional hair-splitters and those who judged with their hearts, saying: "That second group always convicted."[2] Notice though, that such "hair-splitting" involves dissection

[1] Quoted in Baker and Lewis (2009).

[2] France (1912/2007). The translation in the quoted passage is from Kundera (2010, 52).

T. Ralli (✉)
Centre of European Law and Politics, University of Bremen,
Universitätsallee GW 1, 28359 Bremen, Germany
e-mail: ralli@uni-bremen.de

L. Huppes-Cluysenaer and N.M.M.S. Coelho (eds.), *Aristotle and The Philosophy of Law: Theory, Practice and Justice*, Ius Gentium: Comparative Perspectives on Law and Justice 23,
DOI 10.1007/978-94-007-6031-8_8, 

of the details, which are ultimately connected with perception, and so, both from the perspective of the unemotional judge Anatole France applauds, and the engaged perspective exemplified by Judge Sotomayor, the legal process hinges upon understanding the situation of the disputants. Although superficially, there appears to be a gulf between the "empathetic-judge" moniker and the unemotional hair-splitter, the understanding of each disputant's situation is common to both, though lacking in those members of the judiciary, mentally insulated against the concerns and necessities of people appearing before the courts.[3] The fashionable term "empathy" failed to distinguish this last group, at which it was aimed.

Sotomayor's specified invocation of the term "empathy" targeted the problem which I address in this chapter – that of seeing the particulars of cases and the view of the other. But the trajectory of the discussion quoted from the American nomination hearings ended up dealing in categories, law, politics, even feeling, which, like the term "empathy" itself, have limited discriminating capacity. Instead, I propose to draw upon the very old scheme of Aristotelian intellectual virtues, where one finds distinctions across a subset of interconnected skills including, among others, intuition and comprehension. Specifically, I shall look at how the judge – in giving just attention to particulars – exercises discernment or sense (*gnômê*) and the comprehension of what others say (*synesis*). As some translations (such as "fellow feeling" for *synesis*) (Gadamer 1960/2004, note 78, 378) and explanations (such as where "having a shared sense" means "sympathetic") (*EN* VI.11.1143a20, trans. Rowe) indicate, the two concepts are not wholly divorced from feeling.[4] One must be aware of the inherent complexity of the Aristotelian discourse, particularly in considering the syntheses between these concepts and the law-application process.

The following dialogue with the Aristotelian text aims to frame a substantive view of the legal process. I portray the judge as being involved in ways encapsulated by the two virtues or "excellences" of discerning what is reasonable (as a supplement to general rules) and comprehending the situation from the views of others. I conclude with a hypothetical about the legal process contributing to a better understanding of the other in a global environment.

Some previous legal-philosophical accounts of a "sense of appropriateness" (Günther 1993) or a perception-based addition to a theory of judging (Solum 1994, 138–139) have touched upon the ways in which we – cross-culturally – and, more

---

[3] On the campaign trail Obama had commented on the "5%" of the United States Supreme Court cases, where the law was not clear and "good intellect" not enough: "[T]he issues that come before the court are not sport. They're life and death. And we need somebody who's got the empathy to recognize what it's like to be young, teenaged mom; the empathy to understand what it's like to be poor or African-American or gay or disabled or old. And that's the criteria by which I'm going to be selecting my judges." A convention of Planned Parenthood, July 17, 2007, quoted in Livingston and Murray (2009).

[4] The term "empathy" (an early-twentieth-century translation of the German *Einfühlung*, or "feeling into") has contested meaning in science, depending on whether authors focus on the direct perceptual aspect or consider empathy a high-level cognitive phenomenon. For the disagreement over the exact nature of empathy, see, for instance, Preston and de Waal (2002, 2–4).

especially, selected experts conceive particulars. But in hindsight, these enhancements seem instances of what Rosalind Hursthouse has, in ethics, called "the current generalist versus perceptual/particularist debate" (Hursthouse 2006, 286). Below, I shall take up Hursthouse's criticism of the second, the only Aristotelian side of this debate – a criticism according to which preciously little can be said regarding how to develop the needed perception or intuition – and discuss those virtues about whose acquisition we might say more. For instance, one might be at a loss to read correctly the small signs in another person's behaviour (unless they are very similar to one's own idiosyncrasies). Hursthouse's approach appends to any such standard advice that discernment requires experience, the further question about the requisite experience: "What sort?" (Hursthouse 2006, 286). Of what species?

Here is the argument. The judge's discernment is developed by the experience of exceptions, as Hursthouse proposes in the context of practical wisdom. I add that the judge will exercise the discernment by suspending the application of principles while listening, aware of the possible relevance of unforeseen particulars. Furthermore, I add that the exceptions include experiences lived through, which Hursthouse's technical view neglects (Sect. 8.3). The judge uses her comprehension to absorb the components of the situation based upon testimony. I elaborate that, for this grasp of the situation, the judge will have to be open to different perspectives, able to move between them somewhat, and yet courageous enough to stand by what she deems right (Sect. 8.4). These ruminations illuminate the diverging qualities of judges as decision-makers compared with public regulators (Sect. 8.5). In order to arrive at an explanation within a framework inspired by Aristotle, the location of the "merely discriminating" virtues of discernment and comprehension in his scheme, and the reason that their "conclusive" cousin, intuition, is not plumbed further in this chapter, have to be outlined first (Sect. 8.2).

## 8.2 "Minor" Intellectual Excellences and Intuition

The main part of all the human virtues analysed by Aristotle is located at the point of conflict between the rational and the non-rational side of the soul. At this point, the virtuous person listens effortlessly to reason, while the less virtuous who have not achieved such harmony either obey reason, or do not, when another side in them fights against reason and resists being persuaded by it (Aristotle's introductory example of the struggle contrasts people with self-control to people lacking it) (*EN* I.13.1102b14–29). As rephrased in the 1970s, the virtuous person will not balance reasons for and against – for the virtuous the reasons for acting in other ways are not outweighed or overridden, but "silenced" (McDowell 1979/1998, 55–56).

Contrary to such virtues of character, other virtues attach entirely on the rational side of the soul. These are the varied intellectual virtues catalogued in Book VI of *Nicomachean Ethics*. The catalogue is difficult to comprehend, unless one reads into it some overarching line it is supposed to exhibit. One possible interpretation is that Aristotle's discourse maps how practical wisdom (*phronêsis*) – "a true disposition accompanied by rational prescription, relating to action in the sphere of what is good and bad for human beings" (*EN* VI.5.1140b5–6, trans. Rowe) – compares with other intellectual excellences.[5] I shall assume this interpretation for the solely instrumental reason that some unifying thread is supremely helpful in penetrating Book VI, and this one at least prepares and fits the present focus on the judge's action and on the "minor" intellectual excellences of comprehension and discernment, which go hand in hand with practical wisdom.

"Minor" is a later editor's term (*EN* trans. Ross), designating those intellectual virtues that do not give truth or rightness in the form of either affirmation or denial. Those five which do, are called "chief." Among the "conclusive" and thus "chief" intellectual virtues, Aristotle distinguishes first scientific knowledge (*epistêmê*), which concerns invariables, things that are of necessity, from the craft of making things (technical expertise, art, *technê*) and the aforementioned practical wisdom, which are excellences concerning variable things (things that could be otherwise).[6] Then, a fourth excellence, intuition or intelligence (*nous*), is paired with scientific knowledge:

> Scientific knowledge is judgment about things that are universal and necessary; and the conclusions of demonstration, and all scientific knowledge, follow from first principles (for scientific knowledge involves proof). This being so, the first principle from which what is scientifically known follows cannot be an object of scientific knowledge [...] (*EN* VI. 6.1140b, trans. Ross).

So the starting-points of scientific knowledge are assigned to intuition. Eventually, the "most finished" form of knowledge is said to be scientific knowledge in combination with intuitively known true first principles, and when this knowledge concerns the highest subjects, such as causes or heavenly bodies, Aristotle calls it intellectual achievement (*sophia*), also translated as theoretical or philosophical wisdom. Assuming that the overarching line of his scheme is to compare practical wisdom with the other capacities, a plausible goal of defining intellectual achievement thus is to contrast it with practical wisdom, and to dispose of it as a false model for the latter (Broadie 2002, 46–47 and 52–53).

At first sight, those virtues that help us grasp details would appear to be intuition, in some sense, practical wisdom, and the three "minor" virtues connected with practical wisdom – deliberative excellence, comprehension, and discernment. Now, the first of the "minor" ones is deliberative excellence (*euboulia*). Taken in itself, deliberation means enquiring during the often long periods when we have not yet reached assertion, so we are, in Aristotle's words, "searching for something and calculating." (*EN* VI.9.1142b, trans. Ross) To give a legal example, if a judge held onto the same resolution from the beginning to the end of a case, that judge would not be engaged in deliberation. A "strong" judge, as some have mistakenly called this type, would not then be a deliberating one at all, but, absurdly, someone

[5] For this interpretation, see Broadie (2002, 46–47 and 53); *NE*, trans. Rowe, Commentary, 357.

[6] The reason for this first distinction is that "where objects differ in kind the part of the soul answering to each of the two is different in kind, since it is in virtue of a certain likeness and kinship with their objects that they have the knowledge they have." (*EN* VI.1.1139a, trans. Ross)

unmoved by reason. As to excellence in deliberation, it requires having the right end, finding the means that produce the end "most easily and best," finding the means no slower than another person would (*EN* VI. 9.1142b27), and, if the end comes about by one means only, looking to see how.[7] A doctrinal legal adaptation is available: the principle of proportionality, spread around the globe in the late twentieth century, reproduces partly the structure of deliberative excellence, though it may additionally involve deciding the "right end," which Aristotle's deliberation seems to take for granted, as he assumes that "[w]e deliberate not about ends but about means."[8] (*EN* III. 3.1112b, trans. Ross)

But the curious thing about the "minor" excellences is that only the remaining two – comprehension and discernment – are said to deal with particulars, or "ultimates" (*EN* VI.11. 1143a, trans. Ross) among facts; in so doing they are said to come together with the "conclusive" practical wisdom and also a kind of intuition. Before concentrating on these two, comprehension and discernment, I want to address briefly the dual task of intuition (*nous*) in Aristotle's scheme. Intuition forms a pair with both scientific knowledge and practical wisdom in the following respects:

> [I]ntuitive reason is concerned with the ultimates in both directions; for both the first terms and the last are objects of intuitive reason and not of argument, and the intuitive reason which is presupposed by demonstrations grasps the unchangeable and first terms, while the intuitive reason involved in practical reasoning grasps the last and variable fact, i.e. the minor premiss. For these variable facts are the starting-points for the apprehension of the end, since the universals are reached from the particulars; of these therefore we must have perception, and this perception is intuitive reason (*EN* VI.11.1143a–b, trans. Ross).

At the end of this passage, intuition is described as perception in relation to particulars. On being introduced as a pair to scientific knowledge in the preceding indented paragraph, intuition was, by contrast, defined negatively, as affirmation or denial that does *not* follow from demonstration. (The negative way of defining intuition is shared by many contemporary attempts, such as Antonio Damasio's definition of intuition as a "mechanism by which we arrive at the solution of a problem *without* reasoning toward it" [Damasio 1994, 188].) The respective two characterisations of intuition – "practical *nous*" as perception, "theoretical *nous*," what is not the object of demonstrations – allow us to say, at most, that some people sense or "see" correctly what is relevant or right in a particular case. This limited approach leads directly to Hursthouse's complaint, which I paraphrase in the paragraph

---

[7] *EN* III. 3.11.12b12–20 except as indicated in the text. (The quoted expression: *EN* III. 3.11.12b18, trans. Rowe)

[8] According to Robert Alexy's reconstruction of German constitutional jurisprudence, the principle of proportionality has three parts: first, a means under judicial scrutiny should be suitable to a legitimate end; second, between equally suitable means, the means interfering least intensively with another legitimate end in the system should be chosen; and third, if legitimate ends conflict, the greater the degree of non-satisfaction of one end, the greater should be the importance of satisfying the competing end. See, for instance, Alexy (2003, 131–140). The difference with deliberative excellence is the last part, called "balancing" or "proportionality in the narrow sense."

immediately following the next subheading. It would obviously be desirable to discover more positive characterisations of intuition, so that the correlated aspects of legal decision could also be understood otherwise than through the neighbouring virtues – those virtues that deal with particulars or those the like of which intuition is "not."

## 8.3 Discernment and Exceptions

In *Practical Wisdom: A Mundane Account* (2006), Rosalind Hursthouse expresses her regret that in contemporary ethics the practically wise person (*phronimos*) is often assumed to have only the perceptual capacity to see correctly what is to be done – a capacity whose recognition signifies, sure enough, an improvement over the un-Aristotelian "generalist model" according to which a person has special propositional knowledge, knowledge *that*, for instance, knowledge of a "system of what we would call 'moral principles'" (Hursthouse 2006, 284). But Hursthouse is interested in the development of excellences, and on that score the "perceptual model," as she calls this alternative, leaves us in the dark about what kind of experience we need in order to develop our capacity to see correctly and what kind of experience we should seek as a consequence. Thus she turns from intuition to discernment (*gnômê*).[9]

Aristotle characterises discernment as the correct discrimination of what is equitable or reasonable (*EN* VI.11.1143a). He is "tantalizingly brief," Hursthouse notes (2006, 288), and she asks what experience the *phronimos* has that a well-habituated but inexperienced adolescent lacks. She proposes that the *phronimos* has learned to discern such particulars as that, exceptionally, abandoning one's claim may be magnanimous or resigning may be courageous:

> The well-brought-up young would tend to suppose that abandoning one's claim was servile or pusillanimous, even though their natural generosity prompted them to do it, and to suppose that it would be cowardly to resign even though 'mildness' (perhaps? or justice?) suggested one should. [...]
>
> It seems plausible to suppose that the well-brought-up but inexperienced tend to think about what the virtues require and the vices rule out in terms of rather conventional generalizations or paradigms. It is only with the experience of exceptions – when an admired figure does what you thought only a pusillanimous coward would do and is widely praised, when the action of someone you respect surprises you until she explains why she did it, when you hear accounts of such examples – that you come to the more sophisticated understanding – the discernment – that the *phronimos* has (Hursthouse 2006, 290 and 292, respectively).

Hursthouse's resulting advice is that one should take good note of exceptions, when coming upon them, and be mindful of other exceptions (2006, 290).

[9] She declines the either-or question of assigning the capacity to judge exceptions (below in the text) to either discernment or "practical *nous*," remarking that, by developing discernment, one develops the practical *nous* (Hursthouse 2006, 289 and 290).

Her advice is targeted against unthinking generalisations about the virtues. But the same experience of exceptions is of course indispensable in the case of applying general rules to interpret difficult conflicts, including legal disputes. The following is, then, what I myself want to add to Hursthouse's remarks.

Surprising particulars can be relevant – and we ought not to apply principles or rules to others as if knowing, without attentively listening to the other, what their situation is. At least where the other is seriously affected, we should not reason that because somebody does *a*, he or she must also be doing *b* or *c*. Nor should we impose on the other a general tendency or a stereotype before knowing more. There is a difference, I suspect, between assuming things about a person with whom we have no interaction, and assuming things about someone with whom we communicate, even formally. In the former case, our assumptions can be harmless, imaginative or trivial. In the second case, the assumptions reveal a lack of respect. If we wish to speak here of "empathy," then the show of respect to the individual is the commonality between that more recent term and the refined discernment.[10]

In this way, I have specified the statement about exceptions by contemplating individual nuances, which is fundamentally antithetical to stereotyping. For example, a judge who assumes that a victim of rape must yell or have bruises would usually be suspected of lacking experience, because the victim may be rendered inert. Potentially the novice, such as judges taken straight from law school, would more likely apply a stereotype. In the Athens of Aristotle's time, a juror had to be at least 30 years old, and Aristotle agrees with the general opinion that discernment like comprehension and intuition depends on age (*EN* VI.11.1143b7–14). Even so, emphasising age alone invites counterexamples, as not all careers facilitate the same extent of integration of these virtues.

In fact, both the truth that surprising particulars can be relevant and the hesitancy required in applying principles to others follow already from the classical, negatively expressed equity argument, which Aristotle's own brief description of discernment leaves us to pursue. According to that famous argument in the *Ethics*, the equitable or reasonable is not only just but, precisely, a supplement to the legally just, made necessary because "all law is universal, and yet there are some things about which it is not possible to make correct universal pronouncements." (*EN*. V.10.1137b14–15, trans. Rowe)

Where the law provides us with background as portrayed in the equity argument, Hursthouse's instruction to take good note of and to watch out for exceptions, seems sensible. The training for the sake of which she connects exceptions with correct discernment will be familiar to lawyers, who vary cases so that they can tell where the turning points are, which render an action impermissible. For instance, at the time when a precedent-setting opinion is given, it can be unclear as to how broad a rule is derivable from it. The lawyers' intellectual activity of variation creates a kind of continuum between a normal case and exceptional cases. The exceptions will,

[10] The subtext of respect in the case of empathy is acknowledged, for instance, in Mnookin et al. (1996, 220).

incrementally, be justified as if on a continuum from the general rule. Likewise, but looking forward, the judge who is deciding the case which may set a precedent has to consider variations of cases that can come up after the present case.

But this lawyerly technique of how to treat the exceptional is quite another matter than meeting the exceptional, which can force a person to think differently about something. We are struck by experiences that can alter our attitude to life – a fundamental change in a friend, a good friend's breakup or illness, a narcissistic superior in the workplace. These trials demand of us less a technique than a kind of involvement, in the course of which we may have to choose, as friends or relatives must, between learning to cope with the thing and cutting contact with someone altogether. Our choice will often be formed by the type of discernment – "merely discriminating" – and attention to the matter in all its complexity, which indeed precede the moment that we get affirmation and act accordingly.[11]

Hursthouse's advice does not take into consideration the life experiences in which we are involved with the exceptional, though these should surely belong within the wider sphere of wisdom with which her ethical work is concerned. Still, the trained eye for exceptions summarises the lawyerly way of going about relevant details, exemplified by the legal and judicial method of varying cases. But in addition, the experiences that require more than a technique are conducive to the process of conflict resolution. To explore how, I consider the virtue of comprehension.

## 8.4 Comprehension, Listening, Courage

The excellence of comprehension or "understanding" (*EN*. trans. Ross) (*synesis*) means discriminating correctly that which others say. It shows in grasping or "seeing" the other's point, just as one can see the point in scientific conclusions,[12] although we are now concerned with questions of practical wisdom, such as judgment. I again draw inspiration from Hursthouse's commentary.

"What is done is always done in a particular situation," she writes (Hursthouse 2006, 290), but a "'situation' which calls for my doing something may not be *facing* me at all, waiting for me to read it, but rather something whose details I have to work out from what other people say about it" (Hursthouse 2006, 291). The author goes on to discuss the related case where someone is faced with a situation and

[11] As yet unsurpassed is Murdoch's description of this ethical process, a progenitor of decision: "[I]f we consider what the work of attention is like, how continuously it goes on, and how imperceptibly it builds up structures of value round about us, we shall not be surprised that at crucial moments of choice most of the business of choosing is already over." Murdoch (1964/1999, 329).

[12] "[J]ust as seeing the point is called 'comprehending' when one is exercising systematic knowledge, so too one 'comprehends' when exercising judgment in order to discriminate about the things wisdom deals with, when someone else is speaking – and exercising it in order to discriminate rightly (for 'excellently' is the same as 'rightly')." *EN* VI.10.1143a13–17, trans. Rowe.

another person describes it to her or him incorrectly as a "dilemma," or as a situation with only one way out (Hursthouse 2006, 292–293). But here I part ways with her account and focus, instead, on addressing and expressing the situation appropriately.

Living creatures who organise their surroundings need, presumably, some framework in their mind. But this structure tends to be confirmed by its successful use – similarly to what happens in the sort of fortification against which it is customary to warn in scientific contexts, namely, that of only taking in evidence supporting one's own theory. As a corrective, we might, in some contexts, be sceptical about our perceptions (again, perhaps, in scientific contexts, but in general, doubt is not good: it has no intrinsic value in personal relationships, for instance). More generally than in the limited contexts suited to doubt, we should try to see things from many angles and, to do this, we may have to change between perspectives. We cannot completely change into another perspective without forgetting our own – but we can give attention to where the other person comes from and at least try to incorporate the other's perspective into the one from which we start. A single perspective, while conceivably sufficient for organising particulars in a coherent manner, limits the way we see them. We see what we are predisposed to seeing, as the aphorism says. But by being open to different perspectives and able, somewhat, to move between them, we can better listen to others, including our peers, as then the others need not be fitted into one and the same frame. It is this openness to multiple perspectives, and an inherent acknowledgment of their existence, which characterises the *phronimos*, I posit, enabling her or him to get the situation right.

To achieve that discrimination of the situation for which openness is a condition, the legal judge can, at the moment of listening, be on a par with the disputants, despite her position of authority. The aspects of cases that are subject to such periods of dialogue vary, no doubt, across legal traditions and areas of law.[13] But anyone, including the judge, put into the position of solving a dispute, will learn the same humility from the experience of hearing different descriptions of a situation. After perceiving the case in a certain way, we hear the differing account of another party, whose perception varies, and the situation has changed, facing us anew. Openness to perspectives may even be necessary for anything related by the other to challenge us. The other's words can be defective: the cries of injustice that underlie many legal disputes, start because something has been done wrong, may be faint or misconstrued. At this phase, a judge's sensitivity is not yet directed towards a question of right or wrong. By listening to the party, who needs to be understood, the judge shows respect. Importantly, the judge cannot stoop to tepid answers and technical explanations, such as those readily available to a bureaucrat.

[13] Positive law portrays examples of judges' opportunities for dialogue. For instance, according to the German Code of Civil Procedure: "To the extent required, the court is to discuss with the parties the circumstances and facts as well as the relationship of the parties to the dispute, both in terms of the factual aspects of the matter and of its legal ramifications, and it is to ask questions." § 139 Abs. 1S. 1 ZPO.

Openness is but a fragment of comprehension. While the conception of comprehension thus elaborated does not go so far as to say that *synesis* is "fellow feeling,"[14] it does have affinities with some concepts of empathy: both because of the common subtext of respect[15] and because certain cognitive versions of empathy connote "perspective taking."[16] More urgent still, two other, broader conceptual connections have to be established, in order to develop the conception of comprehension towards a fuller account of what immersion in the situation involves.

First of all, openness is necessary, but so is the courage to stand up for that which we are choosing – and to discard those alternatives not chosen. Admittedly, Aristotle's own depiction of the virtue of courage (*andreia*, literally "manliness") is confined to withstanding what is painful, death and wounds, for the sake of what is fine, in the battlefield (*EN* III. 6–9). We need, instead, a wider notion (so that, for instance, letting go of an option, while painful, can be done for the sake of – say – justice). This wider notion is needed, because openness without the courage is empty, eventhough the two may certainly seem incongruent with each other. That last concern is the "tension between empathy and assertiveness" discussed in negotiation literature. There, the tension has been described as a challenge for an effective negotiator; and a mediator has also figured as a model, who might both show an empathetic understanding of all the disputants and, simultaneously, work to assure that each disputant holds onto and asserts their interests (Mnookin et al. 1996; on mediator, 228). If the same model were applied to the judge, she could, in her different role, show an understanding of all the parties while, concomitantly, working out the substance of the dispute and, if needs be, standing up for what she knows to be right. According to Aristotle's distinctions, such an ideal would combine the character virtue of courage – gained by doing courageous things (habituation) – with comprehension (*synesis*) which, as an intellectual virtue, results from experience and teaching (*EN* II.1.1103a17).

Secondly, along with being dimensioned by courage, the judge's comprehension is guided by the quintessentially judicial principle among the ethical doctrines which a "generalist" would consider "knowledge that": the rule of hearing both sides, *audiatur et altera pars*.[17] This rule indicates a method of getting the situation right by listening. But it is equally a precept that the judge can be said to be "reciprocally" following – that is, reciprocally with the above-described active efforts – while listening. I conceive of the judge as distributing attention to all relevant parties in harmony with this rule and gaining an understanding of the situation from their perspectives on facts. But also from a "generalist" perspective, concentrating on principles whenever possible, the judge may hardly be expected

[14] A vicarious experience or understanding for someone (even if the other were a fellow only in the sense of speaking the same language).

[15] As noted, in the case of empathy, in Mnookin et al. (1996, 220).

[16] For instance, Preston and de Waal (2002, 4) offer "perspective taking" as a synonym of "cognitive empathy."

[17] For a "generalist" understanding of this rule, see, for instance, Duff (2003, 219).

to deduce what to do at every level of detail – when to stop listening to defences, for instance – and so needs, in the un-codified region, what should anyway occur, interlaced with deductions – namely, comprehension.

## 8.5 Global Implications and Concluding Remarks

When thinking about institutions in an era of continuing globalisation, besides the ever-present calls for better regulation and better cooperation, a further desideratum often mentioned is contestation.[18] Legal disputation, or even the American legal and regulatory culture of "adversarial legalism" (Kagan 2001), can be cited as examples from the formal end of the contestational spectrum. But one should ask what the value of contestation is. And the foregoing analysis supplies an answer: That value lies in making the decider more involved in the situation, which may bring better understanding. The judge cannot just refuse to listen, and here the character of the decision-maker's role differs compared to a regulator or other official, who is less compelled to listen to and to think about the exact concern of the individual. Discernment enjoys perfect placement within the judge's role, and as previously noted, the experience of exceptions should improve the judge's ability to listen to the individual, recalling the former assertions concerning stereotypes.

Indeed, we might expect judges to exercise similar virtues across legal borders – simply as they "hear and engage," in the ways detailed, the evidence and arguments of both sides, concerning injustices abroad.[19] The American Judge Richard Posner opines that "our adversarial system [...] forces a judge to give a hearing to someone who will challenge the judge's intuition" (Posner 2008, 107). The above Aristotelian analysis may allow us to see the legal process reinforcing the implementation of such virtues beyond the confines of the nation state. The judge is a special kind of expert, one well characterised by the two excellences, discernment and comprehension.

The substantive features of the legal process highlighted make such procedures a distinct model for our own learning of intellectual virtues. Doubtless, no amount of studying the details of legal cases will automatically provoke in us a better understanding of the other, or the aspiration to articulate specific reasons. While the intellectual virtues develop partly as a result of teaching, one cannot underestimate, for this development, the importance of being near a person who exemplifies these virtues.[20] Ongoing discussions of this kind must be characterised by careful

---

[18] For just one example, see, in the context of the World Trade Organisation, Howse and Nicolaïdis (2008, 163–191).

[19] A similar argument is put forward, in the case of private-law claims, in Wai (2005; the quoted expression within the sentence in the text, 480).

[20] On the *phronimos* as a living example of one who is listened to with respect, see, for instance, Broadie (2006, 348).

attention to the subject matter and a lack of self-serving ambition. I hope to have shown how the judge's task differs from those of a public regulator or other official, including an academic observer, for whom the intellectual virtues can mainly be gifted as a consequence of such contingent life experiences, meaning the reciprocated impartation of learning, as encapsulated within the figure of the *phronimos*.[21]

## Bibliography


Alexy, R. 2003. Constitutional rights, balancing, and rationality. *Ratio Juris* 16(2): 131–140.

Aristotle. 1998. *The Nicomachean ethics*. Trans. David Ross. Revised J.L. Ackrill, and J.O. Urmson. Oxford/New York: Oxford University Press.

Aristotle. 2002. *Nicomachean ethics*. Trans. Christopher Rowe. Oxford/New York: Oxford University Press.

Baker, P., and N.A. Lewis. 2009. Sotomayor vows 'fidelity to the law' as hearings start. *The New York Times*, July 14, 2009. http://www.nytimes.com/2009/07/14/us/politics/14confirm.html.

Broadie, S. 2002. Philosophical introduction. In *Nicomachean ethics*, 9–91. Trans. Christopher Rowe. Oxford/New York: Oxford University Press.

Broadie, S. 2006. Aristotle and contemporary ethics. In *The Blackwell guide to Aristotle's Nicomachean ethics*, ed. R. Kraut, 342–361. Malden/Oxford: Blackwell Publishing.

Damasio, A.R. 1994. *Descartes' error: Emotion, reason and the human brain*. London: Picador.

Duff, R.A. 2003. The limits of virtue jurisprudence. *Metaphilosophy* 34(1–2): 214–225.

France, A. 1912/2007. *The gods are athirst*. Project Gutenberg EBook. http://www.gutenberg.org/ebooks/24010.

Gadamer, H.-G. 1960/2004. *Truth and method*. London/New York: Continuum.

Günther, K. 1993. *The sense of appropriateness: Application discourses in morality and law*. Albany: State University of New York Press.

Howse, R., and K. Nicolaïdis. 2008. Democracy without sovereignty: The global vocation of political ethics. In *The shifting allocation of authority in international law: Considering sovereignty, supremacy and subsidiarity*, ed. T. Broude and Y. Shany, 163–191. Oxford/Portland: Hart Publishing.

Hursthouse, R. 2006. Practical wisdom: A mundane account. *Proceedings of the Aristotelian Society* 106: 283–307.

Kagan, R.A. 2001. *Adversarial legalism: The American way of law*. Cambridge, MA/London: Harvard University Press.

Kundera, M. 2010. *Encounter*. New York: Harper.

Livingston, A., and M. Murray. 2009. Context of Obama's 'empathy remark.' First read from *NBC News*, May 1, http://firstread.msnbc.msn.com/_news/2009/05/01/4430634-context-of-obamas-empathy-remark.

McDowell, J. 1979/1998. Virtue and reason. In McDowell, *Mind, value, and reality*, 50–73. Cambridge, MA/London: Harvard University Press.

Mnookin, R.H., S.R. Peppet, and A.S. Tulumello. 1996. The tension between empathy and assertiveness. *Negotiation Journal* 12(3): 217–230.

Murdoch, I. 1964/1999. The idea of perfection. In Murdoch, *Existentialists and mystics: Writings on philosophy and literature*, ed. P. Conradi, 299–336. London: Penguin.


[21] I would like to thank Liesbeth Huppes-Cluysenaer for her thoughtful comments on an earlier draft of this paper.

Posner, R.A. 2008. *How judges think*. Cambridge, MA/London: Harvard University Press.
Preston, S.D., and F.B.M. de Waal. 2002. Empathy: Its ultimate and proximate bases. *Behavioural and Brain Sciences* 25: 1–72.
Solum, L.B. 1994. Equity and the rule of law. In *The rule of law: NOMOS XXXVI*, ed. I. Shapiro, 120–147. New York: New York University Press.
Wai, R. 2005. Transnational private law and private ordering in a contested global society. *Harvard International Law Journal* 46(2): 471–488.

# Chapter 9
# Justice *Kata Nomos* and Justice as *Epieikeia* (Legality and Equity)

**Samuli Hurri**

## 9.1 Introduction

I will discuss certain points in Aristotle's work from the general perspective of law as practice. The focus is on the relationship between truth and justice as that relationship is actualised in legal practice. The background from which the paper stems is not ancient philosophy but modern social theory, notably the works of Michel Foucault. The efforts of an on-going background project aim at elaborating a general theory of legal practice. To do so, Aristotle has provided certain elements to build on. This paper will focus on these elements, but let us first make a note of how truth and justice form the context in the background project.

The theme of the truth-justice relationship lies in Michel Foucault's analyses of the historically changing relations and practices of power. These are modified in intersections between systems of thought and regimes of practice. For example, penal psychiatry is an intersection between psychological knowledge and delinquency control. At one point Foucault depicted the themes of his research as forming in connections between *veridiction* and *jurisdiction*: truth-locution and legal locution. A common feature of the historically occurring intersections between systems and regimes is that they conjoin these two types of discourse, jurisdiction and veridiction, so that their resulting interplay constitutes novel types of power-relations. Power-relations, in turn, are constitutive of everything else in Foucault's world. In an interview in 1978, Foucault explained his research projects in the following way:

> To analyze 'regimes of practices' means to analyze programs of conduct that have both prescriptive effects regarding what is to be done (effects of 'jurisdiction') and codifying effects regarding what is to be known (effects of 'veridiction') (Foucault 2000, 225).

S. Hurri (✉)
Centre of Excellence in Foundations of European Law and Polity,
University of Helsinki, P.O. Box 4, Helsinki, 00014 Finland
e-mail: samuli.hurri@helsinki.fi

L. Huppes-Cluysenaer and N.M.M.S. Coelho (eds.), *Aristotle and The Philosophy of Law: Theory, Practice and Justice*, Ius Gentium: Comparative Perspectives on Law and Justice 23,
DOI 10.1007/978-94-007-6031-8_9, 

On the basis of the analytical distinction between truth-locution and legal locution, Foucault declared his next year's lecture course (*The Birth of Biopolitics*) as "undertaking a history of truth coupled with a history of law." (Foucault 2008, 35) In that history Foucault was interested in a variety of social practices that – unlike law – base their authority on knowledge, on something resembling science. In those practices, however, knowledge is invested with something resembling law's jurisdictional power. The result is subtle coercion, *potentia* in the guise of *scientia*, invisible power that is exercised over ignorant individuals. For that power, Foucault often used the term *normalisation*, meaning empirical knowledge having become normative. Normalisation forms human character on the basis of the categories of 'normal' and 'pathological' instead of the juridical categories of 'permitted' and 'forbidden'. For instance, economic governing is normalisation in that it imposes on subjects a behavioural model according to which the rational pursuance of their best interest is normal for them rather than juridically permitted or forbidden.

It is often thought that law for Foucault was a misguiding language of power, and that otherwise his interest in the law was rather limited. Beyond symbolism, the law figured for him in two basic ways: firstly, it was an institution to which other practices (knowledge-practices) conjoined their functioning; secondly, it was a specific technology whose elements other practices might borrow and modify for their own use. Yet the law was highly central to Foucault in another way: it served him as an ideal-typical practice of power for which empirical knowledge is fundamentally irrelevant. He used the juridical ideal type for analysing veridictional systems and regimes based on empirical knowledge. In the history of these regimes and systems Foucault was looking for the silently working jurisdictional element of all empirical knowledge: exclusion of the abnormal. Moreover, this history – the political history of truth – in Foucault's view develops by the logic of strategy. Strategic logic means that mechanisms and institutions developed at one time for a definite purpose may later be reused in a completely different context and for completely different purposes. For strategic logic, too, the ideal-typical real staging is lawyers' practice, where rules that are empty in themselves can be bent to any purpose (Foucault 1998, 378).

Nonetheless, Foucault was not genuinely interested in the practice of law as such. He never looked into law as a regime of practice capable of confronting other regimes of practice working on the basis of knowledge, science and truth. In other words, he never looked at the confrontation between juridical normativity and normalisation from the perspective of legal practice. Now, that is precisely the theme of this paper: it searches for a juridical paradigm mechanism, by which the law first confronts veridiction and then transfixes it into its own jurisdiction. This is where we meet Aristotle and his presentation of the interplay between justice as equity (*epieikeia*) and justice as legality (*kata nomos*, through positive laws). The goal of the following analysis is to show how this interplay transforms truth-locution into legal locution and legal locution into truth-locution.

Among fields, traditions and schools of research that are perhaps more outspokenly based on the work of Aristotle, there is also a clear and recognised basing of the theory of practice on the work of Aristotle. As is well known, in Aristotle's division of intellectual activities (*dianoia*) into theoretical, productive and practical, practical activity is defined by having as its beginning and principle (*arche*) the actor's performance of choice (*prohairesis*) (*Met.* VI:1, 1025b). Moreover, practical activity is distinct from productive activity (*poiesis*) in that it is self-contained: good conduct (*eupraxia*) itself is the objective of the conduct, not creation or production of something else (*heteron*) (*EN* I.1.1094a; VI.5.1140b).

Hannah Arendt built her political theory on Aristotle's two ideas of practice: self-contained action where performance is accomplishment; the actor as the ultimate initiator of movements and events (Arendt 1989). On the whole, other aspects of Aristotle's political theory (such as the relationship between the community and the individual; arithmetical and geometrical formalisations of justice; just and corrupt forms of government; the nature of the slave and the nature of the citizen) could be seen as overriding the elements that Arendt wished to focus on. Aristotelian ethics of *eudaimôn* and virtues nevertheless implies that practice is self-contained: virtue exists only in actions and is born (*gignomai*) when it is exercised (*energeô*) (*EN* II.3.1105a). A 'happy life' is an end-in-itself with Aristotle, but it is not an end-state where one can put one's feet up and enjoy life. On the contrary, life puts new tests before us every day and 'happiness' must be achieved each time anew by practicing virtue. Practicing virtue is a means to consummate the final cause, yet this final cause is a creation internal to practice. Practice is a means, but it is also an end, insofar as not mere contemplation but actualisation of a good life is an end.

Can one build a legal theory on Aristotle's ideas of practice? To pave the way for accomplishing that, let us isolate from the work of Aristotle a sketch that presents *justice* as something actualising in the practice of law. Some readers would rather hold justice as the final cause (*telos*) given externally and objectively to the law. The present reader leaves that idea of justice to Plato and pursues a faithful understanding of Aristotelian legal justice as actuality (*energeia*). That justice exists only in practice. According to Arendt, *energeia* is defined as follows:

> [...] Aristotle's notion of *energeia* ('actuality'), with which he designated all activities that do not pursue an end (*ateleis*) and leave no work behind (*par' autas erga*), but exhaust their full meaning in the performance itself (Arendt 1989, 206).

My thesis is that the choice (*prohairesis*) that actualises justice every time anew in legal practice is a choice between legality and equity. This choice involves discursive operations that transform jurisdiction into veridiction and veridiction into jurisdiction. This *prohairesis* forms the heart of legal practice in that it brings justice into existence. In legal practice justice is a matter of actuality (*energeia*) so that justice as *telos* is subsumed to justice as *prattein*. In other words, the practice of law reproduces justice irrespective of final causes, that is, through employment of its own structures, legality and equity.

To recap, the objective in the following is to present and discuss two problems. (1) How does legal practice transform veridiction into jurisdiction and jurisdiction into veridiction? (2) How does legal practice bring about justice as actuality?

## 9.2 Legislator, Judge, Litigant

The concrete setting in which legal adjudication took place in a Greek city state at the time of Aristotle was of course wholly different from the modern setting, but what he tells about its elementary structures and principles without doubt applies to the present. Let us first have a look at these structures and principles at the beginning of *Rhetoric* (Book I. 1) where Aristotle defines the functions of legislator, judge, and litigants as parties to the conflict.

In a trial, "the only business of the litigant is to prove that the fact in question is or is not so, that it has happened or not". That is, litigants must stay with presentation of facts (*pragma*, that which has been done). The work of judges, in turn, is to evaluate the facts put forward by the litigants: "whether it is important or unimportant, just or unjust [...] is a matter for the dicast [judge/juryman] himself to decide." Hence, "it is not the business of the litigants to instruct" the judge in legal evaluation of the facts, but only try to prove what has taken place. So litigants present the facts and judges evaluate them (*Rhet.* I.1.1354a, trans. Freese).

What is the division of tasks between the legislator and the judge? "First of all," Aristotle stated, "it is proper that laws, properly enacted, should themselves define the issue of all cases as far as possible, and leave as little as possible to the discretion of the judges." (*Rhet.* I.1.1354a, trans. Freese) So, the legal evaluation of facts should already be there in the legislation. Yet the nature of legislation must be general, legislators must not intrude on the business of assessing individual cases: "what is most important of all is that the judgement of the legislator does not apply to a particular case, but is universal and applies to the future." What is "present and definite" must be left to the judges. So the legislator provides general norms, whereas judges apply those norms to individual cases. Legislators cannot decide on everything, of course. At the very least the "question of a thing having happened or not, of its going to happen or not, of being or not being so; this must be left to the discretion of the judges, for it is impossible for the legislator to foresee such questions." (*Rhet.* I.1.1354b, trans. Freese)

The foregoing means that matters of *veridiction* are to be settled in the relationship between judges and litigants, whereas matters of *jurisdiction* are to be settled in the relationship between judges and legislators. Today, litigants are expected to participate in legal argumentation as well, but that division of labour is not important for the purposes of this paper. What is important is the analytical separation between questions of fact and questions of law, which belongs to the basic design of the practice of law even today.

Yet this is only the basic design. Problems already arise from use of language in both directions, that is, in the litigant's presentation of facts and in legislators' preset models designating the facts in the abstract: "a man, while admitting the fact, often denies the description of the charge or the point on which it turns" in the law (*Rhet.* I.13.1373b–1374a, trans. Freese). Whereas the prosecutor undertakes to present a fact in language that shows it to constitute a definite act of a definite crime, defendants, in turn, would describe the same fact so that their act would not constitute that crime. When all the parties to the case have come to a conclusion and said: "I have spoken; you have heard; you know the facts; now give your decision," (*Rhet.* III.19.1420b, trans. Freese) the judges should be apprised of the facts. It is their job then to choose the proper language applicable to their description and then compare that language to the language of legislation.

A more difficult problem arises when litigants present facts that are not, or seem not, foreseen by the legislator at all. Judges have nothing else as given but these facts, so they must themselves figure out the norm. At this point, the notion of equity, as one of the judge's two basic concepts of justice, enters into the structure and creates a critical tension in relation to legality, the other basic concept of justice. This is the way in which Aristotle presents the problem of equity and legality in *Nicomachean Ethics*:

> The source of the difficulty is that equity (*epieikes*), though just, is not legal justice (*dikaion kata nomos*), but a rectification of legal justice (*nomimou dikaiou*) (*EN* V.10.1137b, trans. Rackham).

Components of this problematic difficulty constitute the basic elements of the following analysis of justice as *energeia*. Practical justice in the shape of adjudication breaks down into justice as *epieikeia* and justice *kata nomos*, equity and legality. Justice as *energeia* will be presented as the way in which these two components collaborate. Finally, this collaboration constitutes a paradigm mechanism through whose application legal-practical jurisdiction confronts veridictional practices and normalisation.

## 9.3 Connections of *Epieikeia* to Legality

Justice as *epieikeia* connects up with justice *kata nomos* from the start. The basic ways in which this connecting-up happens with Aristotle are two. The first of these he puts forward by the point about *correction* that is "the essential nature of the equitable: it is a rectification of law where law is defective because of its generality." (*EN* V.10.1137b, trans. Rackham) This statement is a remarkable event in the history of jurisprudence, in that it depicts the fact that the legal system can look upon itself critically. In other words, the practice of law is self-reflective in its business of administration of justice and injustice. The injustices to be considered at this level are no longer the wrongs that the subjects summoned before the court may have committed. The injustices here are wrongs that the legal system (the practice of law) itself might commit by too rigid, imprudent and indiscrete an application of general legislation. This is the first way in which justice as *epieikeia* connects with justice kata nomos: the working of equity presupposes that a state of legality exists. Without legality on the ground, there would be no *epieikeia*. The type of prudence that has no laws at all is not *epieikeia*.

The second way *epieikeia* connects to nomoi is the basic precept by which judges should handle a situation calling for equitable justice. How to find the maxim of action for this particular judgement? Aristotle laid down the following method for construction of judge-made norms in *Nicomachean Ethics*:

> [...] it is then right, where the lawgiver's pronouncement because of its absoluteness is defective and erroneous, to rectify the defect by deciding as the lawgiver would himself decide if he were present on the occasion, and would have enacted if he had been cognizant of the case in question (*EN* V.10.1137b, trans. Rackham).

In *Rhetoric*, Aristotle stipulates further that to be equitable is "to look, not to the law but to the legislator; not to the letter of the law but to the intention of the legislator; not to the action itself, but to the moral purpose; not to the part, but to the whole." (*Rhet.* I.13.1374b, trans. Freese) This is not only the beginning point of legal hermeneutics, but certainly something very much alive and well in our own time as well. In any event, one perceives that here, too, justice as *epieikeia* connects with justice *kata nomoi*, namely by its fantastic construct of a legislator that is not the real legislator, but only a device of interpretation.

To these two ways in which equity connects with legality, one can add a third: at the moment of the judge's decision, to invoke *epieikeia* means a departure from legislation, but *after the decision* a new law (judge-made law) has been established. Aristotle refers to this effect only in passing, when speaking of the different arguments ('witnesses') that can be drawn upon in trial proceedings: "By recent witnesses I mean all well-known persons who have given a decision on any point, for their decisions are useful to those who are arguing about similar cases." (*Rhet.* I.15.1376a, trans. Freese) Aristotle says here that precedents are arguments, not binding law.

The stricter requirement of consistency of practice can be derived from Aristotle elsewhere, namely, from his elaborations of justice as equality (equal treatment). Aristotle's context for that is not adjudication, but political justice, more precisely distributive justice. In *Nicomachean Ethics*, Aristotle sets out from the following definition: "not everything unlawful (*paranomon*) is unfair (*anison*), though everything unfair is unlawful."' (*EN* V.2.1130b, trans. Rackham). Unfair as *anison* means 'unequal'. Unlike the situational fairness of the equitable, fairness as equality is reducible to a formalistic principle: *anison* as a wrong exists when unequals are treated equally and equals unequally. This must be unlawful, *paranomos*, even in the absence of explicit law. To adapt this to the adjudication setting, one could say that in the absence of legislation on a particular issue one must look to previous cases and be consistent. The idea is that the norm was established when the issue was decided for the first time. If in the next case the same issue were decided differently, that would be unfair discrimination, unlawful despite the non-existence of legislation. In *Politics*, Aristotle stipulates the following:

> For instance, it is thought that justice is equality (*ison*), and so it is, though not for everybody but only for those who are equals; and it is thought that inequality (*anison*) is just, for so indeed it is, though not for everybody, but for those who are unequal (*Pol.* III.9.1280a, trans. Rackham).

The context of this statement, too, is political justice – the right distribution of honours and riches that requires an assessment of the merits of individuals. *Mutatis mutandis*, however, the same precept – according to which *equals should be treated equally, but unequals unequally* – underlies the requirement of consistency in the practice of law. This requirement stands out as the third way in which *epieikeia* connects with legality. Even in the absence of legislation, or in cases where legislation is departed from, a new state of legality will emerge from any application of *epieikeia* in that it constitutes a model (a precedent) for future cases.

## 9.4 Extra-Positive Registers of *Epieikeia*

Despite the above three ways in which *epieikeia* is connected to legality and legislation, it is still something that one uses for the purposes of departing from the letter of law or for the purposes of establishing a norm where justice *kata nomos* is silent. To achieve this, the strategy of *epieikeia* plays two registers, both of which are external to *nomoi*. The first register is that of facts and veridiction. Equity is activated by an account of facts, and its work is adapting the normative system to facts. Aristotle illustrated this basic function of *epieikeia* by likening it to the technique of measuring stone-blocks of indefinite shapes by construction engineers:

> For what is itself indefinite can only be measured by an indefinite standard, like the leaden rule used by Lesbian builders; just as that rule is not rigid but can be bent to the shape of the stone, so a special ordinance is made to fit the circumstances of the case (*EN* V.10.1137b, trans. Rackham).

The other register played by the strategy of *epieikeia* is that of the so called *unwritten laws* "established by each people in reference to themselves" as written laws are, and even beyond that, the *general laws* that are "those based on nature." (*Rhet.* I.13.1373b, trans. Freese) As for general laws, Aristotle draws on Sophocles' *Antigone*: "For neither to-day nor yesterday, but from all eternity, these statutes live and no man knoweth whence they came." (*Rhet.* I.13.1373b, trans. Freese). Thus Aristotle's *epieikeia* has at its disposal not only the register of facts and veridiction, but also the register of supra-statutory law and extra-legal justice: "equity is justice that goes beyond the written law." (*Rhet.* I.13.1374a, trans. Freese)

An *epieikeia* case normally occurs when judges come up against circumstances which have simply escaped legislators' notice. Further complexity is generated to this structure in that sometimes the foresight of legislators covers a possibility that particular circumstances may arise in which legislation would not be applicable. Briefly put, legislators prepare for surprises. As surprises are inevitable, legislators would insert an element of *epieikeia* into the *nomoi*, knowing that they will be "unable to define [the law] for all cases". Then the legislator would employ "a universal statement, which is not applicable to all, but only to most, cases." (*Rhet.* I.13.1374a, trans. Freese)

Let us quickly summarise so far. In the basic design, questions of fact were distinguished from questions of law and general legislation was distinguished from individual decisions. Then the problem of justice *kata nomos* and justice as *epieikeia*, of legality and equity, was introduced. Equity connects with legislation in that it corrects deficiencies in legislation, in that it uses the methodical construct of legislator for the purposes of interpretation, and finally in that every decision based on equity creates new law. At the time of the decision, however, equity disconnects the practice of law from strict legalism by drawing on two registers: discourse on facts and discourse on unwritten law and justice.

The above abstract analytic constructed on the basis of Aristotle seems to suggest something like a universal structure of the legal field. It should be noted that in the reality of legal practice *epieikeia*-legality is a scheme of action rather than of analysis. There it is all about movement, not stability. When the legal struggle (rhetorical or dialectical) begins, the analyst must be prepared for an endless process of re-conceptualization. Legality and equity will disband and come together again in as many ways as there are action situations. For the participants the question is "what use should be made of them when exhorting or dissuading, accusing or defending." (*Rhet*. I.13.1375a, trans. Freese)

> For it is evident that, if the written law is counter to our case, we must have recourse to the general law and equity, as more in accordance with justice; and we must argue that, when the dicast takes an oath to decide to the best of his judgement, he means that he will not abide rigorously by the written laws; that equity is ever constant and never changes, even as the general law, which is based on nature, whereas the written laws often vary (*Rhet*. I.15.1375a, trans. Freese).

If legality would not support one's objectives, "it is necessary to see whether the law is contradictory to another approved law or to itself". Furthermore, if "the meaning of the law is equivocal, we must turn it about, and see in which way it is to be interpreted so as to suit the application of justice or expediency, and have recourse to that." Finally, if "the conditions which led to the enactment of the law are now obsolete, while the law itself remains, one must endeavour to make this clear and to combat the law by this argument." All of these are tactics of justice as *epieikeia* for getting round the type of justice that is established *kata* nomoi (*Rhet*. I.15.1375b, trans Freese).

> But if the written law favours our case, we must say that the oath of the dicast 'to decide to the best of his judgement' does not justify him in deciding contrary to the law, but is only intended to relieve him from the charge of perjury, if he is ignorant of the meaning of the law; that no one chooses that which is good absolutely, but that which is good for himself; that there is no difference between not using the laws and their not being enacted; that in the other arts there is no advantage in trying to be wiser than the physician, for an error on his part does not do so much harm as the habit of disobeying the authority; that to seek to be wiser than the laws is just what is forbidden in the most approved laws (*Rhet*. I.15.1375b, trans. Freese).

## 9.5 Transformations: Jurisdiction and Veridiction

Aristotle's model assumes that legality should always be the starting point whereas equity only corrects or pre-empts the mistakes that might ensue from indiscriminate application of written laws. In spite of equity's connections to legality (correction, operation with the construct of legislative will, feedback), the point of equity is that it plays extra-legal registers: facts and non-positive justice. This brings us to the interplay between veridictional and jurisdictional discourses that begins at once when any practical legal strategy makes its choice between the two available tactics: legality or equity.

*Epieikeia* creates an aperture in prevailing legal texture for insertions of veridictional discourse that is not simple presentation of facts. Normative argumentation building on the particular circumstances of the case at hand is the basic and most explicit way in which juridical discourse (jurisdiction) embraces and integrates veridictional discourse. Complexity exists at levels that are less explicit. In the filters of subjective strategies – that is, at the level of individual legal confrontations – one sees nothing but individual facts and general laws that correspond more or less well with each other. In the objective structures of the legal field this same practice nonetheless sets in motion (*archein*) detectable interactions between jurisdictional and veridictional discourses. The two discourses emerge from each other when passed through (*prattein*) the vehicle of basic legal concepts of justice. Two tacit transformations take place: jurisdiction becomes veridiction and veridiction becomes jurisdiction.

The first transformation is exposed by understanding the true nature of the legality-argument. Drawing on the formalistic idea of justice as legal certainty, the legality-argument implicates a more or less hidden, but all the same enormous change. The whole system of law is transposed in the realm of facts. The legal system becomes a *factual state of legality*. Aside from the fact that it is a fact, nothing makes this state just. Legality must be a fact upon which people may rely, so that the justness of law is identical to its factness. Arguing from legality, one really draws on no other value but the value of ontological security. This insight has its liberating effects as well. Insofar as one considers the legal system as a mere fact, one would there have a negative constitution for one's moral autonomy and freedom. Who can really bear this freedom is another matter. Both of these things (factuality of law, morality of the person) are nevertheless conceived as effects or functions of legality. What happens is a tacit transformation: jurisdictional discourse on norms turns out to be veridictional discourse on facts in the end.

To develop this type of ultra-critical legal positivism further, one could use some of Slavoj Žižek's ideas in his *Sublime Object of Ideology*. To begin with, the realization that the law has a 'constitutively senseless character', which means that it is nothing but a fact of brute force, is 'traumatic' to subjects (Žižek 1999, 37). Speaking of law as an ideological state apparatus, Žižek suggests that "it is precisely this non-integrated surplus of senseless traumatism which confers on the Law its unconditional authority." (Žižek 1999, 43). He concludes, however, that "through this acceptance of the customs and rules of social life in their nonsensical, given character, through acceptance of the fact that 'Law is law,' we are internally freed from its constraints [...] we render unto Caesar what is Caesar's, so that we can calmly reflect on everything." (Žižek 1999, 80) Therefore, what he calls 'obscenity' in the foundation of law may have liberating effects on subjects. In Žižek's view this is a modern phenomenon connected to Enlightenment:

> In the traditional, pre-enlightened universe, the authority of the Law is never experienced as nonsensical and unfounded; on the contrary the Law is always illuminated by the charismatic power of fascination. Only to the already enlightened view does the

universe of social customs and rule appear as a nonsensical 'machine' that must be accepted as such (Žižek 1999, 80).

The other transformation is the reverse of the first and occurs in application of the equity concept of justice. As noted, the equity-argument plays two registers that disconnect it from legality. First, facts on the ground: veridictional discourse on special circumstances that the legislator has not foreseen. Second, extra-legal justice: jurisdictional discourse on norms that supplement and correct the lawgiver's omissions and faults. These two steps belong to one single tactical move, but a change in the mode of discourse occurs there at some point. What first was speech on facts will in the end become speech on norms ("This change is imperceptible"; factual "observations concerning human affairs" connect "with an ought," but "how this new relation can be a deduction from others, which are entirely different from it?" Hume 1978, III.I.I.469). It is like walking in a city and suddenly realising that the name of the street has changed. There must have been a crossroads where the road forked, but the junction passed unnoticed.

Let us stroll back to the corner where the turning occurred. So much is clear that the two ends of equity discourse are different in nature (facts and norms), but what is the nature of the junction between them? Once again: the legislator's activity and justice kata nomoi must be based on what exists generally, which is fine as far as it goes. Yet indefinite situations exist where general legislation must be departed from and replaced by justice as *epieikeia*. Special circumstances make one turn away from legislation and invoke extra-positive justice. At this point one must be extremely cautious: if one now simply turns to the precepts of extra-positive justice, the junction will be left behind once again. Therefore, let us not hasten to these precepts of which 'no man knoweth whence they came', but concentrate on the antecedent turn-away from legality. The key to our problem may be this negative act of rejection.

For making that act intelligible, the notion of *normality* can be introduced. What is normality in laws? In a certain way, normality has always already infiltrated into legal norms insofar as they must reflect concrete types of social relations. One can perhaps say that these types lie in law's abstract models at each instance of application, because one has to have in mind something concrete that makes words in laws meaningful. Indeed, rights and duties of spouses build on received wisdom on the family household; regulation of entrepreneurship builds on certain ideas of production and exchange; limitations to government competences build on one's perceptions of high-powered offices; definitions of crimes, in turn, on stereotyped motives for human action. Legislation works and is workable to the degree to which particular situations correspond to the normal types presupposed by legislation. But are these background 'normal types' and 'normality' factual or normative? One quickly realises that they are at once both factual and normative. They constitute our social conceptions of external reality, yet they exert formative power on us and our society through these same conceptions. Conceptions conceived as factual become – or perhaps were all the time – normative.

Returning to the fact-norm junction, one notices that a kind of *norm* was wrapped up in the fact from the start: normality that tacitly imposes models on our lives. The double nature of normality allows it to stand at the junction, as a hinge, between the two ends of equity discourses. An act of rejection is really directed at the normative power of something that the *nomoi* take for granted. This power is a matter of factual assumptions that regulators had meant neither to change nor not to change. In the practice of *epieikeia*, the rejection of factual assumptions is where the transformation of discourses really begins. Yet the source of equity's extra-legal norms is not the factual description of special and normal circumstances as such, but the normative counteraction this description is invested with. Strictly speaking, this counteraction is not repudiation of the validity of general legislation, but repudiation of the normality presupposed by general legislation. Clearly, normality regulates our lives everywhere in society, but it also does so in the presumptions of legal laws. Yet the equity mechanism shows that in the field of law one can also escape from the sway of normality. Indeed, Aristotle's *epieikeia* is meant to destroy the phenomenon it has revealed: the undercover jurisdictional power of factual conceptions, normalisation.

## 9.6 Conclusion: Justice as *Energeia*

Let us in the end return to the question of justice as *energeia*, actuality. Justice as *energeia* should be justice that exists only in practice and practice should bring this justice about by employing its own structures. This paper has presented the collaboration between the jurisprudential principles of legality and equity and the discursive transformations this collaboration involves. Are these able to bring justice of this specific kind into existence?

Practice of law requires legality from other practices in several ways, but when it directs this requirement to itself, it grows into a merciless introspective gaze. Legal practice is part of the legal system that it upholds, but with legality it may objectify this system and thus look upon itself as something external and factual. This brings us to the old question: How can it be that a practice may look upon its norms and rituals as elements of the external word – or of a 'senseless machine', as Žižek put it – but yet hold these norms and rituals normatively binding on it? A normative binding is an internal feeling of 'ought'; it is a subjective experience that every appeal for legality wishes to generate in its addressees. This experience constitutes the goal of the game of law, in the same way as scoring is the goal of a football game and neutralization of enemy forces is the goal of war. Legality's ought-experience should make one genuinely feel that whatever is to be done will in one way or another be done in the name of justice. Yet the perverse truth of legality is that at its heart lies a stiff requirement of lifeless automatism, a truth which legality imposes on the reality of legal practice. This makes the law appear like a sad machine that has a consciousness, so that it can feel sorry for itself, but no will, so that it knows that its mere functioning is its raison d'etre.

A functionally associated, but nevertheless different notion of justice is *epieikeia*. In *epieikeia*, legality's ordinary justice appears to be overridden by a godly rumble of cosmic and immemorial justice. Yet necessary for the practice of *epieikeia* is not some comprehension of universal and total morality, but only sensitivity to what is singular in events. *Epieikeia* concerns extraordinary circumstances coming and going in time and place. With that one understands that the source of this justice is not to be searched from the galactic spheres of eternal rightness. The source of justice in *epieikeia* is counteraction. It is counteraction against normalisation, that is, against the normative power of empirical knowledge tacitly incorporated in law's factual presumptions. *Epieikeia* instigates a discursive process that encounters prejudices that have quietly gotten under the law's skin. This process brings to the foreground what until then has worked unproblematically in the background. The heart of equity is a fight against the silent and self-nourishing force of normality. Yet it is evident that the force of normality constantly overflows critique; normality's expanse stretches far beyond the horizon of our intentionality and imagination, whereas the situational and momentary critiques of *epieikeia* appear there like small islands projecting out of the ocean. Were *epieikeia* fighting all alone, its inevitable fate would be to run from defeat to defeat, and every single battle would simply vanish into the thin air of desperation.

*Epieikeia* is not fighting alone, however, but in alliance with *nomoi* that take the place of factual normality in the practice of law. *Nomoi* require *epieikeia*, because otherwise, with law's gaze upon itself through legality, they would go lame as in one of those dreadful dreams, where one cannot move one's limbs but must simply go through the dream paralyzed by fear. *Epieikeia* requires *nomoi*, because facts that hide norms must be found in the folds of legislation. Without legislation, veridiction as veridiction can only be fought by way of veridiction. Pure and simple veridiction is entirely immune to *epieikeia*'s critique. But veridiction tends to have jurisdictional effects of normalisation, no less than jurisdiction tends to effect things in the factual world. Both of these processes are staged in legislation, so that the practice of law is able to engage with them. This engagement, I think, is justice as *energeia*, where its game is one of combinations and dissociations between 'is' and 'ought'. The practice of law leaves behind no work in the nature of ought, but exhausts its full meaning of this nature in the performance itself. What it leaves behind is ever new institutionalisations of the law that are always and immediately in the nature of 'is'.

The world's most famous scrutiny of the problem of 'is' and 'ought' appears in David Hume's *Treatise of Human Nature*. In the midst of this scrutiny, a peculiar episode occurs. All of a sudden, the book asks the reader to imagine a person, who watches the author, Hume, committing carnal sins. Then, says Hume, this person, who through "a window sees any lewd behaviour of mine with my neighbour's wife, may be so simple as to imagine she is certainly my own." (Hume 1978, III.I.I.461) The facile philosophical problem here is whether Hume's lustful pleasure in this woman is wrong because someone watching the scene can be led astray from what is the reality of the situation, or because of some other conditions that have nothing to do with facts and their comprehension. In other words, is Hume's libertinism offending sexual *normality*, or is it offending sexual *morality* in some other sense? Perhaps it is offensive, because the woman belongs to another man? Perhaps it is not offensive, because it is done out of love?

Be that as it may, I think the example has a less explicit but much more interesting moral message to tell. This one is conveyed by its picture of voyeurism, the image of Hume's hidden observer. One might ask whether the secret viewing of an intimate act between others is not just as obscene as anything that provokes it. It surely is not nice, but there is even more in this gaze. I feel that Hume's real point here about morality is not about sexual morality ('only married couples may have sex'), but about social conventions that block one's mind as they have blocked this observer's mind. He is a victim of anonymous coercion through social patterns that dominate his perception. What stands outside this perception does not stand as wrong in this little portrayal of a moral situation. Social patterns that determine perception generate capillary social control of individual life; perhaps this is thought by Hume as wrong. Behaviour that is irregular up to the point of *unimaginability* is a huge transgression of normalisation. Exclusion of what is not in the range of social regularity is an effect of the anonymous gaze of normalisation. Insofar as legal practice knows how to root this gaze out from its hiding places, from its insidious silence, legal practice can also unlock the reserves of dissentience and uphold justice as *energeia*.

**Acknowledgement** The author wishes thank the two anonymous referees for their comments and Mr. Christopher Goddard for revising the language of the text.

## Bibliography

Arendt, H. 1989. *The human condition*. Chicago/London: The University of Chicago Press.

Aristotle. 1926. *Rhetoric*. Aristotle in 23 Volumes, Vol. 22. Trans. J. H. Freese. Cambridge, MA/London: Harvard University Press/William Heinemann Ltd.

Aristotle. 1933,1989. *Metaphysics*. Aristotle in 23 Volumes, Vols.17, 18. Trans. Hugh Tredennick. Cambridge, MA/London: Harvard University Press/William Heinemann Ltd.

Aristotle. 1934. *Nicomachean ethics*. Aristotle in 23 Volumes, Vol. 19. Trans. H. Rackham. Cambridge, MA/London: Harvard University Press/William Heinemann Ltd.

Aristotle. 1944. *Politics*. Aristotle in 23 Volumes, Vol. 21. Trans. H. Rackham. Cambridge, MA/London: Harvard University Press/William Heinemann Ltd.

Foucault, M. 1998. Nietzsche, genealogy, history. In *M. Foucault: aesthetics, method, and epistemology. Essential works of Foucault 1956–1984*, Vol. 1, ed. James D. Faubion, 369–391. Trans. Donald F. Brouchard, and Sherry Simon. New York: The New Press.

Foucault, M. 2000. Questions of method. In *An interview in Michel Foucault: Power. essential works of Foucault 1954–1984*, Vol. 3, ed. James D. Faubion, 223–238. Trans. Robert Hurley. New York: The New Press.

Foucault, M. 2008. *The birth of biopolitics. Lectures at the Collège de France, 1978–1979,* ed. Michael Senellart. Trans. Graham Burchell. Basingstoke/New York: Palgrave Macmillan.

Hume, David. 1978. *A treatise of human nature*. 2nd ed, ed. L.A. Selby-Bigge. Oxford: Clarendon Press.

Žižek, Slavoj. 1999. *The sublime object of ideology*. London/New York: Verso.

# Chapter 10
# Legality and Equity in the *Rhetoric*: The Smooth Transition

**Miklós Könczöl**

## 10.1 Introduction

In his paper 'Justice *kata nomos* and Justice as *epieikes* (Legality and Equity)', Samuli Hurri seeks to base a practical notion of justice on Aristotelian conceptions of legality and equity. This he does both by a clear argument and with considerable respect to Aristotle's text. Indeed, it would be hard to challenge his basic idea of justice as *energeia*, even though one wonders if it is really necessary for Hurri's project to conceive of justice in decidedly un-Aristotelian terms, "as something that *only* exists in practice," as opposed to "justice [...] considered as the externally and objectively given final cause of the practice of law" (Hurri 9.1, emphasis added). Accordingly, the following remarks will focus on the Aristotelian grounds rather than the consequences Hurri quite convincingly draws. First, I am going to offer an interpretation of *Rhet.* 1.1.1354a 26–30 somewhat different from the traditional (and Hurri's) way of understanding Aristotle's explanation of the roles of the orator, the judge, and the legislator (10.2). Then I shall turn to the functioning of equity as described in *Rhet.* 1.13.1374a 18–b 23, trying to highlight the close links between *epieikes* and the legal text (10.3). I hope to conclude by showing that a less strict opposition of facts and qualities, and a more limited understanding of arguments from equity, which I think both are more faithful to Aristotle, actually provide more firm grounds for a practical notion of justice (10.4).

M. Könczöl (✉)
Department of Classics and Ancient History, Durham University,
38 North Bailey, DH1 3EU Durham, UK
e-mail: miklos.konczol@durham.ac.uk

L. Huppes-Cluysenaer and N.M.M.S. Coelho (eds.), *Aristotle and The Philosophy of Law: Theory, Practice and Justice*, Ius Gentium: Comparative Perspectives on Law and Justice 23,
DOI 10.1007/978-94-007-6031-8_10, 

## 10.2 The Possibility of Legal Argumentation

In the introductory chapter of his *Rhetoric*, Aristotle seems to draw a sharp line to demarcate the roles of the parties and the judges of a legal debate:

> It is clear (he says) that the opponents have no function except to show the facts (*deixai to pragma*): whether something is or is not true or has happened or has not happened; whether it is important or trivial or just or unjust, in so far as the lawmaker has not provided a definition, the judge should somehow decide (*gignôskein*) himself and not learn (*manthanein*) from the opponents (*Rhet.* 1.1.1354a 26–30, trans. Kennedy).[1]

The following passage (1.1.1354a 31–1354b 16) then opposes the responsibilities of the lawmaker and the judge. Highlighting some structural differences between the situations in which laws and judgements are passed, Aristotle concludes that:

> in other matters […] the judge should have authority to decide as little as possible; but it is necessary to leave to the judges the question of whether something has happened or has not happened, will or will not be, is or is not the case; for the lawmaker cannot foresee these things (*Rhet.* 1.1.1354b 11–16, trans. Kennedy).[2]

According to Hurri, "[w]hat this means is that matters of *veridiction* are to be settled in the relationship between judges and litigants, whereas matters of *jurisdiction* are to be settled in the relationship between the judges and the legislators." He then adds that "[t]oday, litigants are expected to participate in the legal argumentation as well," suggesting a contrast between the practice of the ancients and moderns (Hurri 9.2). This interpretation follows a respectable tradition of Aristotelian scholarship (its most recent exponents are e.g. Schütrumpf 1994; Kennedy 2007), which seems to be based on two assumptions, namely, (1) that 1354a 26–30 is basically about the distribution of roles within a court trial, and (2) that the task of the orators is opposed to that of the judge. I shall first elaborate on these two points, reflecting at the same time on the two questions that are in the focus of Hurri's investigations, i.e. "[h]ow the practice of law transforms veridiction into jurisdiction and jurisdiction into veridiction" and "[h]ow the practice of law brings about justice as actuality."

The first thing to do is to have a look at Aristotle's terminology, which I think allows for a different approach to his 'distribution of roles'.

Let us begin with the verbs. Firstly, what the orators do with the 'facts' is *deixai*, they 'show' them to the judge.[3] Facts sometimes do not need formal logical proof, but they always have to be 'shown' or explained to the judge, as they serve as the basis of the final decision on lawfulness. One may think of non-technical proof (*pisteis atechnoi*, cf. *Rhet.* 1.15.1375a 22–1377b 12), like documents or testimonies, which may prove that something happened without any additional support, whereas their validity or reliability may still be challenged. Thus, the participants of the debate certainly need to offer a narrative of what happened in order that the

[1] I have modified the translation of Kennedy (2007) on the following points: 'show the facts: whether etc.' instead of 'show that', and 'judge' instead of 'juror'.

[2] I have slightly modified the translation of Kennedy (2007), rendering *kyrion* as 'have authority to decide' rather than 'have authority to determine'.

[3] On the meaning of *deixai*, see Grimaldi (1980, 12).

judge may formally adopt one of the competing narratives (cf. Eden 1986, 72), and they have to support their narrative with adequate proof.

Next comes the contrast between *gignôskein* and *manthanein*. *Gignôskein* here may be interpreted in two ways, and it seems that we do need both senses of the word to understand the passage correctly. On the one hand, *gignôskein* refers to the act of acquiring knowledge, i.e. to the active efforts of the judge to see what actually happened from the perspective of law. That contrasts it with *manthanein*, which would mean that the judge accepts what the litigants say as truth.[4] On the other hand, *gignôskein* is often used as the term of judicial decision-making. What follows from this twofold meaning of *gignôskein* is that intellectual efforts are needed from the part of the judge if he is to live up to his function of deciding the case.

It is important to note that throughout Aristotle's review of earlier textbooks at the beginning of Book I, the focus is on the opposition of two means of persuasion: manipulation and proof (cf. Dow 2007). Aristotle seems to distinguish between two groups of things that can be possibly said by the orators before the court: something is either relevant to the case and thus qualifies as (rational) proof, or irrelevant and may serve only to (irrationally) manipulate the judges. Understood in light of that, the task of the orator (i.e. *deixai to pragma*) is opposed not only, and not even primarily, to that of the judge but rather to speaking *exô tou pragmatos*.

*Deixai to pragma*, then, is something that has to be done by the litigants. Moreover, it can only be done by the litigants and no one else. Decision on the case, in turn, is to be made by the judge. This distribution of roles does not necessarily prevent the parties of the debate from participating in legal argumentation. Those arguing for a certain decision can legitimately present their narratives concerning the facts, which may or may not include statements concerning the lawfulness or importance of the acts described. The judge has, in turn, to weigh and evaluate their arguments and the proof which these are based upon, before passing the final decision. The opposition of *gignôskein* and *manthanein* serves to emphasise the importance of this function.

The above remarks further suggest that *pragma* is best understood as the narrative concerning the case presented by the litigants, and that the limitation 'whether something is or is not true or has happened or has not happened' is not intended to eliminate legal arguments on their part, but to indicate what is inevitable for them to do, i.e. to furnish (factual) proofs. Now I shall examine the role of the 'facts' in establishing the legal 'quality' of the case (i.e. deciding the question of lawfulness), on the basis of Aristotle's discussion of the rhetoric of *epieikes*.

---

[4] A parallel passage can be found in Plato's *Laws*, where the Athenian explains the duties of the judges at artistic competitions. The judges, he says, need more than just taste: they have to be able to decide against the opinion of the audience. The judge should not listen to the audience (*para theatrou… manthanonta*), nor should his decision (*gignôskonta*) be influenced by fear and unmanliness (*Lg*. 2.659a 2–7). The passage is also mentioned by Schütrumpf (1994, 106–107 note 47) as a possible sign of Plato's influence on the terminology of *Rhet*. 1.1.1354a 30. Contrary to his argument, however, it seems that Plato's view of the desirable assessment of artistic competitions does not say anything about either the participants' or the audience's licence to give their opinion. Rather, it is the task of the judge to disregard these opinions and decide according to the actual merits of the presentations.

## 10.3 Equity and Definition

Hurri actually comes quite close to a broader understanding of the orator's task as he emphasises the problems resulting from the use of language for describing a certain act, be it the "litigant's presentation of facts" or the "legislator's pre-set models designating the facts in the abstract" (Hurri 9.2). He also points out that Aristotle's description of legal definitions (*Rhet.* 1.13.1373b 38–1374a 17) shows awareness of these problems. His subsequent discussion of the links between *epieikes* and legality, however, does not examine the role of definitions in the functioning of arguments from equity.

According to the well-known passages of the *Rhetoric* (1.13.1374a 18–b 23) and the *Nicomachean Ethics* (5.10.1137a 31–1138a 3), *epieikes* seems to work by way of supplementing the normative text, without denying its validity (see Harris 2004), apparently in order to make it conform to the legislator's intent. Yet the situations suggested by both of these texts, in which equity is likely to have a role, are pointed at the non-application of a certain rule.

Arguments from the legislator's intent have a twofold character. On the one hand, they exemplify what are often termed 'consequentialist arguments'. As Jacques Brunschwig (1996, 139) puts it in his interpretation, "there exists a perfectly applicable law, but [...] a mechanical or blind application of it would be too severe according to the moral intuitions of the judge and those of the society in which he works." Consequently, the argument is based on the assertion that the legislator would not have intended the law to lead to such a verdict. On the other hand, the legislator's intent is still something referred to in 'rule-based reasoning', where it appears as a means of interpretation, which is intended to help establishing the meaning of a normative text, by explaining how the legislator actually meant what he put into words. What is important for us to see here is that this method of reasoning—i.e. advocating equity by way of interpreting the text—makes it possible for the orator to avoid putting the validity of written law to question.

What, then, remains of *epieikes* for arguments that can be safely used in a speech without appearing to be seeking, in Aristotle's words, "to be wiser than the laws" (*Rhet.* 1.15.1375b 23–24)? It may be a good idea to have a look at the example Aristotle gives for using equity in a particular case of judging an offence: "[I]f someone wearing a ring raises his hand or strikes, by the written law he is violating the law and does wrong." Here the discrepancy between the law and the truth is due to the fact that the law does not define "how long and what sort of iron has to be used to constitute 'wounding'" (*Rhet.* 1.13.1374a 32–36, trans. Kennedy).[5] The law, as far as it can be reconstructed from Aristotle's words, forbids and punishes assault with 'iron'. In case someone strikes another person with an iron ring on his hand, the conceptual requirements for applying the law obtain and the action qualifies as 'wounding'. In such a case, applying the sanctions attached to wounding would lead

[5] Here I have modified the translation of Kennedy (2007) to keep Aristotle's 'iron' (*sidērōi*), which illustrates the point much better than Kennedy's 'weapon'.

to injustice, as it would mean treating different actions (e.g. deliberately using a weapon and wearing a ring) in the same way. This is, in the words of the *Nicomachean Ethics*, an example of "things about which one cannot speak correctly in universal terms" (5.10.1137b 14, trans. Crisp).

In such a case, the defendant can suggest that further qualification has to be added by the judges, saying "what the lawgiver [...] would have included" (*EN* 5.10.1137b 23–24, trans. Crisp). For example, further details concerning the characteristics of the object made of iron can be described, in order to make the difference between a ring and a weapon appear in the judgement. Or the intention of the person 'raising his hand or striking' can be taken into consideration, in order to distinguish between the deliberate use of a weapon and the wearing a ring on one's hand: looking "not to the action but to the deliberate purpose", as Aristotle puts it (*Rhet.* 1.13.1374b 13–14, trans. Kennedy). These qualifications would then concern the concept of 'wounding' as defined by the law. The defendant would argue that he 'raised his hand' or 'stroke' but did not 'wound', denying not the fact itself but its legal qualification.

This way of reasoning would then be the same as what is described in the paragraphs immediately preceding the discussion of equity. "People often admit having done an action and yet do not admit to the specific terms of an indictment or the crime with which it deals," Aristotle says (*Rhet.* 1.13.1373b 38–1374a 2, trans. Kennedy), introducing the topic of definition (*epigramma*).

Having accepted such an argument, the judges have to supplement the text of the law interpreted, thereby making it irrelevant for judging the action under dispute. The intention of the legislator is thus referred to in order to make it clear that it would be contrary to this intention to punish the defendant for having committed the crime he is charged with.

In this sense, we may agree with Jacques Brunschwig (1996, 151–152), who observes that the phrase used by Aristotle in the *Nicomachean Ethics*—"what the lawgiver would himself have said had he been present, and would have included within the law had he known" (5.10.1137b 22–24, trans. Crisp)—refers to two different things. Supplementing the definition by adding further qualification of the action in terms of facts (including intention) is done by reconstructing the abstract and general will of the legislator, while deciding that the rule thus obtained is not relevant for the facts of the case is "what the lawgiver would himself have said had he been present."

Yet these are two consecutive steps of the same line of reasoning: the consequentialist part of the argument, which leads to the decision not to apply the law, needs the backing of the interpretive or rule-based part in order to make the judges feel safe in deciding the case apparently 'according to the laws and decrees of the Athenian people', as contained in the 'dikastic oath' they have sworn (cf. Harris 2006).

On the other hand, Aristotle's final clause 'had he known' highlights the interdependence of the two steps. It is on the basis of the knowledge of the particular circumstances of the case and pondering the consequences of their judgement that the judges can decide where the text says less than what is necessary for a just decision.

Taking into account the particular situation and offering a corresponding interpretation of the general rule of decision, the topic of definition can serve the aims of equity, so that "we will be able to make clear what is a just verdict." (*Rhet.* 1.13.1374a 9, trans. Kennedy)

It may be objected to the above argument that it downplays the importance of one of the 'extra-positive registers of *epieikes*'. That is, in fact, a third point where my interpretation departs from Hurri's.

Hurri regards the arguments based on an opposition of positive law and unwritten laws as "the other register played by the strategy of *epieikes*" (Hurri 9.4). What speaks for this interpretation is the fact that Aristotle does use the terms *epieikesterois* and *epieikes* when describing arguments against the application of a written law:

> [I]t is evident that if the written law is contrary to our case (*tôi pragmati*), one must use common law and what is more equitable and just (*epieikesterois kai dikaioterois*); [say] that to use 'best understanding' is not to follow the written laws exclusively; and that equity (*to epieikes*) always remains and never changes nor does the common law (for it is in accordance with nature) but written laws often change (*Rhet.* 1.15.1375a 27–33, trans. Kennedy).[6]

It should be noted, however, that these sentences are not part of Aristotle's presentation of equity, but items of a list of arguments concerning *pisteis atechnoi*, which focus on the applicability of a (written) rule. Here, then, *to epieikes* is not referred to in order to serve as the basis of a particular solution, but as a topic that can be exploited by a speaker arguing against the use of a specific written rule. The first argument rests on the assertion that 'the written law is contrary to the facts', which does not seem to add much to what we have seen so far: it is not its being written that makes the law irrelevant for the case. The second case is quite different, as the argument there is based on the general mutability of written law, as opposed to the stability of *to epieikes* and the common law. Yet this stability does not mean that arguments from equity would go beyond 'the register of facts and veridiction': it is not *to epieikes* that "has [...] the register of supra-statutory law and extra-legal justice at its disposal" (Hurri 9.4). Quite the contrary, it is the arguments from equity that allow general justice to be accomplished, by making corrections to legal justice from within.

## 10.4 Conclusion

Let me now summarise, by way of conclusion, the differences between Hurri's and my interpretation of Aristotle, to see why I think the latter may help to support Hurri's practical conception of justice.

[6] I have modified the translation of Kennedy (2007) on several points. He renders 1375a 27–31 as 'for is evident that if the written law is contrary to the facts, one must use common law and arguments based on fairness as being more just. [One can say] that to use [the jurors'] "best understanding" is not to follow the written laws exclusively; and that fairness always remains'.

In terms of the distribution of roles between judges and litigants, I argued against Hurri (and what seems to be the *communis opinio* in scholarship) that the latter are not excluded from legal argumentation by Aristotle. On the one hand, their narratives may contain elements pertaining to the lawfulness of the action at issue, and it is only the acceptance or rejection of these elements that has to be decided by the judge. On the other hand, the legal quality of the 'main' facts is usually decided upon with regard to other facts. This suggests that while it is the judge who performs jurisdiction (in the sense of the moment of legally binding decision), veridiction does overlap with legal argumentation (furnishing the epistemic sources of that decision).

The way Aristotle describes arguments from equity is, it seems to me, illuminative of that overlapping. Here, the difference between Hurri's view and mine lies in the assessment of the 'registers' of *epieikes*. On the basis of Aristotle's example of equity-based reasoning and my interpretation of the topics against written law, I argued that justice as *epieikes* goes beyond the law by interpreting its text rather than by questioning its validity. It is by an addition to the legal definition that the irrelevance of a written rule can be shown. The addition, to be sure, belongs to the sphere of jurisdiction, yet it is based on the knowledge of the particular factual circumstances of a given case. The transition between facts and law, veridiction and jurisdiction is concealed by a reference to the legislator's intention.

Now, if equity always has to travel the same route from facts to qualities through complementing the definition of an act, it becomes clear why justice as *epieikes* cannot be conceived of otherwise than as an activity. Hurri is perfectly right in saying that "[*e*]*pieikes* creates an aperture in the preset legal texture of taxonomies for insertions of veridictional discourse, a space for the argumentation from the particular circumstances of the case at hand." (Hurri 9.5) It has to be added, though, that this aperture has to be created by way of legal argumentation. Justice as *epieikes* works through gaps in the texture of written law, and it is hard to tell whether these gaps are just shown or actually made by the arguments.

**Acknowledgement** I am indebted to Péter Cserne, Edward M. Harris, and the editors of this volume for their comments and advice. My participation at the workshop was supported by an Ustinov Travel Award I received from Ustinov College, Durham University.

## Bibliography

Brunschwig, J. 1996. Rule and exception: On the Aristotelian theory of equity. In *Rationality in Greek thought*, ed. M. Frede and G. Striker, 115–155. Oxford: Clarendon.

Aristotle. 2000. *Nicomachean ethics.* Trans. & ed. R. Crisp. Cambridge: Cambridge University Press.

Dow, J. 2007. A supposed contradiction about emotion-arousal in Aristotle's *Rhetoric. Phronesis* 52: 382–402.

Eden, K. 1986. *Poetic and legal fiction in the Aristotelian tradition*. Princeton: Princeton University Press.

Grimaldi, W.M.A. 1980. *Aristotle, Rhetoric I: A commentary*. New York: Fordham University Press.
Harris, E.M. 2004. Le rôle de l'*epieikeia* dans les tribunaux athéniens. *Revue historique de droit français et étranger* 82: 1–14.
Harris, E.M. 2006. The rule of law in Athenian democracy: Reflections on the judicial oath. *Dike* 9: 157–181.
Kennedy, G.A. 2007. *Aristotle, On rhetoric*: *A theory of civic discourse*, 2nd ed. New York/Oxford: Oxford University Press.
Schütrumpf, E. 1994. Some observations on the introduction of Aristotle's *Rhetoric*. In *Aristotle's Rhetoric: Philosophical essays*, ed. D.J. Furley and A. Nehamas, 99–116. Princeton: Princeton University Press.

# Chapter 11
# Legal Rules and *Epieikeia* in Aristotle: Post-positivism Rediscovered

**Jesús Vega**

> *Some persons in fact believe that Solon deliberately made the laws indefinite, in order that the final decision might be in the hands of the people. This, however, is not probable, and the reason no doubt was that it is impossible to attain ideal perfection when framing a law in general terms; for we must judge of his intentions, not from the actual results in the present day, but from the rest of his legislation (Const. Ath. 9).*

## 11.1 Introduction: Aristotle and a Current Debate

Much of the discussion in contemporary legal theory addresses the topic of the function and the scope of rules in the law. The two main conceptions on the field argue on whether rules, which are necessary ingredients of the law, are also sufficient criteria for the justification of legal decisions. It is a discussion on the normative structure of the law: shall we understand the law as including certain groundings going beyond legal rules (principles, in general)? Are the limits of the law tantamount to the limits of its rules? Or does it instead need to be expanded to comprise justificatory contents belonging to other different normative spheres (moral-political ones, in general)? The answer to these questions is always philosophically committed to a given general conception about the law.

This paper has been developed within the framework of the research project *Constitucionalismo y argumentación*. (DER2010-20132) held by the Spanish Ministry of Science and Innovation. I am most grateful to Jenny Porter and Victoria Roca for their work of language revision and correction.

J. Vega (✉)
Dpto. Filosofía del Derecho, Facultad de Derecho, Universidad de Alicante,
Campus de San Vicente del Raspeig, 03690 Alicante, Spain
e-mail: jesus.vega@ua.es

L. Huppes-Cluysenaer and N.M.M.S. Coelho (eds.), *Aristotle and The Philosophy of Law: Theory, Practice and Justice*, Ius Gentium: Comparative Perspectives on Law and Justice 23,
DOI 10.1007/978-94-007-6031-8_11, 

This sketch, oversimplified as it might be, brings before us the current dilemma between positivist conceptions of the law (Hart, Raz, *soft positivism*), on the one side, and "antipositivist" (Dworkin) or "no positivist" (Alexy) conceptions of the law, on the other—it should be more accurate to label them "post-positivist" rather than "natural law" conceptions. The former position emphasizes the importance and self-sufficiency of rules in the law. In fact, the study of the logical and conceptual structure of legal rules has been elaborated mainly by legal positivism. The fact that current legal positivism, especially after the Dworkinian critique to the "model of rules", has *nolens volens* taken principles into account, did not imply any retrieval at all from the centrality of rules to the law. Rules are still conceived as the main components of the legal institutional system, establishing its own conventional, immanent limits—principles themselves are understood as being conventional. The permanent risk that threatens this position is formalism, that is, the substantivization of the regulative or authoritative dimension of the law as derived from a self-referent legal normativity—its ultimate foundation and justification is to be found in its very self, rather than beyond. Contrariwise, the latter position—post-positivist theories—approaches rules from normative frames that are external to the institutional legal-positive machinery (thus, theories of this kind are inertially labelled as "natural law" theories) That framing is fundamentally of a moral, yet political, nature. The main consequence of this thesis is that rules are pushed to the back row whereas principles come to the front one. Rules are subordinated to principles for those are the genuine normative grounds on which legal conventions lie, thus setting wider limits to the law. Now the risk at stake under this conception is the superfluity of legal rules, reduced, as they seem, to a purely declarative, non-constitutive role.

In this paper I defend that the Aristotelian reflections on the nature of law might throw a new light on this discussion. Actually an old light being shed anew for, as a matter of fact, Aristotle's theory of law has historically been the doctrinal starting point for the two above-mentioned basic positions (the debate is not so modern after all!). Not only traditional iusnaturalism, with its dualist theory of law, valid as it was all through the Middle Ages, and modern rationalism were both elaborated under the Aristotelian inspiration—but also legal monist positivism coming afterwards (Conklin 2001: 13–36). Yet, having prefigured them, it is important to note that the Aristotelian theory itself is neither iusnaturalist, nor positivist. Its importance largely exceeds these historical developments. Rather it represents a true "niche" of philosophical ideas on law. A place to return to in a moment of transition like the one we are currently living, now that the positivist cycle is coming to an end (as the natural law one did before). The new philosophical conception of the law and its normative structure that this post-positivist stage requires is barely coming to life. A new basis for the conceptual reconstruction of the relations between the authoritative and the justificatory dimensions of law is required.[1] It is here where the Aristotelian ideas about law and legal rules can prove their great relevance. I will try to show how this is particularly the case with his reflections on *epieikeia* or equity.

[1] See Aguiló-Atienza-Ruiz Manero (2007: 16 ff.).

## 11.2 Law and Rules in Aristotle

Rules play a substantial role in Aristotle's philosophy of law. As it is widely known, their importance comes above all from a fundamental thesis about the "rule of laws, not of men" (*EN* V.6.1134a35)[2] which is based upon the postulate of law as a system of general, positive rules conceived as the essential instruments for the public organization of the *polis*. The very idea of justice is defined in Aristotle in terms of rules or *nomoi*. *Dikê* is nothing but "the order of political community" (*politikês koinônias taxis*) resulting from *nomoi*, and *dikaiosunê*—the virtue of justice—is "the collection of statements determining that what is just [*dikaiou krisis*]" (*Pol.* I.2.1253a37-39)[3] by means of the recursive application of positive law rules or "legal justice" (*nomikon dikaion*). Justice is identified in a sense—its formal one, yet not purely formal—with the political implementation of a system of general legality: justice is that which is "legal and equal [*to nomimon kai to ison*]"[4] (*EN* V.1.1129a34). So, Aristotle thinks of these rules as universal (*katholou*)[5] despite their particularity and variability for they are products of each and every single, politically circumscribed, legal system.

This universality is closely attached to the "architectonic" practice of ruling social conduct by means of general rules that are applicable at the scale of a political society. An institutional apparatus actually able to impose habits and to keep them constantly in force against social customs is required, this being the primary function of legislators (*EN* I.1.1103b3-4). The latter are, in turn, the main interlocutors of the *Nicomachean Ethics* (Bodéüs 1993: 60–61). Law is a practical, regulatory institution which is oriented to produce rules in the social sphere, whether newly enacted rules or already settled ones to be incorporated into the legal practice and redefined by it. Institutionalized coercion is necessarily attached to legal rules all along the extended processes of application thereof, where they are inculcated by means of singular compulsive acts. This is the authoritative (*kyrios*) aspect of legal rules (Schroeder 2003: 46 ff.), which explains the impossibility of law without

---

[2] Unless otherwise indicated, all the translations are taken from the revised Oxford edition of Aristotle's works.

[3] I have modified the Oxford's translation: "the principle of order in political society" and "the determination of what is just."

[4] I have modified the Oxford's translation: "the lawful and the equal".

[5] "Of things just and lawful [*dikaion kai nomimon*] each is related as the universal to its particulars; for the things that are done are many, but of them each is one, since it is universal." (*EN* V.7.1135a5-8). Although it is common to use "generality" to refer to the general or generic character of rules, in this paper I choose to use "universality." The first reason for this preference lies in the original Aristotelian *katholou*, whose epistemological meaning indicates a truly universal knowledge. The second, as my argument will hopefully make clear, is that the "generality" of rules pertaining to the practical domain is a semantic or logical property whereas their "universality" is a pragmatic or axiological one, thus belonging to a deeper, justificatory level of rationality. See MacCormick (2003: 78, 97 ff.).

judges or, in general, without officials as authorities in charge of the enforcement of the legal system (*EN* X.9.1179b5-1180a5).

Thus the authoritative dimension of the law is sharply emphasized by Aristotle. Actually, here we find the first historical formulation for the positivity of legal rules in terms of *practical* authority. Whether formalized or not—including therefore customary rules: *nomos* covers both meanings—the law necessarily consists of rules that are told to be universal and coercive. This is a necessary feature of the law as a specific social technique, i.e. an instrument for the *generalization* of social habits within the *polis*. There is no doubt about the conventional origin and nature of these rules either: in the famous first paragraph of *EN* V.7 the *nomikon dikaion* is defined in terms of the production of positive legal rules by means of the constitutive decisions of legal authorities (1134b20-22). In other words: once they have been established, legal rules open a new course of subsequent decisions that they intend to *generally* govern thereinafter. So, Aristotle—adopting now not the point of view of the production of rules but that of adjudication (*krisis*) by judges—makes it clear in a number of occasions that the judiciary should be a power strictly submitted to the authority of such legal rules. The philosopher left us some statements along those lines which would certainly be endorsed by any modern positivist-legalistic jurist of the post-codification era.[6]

Now, justice, in Aristotle's terms, includes a material dimension too. The universality of legal rules has not merely a descriptive or structural sense, but also a *justificatory* one. It is at this point where Aristotle introduces the notion of "natural justice" (*physikon dikaion*): that "which everywhere has the same force and does not exist by people's thinking this or that" (*EN* V.7.1134b19-20). This is a *praxis*-circumscribed universality, not a proper naturalist one anyway ("as fire burning both here and in Persia", states Aristotle critically, both against Sophists and Platonists in 1134b26), for natural justice is an *internal* part of political justice.[7] It is not then about the *physis* in its non-practical meaning, which implies invariable relations of strict necessity and universality. Natural regularities cannot provide a *normative* grounding for the law since it is a product of human practice (*anthrôpina dikaia*, 1135a4), 'something human" (*anthrôpinon*, 1137a31). Hence, the relevant foundations here are not naturalist or pre-positive but immanent to *praxis* (against iusnaturalism); and they are normative but not immanent to legal justice (against positivism), in other words, they are moral and political foundations—thus, always changeable (*kinêton*, 1134b29). The distinction between "legal justice" and "natural justice" is to be properly interpreted in terms of the distinction between law as a system of rules and the values which (universally) justify its existence and functioning.

The requirement of the existence of a set of general rules for the construction of the state is a universal feature of every political system. This justifies that

---

[6] For instance this celebrated one: "Now, it is of great moment that well-drawn laws should themselves define all the points they possibly can and leave as few as may be to the decision of the judges" (*Rhet.* I.1.1354a32).

[7] See Nussbaum (1985: 212); Yack (1993: 128ff, 194 ff.); Bodéüs (1993: 71 ff.); Burns (1998: 155).

"all regimes adhere to a certain justice" (*Pol.* III.9.1280a9). Besides, the central *normative* criterion in Aristotle's political philosophy is also here stated: "there where the *nomoi* do not rule there is no *politeia*" (IV.4.1292a32, my translation; see 5.1292b5ff.). Positive constitutions may have different contents for each may incorporate different, even incommensurable, material values,[8] "there is but one which is everywhere and by nature the best" (*EN* V.7.1135a5). Monarchy, aristocracy and democracy are all dependent on different principles of distributive justice. However, for Aristotle they are not equivalent from an evaluative point of view, there is rather an internal order among them. The criterion for that arises from the fact that laws, as general rules, are meant to promote the common good of each and every citizen, not the good or interest of a part (a group or faction). The distinction between *correct* and *deviate* constitutions (*Pol.* III.9-13) is here derived, directly based then on the common good that general legal rules are able to introduce into the *polis*: "And "right" [*orthon*] must be taken in the sense of "equally right," and this means right in regard to the interest of the whole state and in regard to the common welfare of the citizens" (*Pol.* III.13.1283b40-42).

Legal rules (*nomikon dikaion*) are then those components by means of which political justice among equal and free citizens living under the rule of law can materialize (*EN* V.61134a28ff.). This kind of justice proves then to be irreducible to a purely rule-formalist or procedural justice: it is a kind of justice based on *principles*. These are, on the one hand, substantive or material principles (that is, moral principles), and on the other, formal or institutional ones (that is, political) They both justify the universal character of legal rules and rule-based decisions. As stated before, *nomos* is a kind of order or *taxis* which, in its turn, organizes a complex set of other pre-existent norms. These are the informal, customary norms shaping the different communities (basically, households and villages), the normative layers on which the *polis* is built (*Pol.* I.2). Starting from this positive morality, law comes into existence as a second-order system of rules, thus making its way from ethics to politics. Progressively, customary law (much of it of a traditional, even ancestral character) needs to be declared and reformulated in terms of central, legislative rules by a political authority. This positivization characterises the transition from *thesmos* to *nomos* as a new type of general or universal rule (Ostwald 1969: 137 ff.). Law is therefore the institution that throughout the process of construction of a political society *totalizes* all the other social norms and posits certain standards that are universalized for *all* the members of the community now *as citizens*.[9] It is due to its *erga omnes* character that legal rules need to be universal.

Conceived as a *praxis*, the law consists of formal rules that turn it into a specific, relatively autonomous institution revealing at the same time its normativity as internally connected to the material values of morality and of the political organization.

---

[8] "The goodness or badness, justice or injustice, of laws varies of necessity with the constitutions of states" (*Pol.* III.11. 1282b8-10).

[9] Laws are "the "works" of the political art", and "in their enactments on all subjects [*peri apantôn*]" they affect all men and aim at "common interest of all" (*EN* VI.8.1141b25; X.9.1181b1).

So the concept of law expands beyond rules to comprise within its realm the grounds or values—principles—that substantively support and justify them. It is only in terms of such principles that justice can be predicated from law in a strong, not an accidental or weak sense. The whole Aristotelian theory of law revolves around this fundamental idea of the axiological unity of the legal, the moral and the political.

The key question, of course, is the extent to which, and exactly how, such a unity is to be attained along the recursive application of *nomoi*. This question can be broken down into the following: How do legal rules, which in their continuous enforcement process represent the impersonal, supra-individual power of the whole political community and its public standards, actually achieve the common good? Assuming that it is just "through rules [*dia nomôn*] that we can become good" (*EN* X.9.1180b24-25), how do the moral and the political good relate to one another according to Aristotle? What sort of relation connects general legal statements that determine the judges' applicative reasoning to each particular situation? Particularly, how should the legislator approach and balance the aggregate—political—consequences of universalizing legal standards? In sum, how are the formal and the material aspects of the universality of law—its authoritative and justificatory dimensions—harmonized as required by justice?

It is widely known that most of the Aristotelian answers to those questions lie in his crucial concept of *epieikeia*—equity—as "the correction of law". As it has been well demonstrated by Shiner, among others,[10] the relevance of this concept for the contemporary discussion within the general post-positivism debate on the role of rules in law is beyond any question. I assume that the theses on legal rules and their underlying moral and political principles presupposed by equity are then a fruitful platform for the critical analysis of the current discussion on the concept and limits of the law. In particular, I attempt to show how, against positivist conceptions of law, a sounder conceptual clarification on legal authority and the structure of legal practice under the rule of law can be obtained from the Aristotelian *epieikeia*. In the following pages, I will focus on one salient contribution in the last decade's debate, that by F. Schauer. In his theory of rules, Schauer (1991) defends a positivist position—called "presumptive positivism"—which I will take as a good exponent of a widespread view on rules in legal methodology and therefore as a good critical-comparative test to highlight the actuality of Aristotle's position. Specifically, I will try to show how a post-positivist—"Aristotelian"—conception of law needs to reject the "asymmetry of authority" characterizing a rigid, formalist rationality of rules in practical decision-making as claimed by Schauer's sort of legal positivism. To that end, I will compare the alternative positions of what I shall call the "Schauerian" and the "Aristotelian" legislators on the insufficiencies of the rationality of rules and the need for their "correction" or defeasibility by their underlying principles or values.

Let us begin with a brief introduction to Schauer's main ideas on rules.

[10] See the analysis of the controversy Hart-Dworkin he makes (Shiner 1994: 1253ff). See Schauer (1997: 287 ff.); Zahnd (1996: 273–274).

## 11.3 The Rationality of Rules: A Positivist Approach

Schauer revisits the topic of the generality of rules. His proposal is to reconstruct them as "entrenched generalizations." As they work by making generalizations, rules are selective of the appropriate properties and simultaneously suppressive of the rest. What becomes generalized in a rule is a "factual predicate" or "hypothesis", to which a "consequent" is added. So, "any rule can be reformulated to take the canonical form of a hypothetical factual predicate followed by a consequent."[11] The rule as a whole, as well as the generalization that it embodies, always serves a purpose. This is what Schauer calls the rule's rationale or underlying justification, that is, the state of affairs attempted to be achieved through the enactment of the rule, i.e. "the evil sought to be eradicated or the goal sought to be served." (26) The properties that are selectively generalized (as well as those others suppressed as irrelevant) by the rule's factual predicate—individuals, actions and circumstances—depend on that purpose. The selected properties are then precisely those that are perceived as causally connected to it, in that they maximize the likelihood of the implementation of the desired state of affairs that justifies the very rule.[12]

The identification and generalization of the relevant properties constituting the rule depends then on the balancing of reasons made by the rule-maker. From that moment onwards, the application of the rule has to be exactly referred to those selected and universalized properties, and not to the justification itself, so that the rule-applier has to leave her deliberation on it totally aside. All she has to do is to identify the selected properties in every particular case and apply the foreseen consequence. Now, since the relationship between the generalization and the justification is only of probabilistic causation—and then the rule's factual predicate is true for the most part but not necessarily for all the cases—an inevitable consequence arises: there shall be cases in which the rule is applicable though it does not promote its justification, and cases in which the justification is applicable but the rule itself is not. In the former cases (for example, the case of a guide dog in the restaurant), the rule is *over-inclusive*, whereas in the latter (the case of annoying disturbances caused by agents or animals other than dogs) it is *under-inclusive* (39). What is to be done in particular cases like these where the generalization fails and the connection between the rule and the justification is broken?

---

[11] Schauer (1991: 23). All numbers with no further reference henceforward indicate pages in this work.

[12] Schauer's example is that of a rule such as "no dogs allowed" appearing in a restaurant. The justification of this rule—preventing client annoyance—guides the rule-maker in his act of generalizing, from the particular ways of behaviour characteristic of dogs and which may disrupt clients in a restaurant (barking, running, jumping, eating), to the general category "dogs" itself. This category or property ("dogness") is selected in light of its causal relevancy to the presence of annoying disturbances, and thus the prohibition of dogs as a type actually contributes to the increasing of the likelihood of achieving the desired, opposite goal (in this case the avoiding of an evil), which is the rule's justification, *viz.* preventing annoyances in the restaurant (28–29).

Here Schauer identifies two different general decision-making modes. The first one is called "particularistic" and its main characteristic is that it gives priority to the justification. The rule is seen as deprived of any autonomous value, so that its generalization appears merely as a *defeasible* instantiation of that justification. The decision of those cases in which the rule shows itself to be over- and under-inclusive is directly guided by its underlying rationale. In the second decision-making mode, by contrast, instead of being continuously malleable, generalizations become *entrenched*. That is, it is the rule itself what guides the decision in every single case, even when its application frustrates the justification behind it either due to its over- or to its under-inclusiveness. Now the decision-maker treats the generalizations as more than mere indicators or guides, but as supplying reasons for decision independent of those supplied by the generalization's underlying justification. Given that in this model the rule is thus imbued with an intrinsic, autonomous force or weight, Schauer calls it *rule-based* model (42 ff., 51 ff., 77 ff.). This autonomous value or "resistance" of the rule presupposes that the meaning of the generalization is not coextensive with that of its background justification. Schauer calls this the *semantic autonomy* of rules, which is in its turn based on the linguistic fact that the meaning of the rule's formulation exists independently of its singular interpretation on every particular occasion, i.e., it is partly acontextual (55 ff., 67). The differentiation between both modes of rule-reasoning somehow runs parallel with the opposition that Schauer formulates between two subtypes among prescriptive rules, viz. "rules of thumb" (which are "transparent" to their justifications) and "mandatory rules" (which are "opaque" or resistant to their justifications and thus rules in the strict sense). Rules of thumb are also redefined as those vulnerable to the inapplicability of their background justifications, while strict rules are those resistant (not absolutely though) to defeat or override not only in the face of their justifications but also of reasons external to them[13] (109–110, 117–118).

The justifications underlying rules are not only *substantive* justifications (the values and purposes the generalization intends to promote), but also *rule-generating* ones, i.e. justifications for having rules themselves. These are formal, institutional justifications, as they refer to the reasons for always instantiating substantive reasons *in the form of rules* (94 ff.). A third decision-making mode is in this respect introduced by Schauer: the "rule-sensitive particularism." According to this model, even when they are suboptimal, rules can prevail against their substantive justifications by consideration of reasons concerning certainty, reliance, predictability, stability, equality, etc. Such reasons bring into play the point of view of the rule-maker and not only that of the rule-applier. Particularly, when we speak of political and legal institutions, those reasons for having rules are at the basis of the very design and "division of work" inside them. Rules operate as a device for the allocation of power, and a decision-making mode that runs entrenching the generalization against their underlying substantive rationale can be entirely justified "without committing the

[13] We find the same idea in Raz's conception of mandatory rules as "exclusionary rules" (Raz 1990: 58 ff.).

sin of rule-worship" (98). In effect, preventing the decision-maker from considering the direct application of the substantive justifications on every occasion may appear as the only way to patrol the boundaries among institutions, thus preventing certain decision-makers from taking certain kinds of decisions (in sum, a way of granting separation of powers and the rule of law).

Schauer relates this point—reasons for imposing rules not only being different but even stronger than reasons for following them—to what he calls the "asymmetry of authority" (128 ff.). The rule-making authority has an aggregate perspective in relation to the whole set of individual future applications of the rule, thus she can consider that the collective consequences of a strict rule-based decision-mode are rationally *better justified* than a collection of particular judgments that apply the rules' underlying justifications, even if, necessarily, following the former may sometimes mean taking *wrong* decisions (i.e., decisions the very authority, were she present, would *not* take). The reason is that from the perspective of the authority there will be more cases of mistaken non-following than mistaken following of the rule. So the authority imposes in advance the obedience to her rules, overcoming the future possible disagreement of the rule-appliers in those cases of over- and under-inclusiveness. The process of rule-making can be seen indeed as the anticipation by the authority of these kinds of situations, thus inducing by means of sanctions or rewards the rule-applier to relinquish his allegedly best judgment even under those circumstances when it *is* best not to follow the rule but its justification. Rules are then the second-best solutions[14] and the price to be paid for such diminution in subject error is suboptimality.

Schauer introduces a fourth decision procedure, the one he finally endorses: the *presumptive positivism*, a rule-based decision-mode that again tries to mediate between blind obedience to rules and the full particularism.[15] Rules are given a "strong but overridable priority", so that "decision-makers override a rule [...] not when they believe that the rule has produced an erroneous or suboptimal result in this case, no matter how well grounded that belief, but instead when, and only when, the reasons for overriding are perceived by the decision-maker to be particularly strong." Rules then have enhanced but no conclusive weight, and whenever the result they indicate "is egregiously at odds with the result that is indicated by [a] larger and more morally acceptable set of values", they should be defeated (204–205). So the only reason why they should not determine the decision and prevail against its underlying justifications should be whenever the error or frustration suffered by these materializes in a particularly severe manner. Otherwise there

[14] "A decision procedure that aims to optimize in every case may be self-defeating, producing worse results in the aggregate than a decision procedure with more modest ambitions" (101–102).

[15] This endorsement is, however, presented in descriptive terms: "[P]resumptive positivism is a descriptive claim about the status of a set of pedigreed norms within the universe of reasons for decision employed by the decision-makers within some legal system, [...] [it] may be the most accurate picture of the place of rules within many modern legal systems" (203, 206).

exists a presumption in favour of the rule's generalization. Unfortunately, Schauer does not elucidate the criteria for the identification of those "particularly exigent circumstances" (196) or "particularly strong" reasons for overriding a rule, nor those for the identification of a result which is "egregiously at odds" with an alternative set of values. In any case, the sources of the presumption are the general ones we have already seen behind any rule, i.e. the entrenched generalization that operates by preventing the decision-maker from doing what she would have done, all things considered, in the light of substantive reasons.[16] But also the presumption in favour of rules is supported by institutional reasons—reasons for having rules—which determine the necessity of rules as second-best solutions in decision-making and particularly in the organization of political power.

## 11.4 Three Contexts of Epieikeia

The significance of *epieikeia* as the "correction of law where it is defective owing to its universality" (*EN* V.10.1137b26-27) lies in the fact that it focuses our attention on at least three principal points: (i) the epistemological dialectic between theoretical and practical reason, (ii) the moral and political functions that legal rules as a whole are serving, and (iii) the articulation of the internal legal method—how law "rules" by means of generalizations that intend to guide actions, and specifically by governing particular judicial decisions.

Let us start with the latter. Here we encounter the connection between the legislature and the judiciary, i.e. between the legislative and the adjudicative practices, both constituents of the legal institution as a regulatory or normative one. Equity presupposes that in the law both the practice of legislating and the application of legal rules refer to each other, but they also differ from one another and the two are deferred both in space and time, thus organized in a second level with respect to the social practices they intend to rule. First, legislation provides "the positive categorization of those actions which people should perform and those from which they should abstain [*nomothetousês ti dei prattein kai tinôn apechesthai*]"[17]—that is, the linguistic formulation of "generalizations" in Schauer's sense. And, second, it is required for those formulae to be generally applied and enforced by means of public coercion, this being the central institutional duty of the judiciary. The way for legal rules to be corrected is by means of the incorporation to the practice of adjudication of the weight of the relevant moral and political values, on which those rules rely, even when this operates against the generalizations of rules themselves. Such values, we already know, are, firstly, those resulting from substantive conceptions

[16] "With respect to any decision-making agent, a rule exists for that agent [...] only when an instantiation provides a reason for action independent of that supplied by the instantiation's background justification" (113).

[17] *EN* I.2.1094b5-7 (I have modified the Oxford's translation: "as to what we are to do and what we are to abstain from"). See *Pol*. III.15.1286a15-16; *Rhet*. I.1.1354a22.

on distributive justice according to every particular constitution,[18] and secondly, those moral values regarding the different virtues (as well as customary norms and uses, traditions, etc. which form positive morality).[19]

On this basis, Aristotle establishes a conclusive limit to any formalist understanding of legal methodology—and so of law itself—along the lines of positivism, inasmuch as he highlights that legal rules are always oriented to the common good. Since this common good "seems at all events something greater and more complete both to attain and to preserve" than the good of single individuals, "for though it is worthwhile to attain the end merely for one man, it is finer and more godlike to attain it for a nation or for city-states" (*EN* I.2.1094b9ff. See X.9.1180a23-24), we can certainly here assume Schauer's considerations about "reasons for having rules" as a main source for the universality of legal rules. Moral disagreements as to the content of virtue and happiness are not to be politically resolved in moral terms any longer but in legal terms, that is, by adopting a constitution considered to be mandatory for all the citizens. Any constitution, though, consists in the incorporation and the articulation in the political practice of a heterogeneous diversity of moral values. Legislative reasoning, a political-architectonic and prudential one (*EN* I.1.1094a14; 2.1094b5), must anticipate the aggregate consequences of universalizing legal standards, given that only by way of imposing equal and general standards in public as well as in private relations is it possible to purport to reach an effective universalization of justice within the *polis*. This, however, provides by no means decisive support to the thesis of the "asymmetry of authority."

In effect, from the point of view of the Aristotelian legislator, the political purpose of *generally* imposing the referred substantive values is just an *instrumental*, formal value. Hence, Schauer's institutional reasons for having rules—"rule-generating reasons"—have *justificatory* force according to Aristotle only to the extent that they are *actually* spreading and propagating the substantive reasons that underlie those rules. Otherwise we would not have a correct understanding of the relation between law and morality. Law, by means of its rules, is an instrumental mechanism for the incorporation and implementation of the moral values in the political practice, rather than a self-referent device. On the other hand, rules being coextensive with their underlying justifications do neither imply they not being "rules" at all nor they being mere "transparent rules of thumb" at most. This conclusion is precisely a formalist one, for it has its premises in an erroneous conception of practical reason mistaking it for a theoretical one. Rules are not generalizations of the same kind as descriptive generalizations, that is, just logical-linguistic or semantic ones.

---

[18] According to Aristotle, "the laws are, and ought to be, framed with a view to the constitution, and not the constitution to the laws." (*Pol.* IV.1289a13-15) Legislative activity is of an interpretive nature and it is developed within the constitutional framework, therefore there can be legal rules that are contrary to the regime's foundation (II.9.1269a32.).

[19] Legal rules have a justificatory, not only a genetic connection to substantive moral values, and this provides them with some kind of universal dimension ("everywhere the same force"), as well as exposing them to the possibility of moral criticisms. That is why justice in its general sense is a synonym for virtue, since "in justice all the virtues are contained" (*EN* V.1.1129b30; see 2.1130b18ff.).

They are *deliberative* generalizations, hence of a distinctive practical, axiological character.

The first deliberative instance is that of the rule-maker. The rule's generalization cannot be taken as a self-referent logical-theoretical construct, but as the result or *effect* of a practical reasoning made by a subject in relation to others with a *purpose* (interfering in their future conduct) and in view of some *values* (an *evil* to be eradicated, a *good* to be achieved). Only as a consequence of it being a practical reasoning is it possible to say that the rule possesses a *rationale*, that is, a *justification* (as different from an epistemic, theoretical *explanation*). Generalizing is here, then, not only a logical but also an *axiological* operation. The selected and suppressed properties at the core of the rule's factual predicate are not so "factual", as both selection and suppression result from an *evaluative* judgment. Of course, there are causal considerations of an objective nature too—i.e. technical or even scientific—involved in the making of the generalization. But this epistemic element is not enough to rule out the practical character of such reasoning, since its instrumental objectives are only to be attained *by means of the action of other individuals*. Here we face the second deliberative level of a rule: that of its appliers, who come into play as practical agents too.

The rule is then indeed constituted by the continuous practice of application of the generalization at stake to particular given situations. This iterated practice is what Schauer calls the "entrenchment" of rules. Still, in my opinion, Schauer manifests his strongly—otherwise widespread—positivist view on legal rules when he is inclined to interpret the entrenchment as primarily a logical or semantic property: in his understanding a rule is essentially an "entrenched meaning" (1991: 72) from which its normative force stems. The entrenchment is associated to the linguistic formulation of the rule, that is, to the logical generalization at its core. According to this conception, entrenchment is about determining the appliers' decision in congruence with the *literal* formulation of the rule. But, above all, it is about *leaving aside their deliberation* when they have to undertake rule application, that is, it is about not allowing appliers to decide otherwise. And "otherwise" means here "in the light of the justification that underlies the generalization." For, it is said, there can only be a "rule-based" decision-making insofar as the rule-formulation, whenever determining solutions to a particular case which happen to be different to those resulting from its justification, is able to oppose a certain level of "resistance"—not necessarily absolute though—to such a justification and therefore to the solutions that the applier should otherwise apply. Entrenching a generalization is therefore, for Schauer, entrenching its meaning in order to make it "opaque" or "autonomous" against its justification.

Now, to consider the propositional content of the rule's generalization as the very key of the rule is just a formalist or logicist prejudice. For it is only through a continuity established by the recursive application of the criteria expressed in the generalization that the rule becomes "entrenched." Rules are not just logical generalizations but, once these latter have been settled, they are general complex practices of application of these generalizations to particular cases. In other words, rules reveal themselves as chains of practical reasoning, and particularly as the *process* composed of the recurrent and persistent practice of uniform application of certain generalization as a deciding criterion. It is more in such a generalized practice than in the general abstract criterion itself that the rule is to be found. It is then about the *justified* generalization of a course of action, which essentially involves an *axiological continuity* between the rule-maker and the rule-appliers. This is the only way in which formal universality can be a sound criterion of rationality in the sphere of *practical reason* too.[20]

In the case of the lawgiver the articulation and implementation of political and moral values implies that, after a deliberative balancing among them, they are translated into courses of action typified in legal, general and abstract statements. Those anticipated courses of action will only generate a rule inasmuch as they become recurrently identified in subsequent practices. This implies that the rule is the practical series of all those particular cases that are covered by the justificatory balance among values that the legislator has translated into the courses of action typified in general and abstract statements. The deontic meaning of a rule goes far beyond both the semantic meaning of its formulation and the logical consistence of its iterated applications: it includes an axiological *coherence* among them. And this is exactly the rationale of *epieikeia*: to remove the possible incoherencies or axiological discontinuities that rules might be generating along the process of their application in social practice. In other words, *epieikeia* is tantamount to restoring the suboptimal scope of rules to its justified scope by means of ruling out—instead of tolerating as the second-best choice—those cases of over- and under-inclusiveness that the literal—"entrenched"—application of them would otherwise provoke. It is then the correction of the rules' logical universality in the name of their practical universality.

## 11.5 The Limits of the Authoritative Dimension of Rules

Therefore, the *epieikeia* has centrally to do with judicial practice.[21] But in fact it is an *institutional*, formal principle governing adjudication from the point of view of the political design that the Aristotelian legislator is attempting to establish. However, as we saw, the Schauerian legislator's rationality is precisely one that

[20] As it is well known, the best philosophical expression of this idea in the history of philosophy is Kant's categorical imperative, which operates by being "constructively" applied to *material* moral maxims of action. The formal, logical universalization is a procedure for ruling out *via negationis* immoral maxims and so for obtaining those that are consistent not only logically but also morally. In this sense, it gives rise to a "teleological", *viz.* practical, not purely "deontological" ethics. See Wimmer (1980: 202 ff.); Schnoor (1989: 78 ff.). Let us remember too, *en passant*, that the categorical imperative is an *a priori* but *synthetic* judgment: that is, the empirical product of a "pure practical reason" which *causes* action in an Aristotelian manner.

[21] As it is known, Aristotle also treats equity, following an ancient Greek tradition, as a moral virtue of the *citizen*. I will leave aside this dimension. See Hamburger (1971: 89 ff.); Georgiadis (1987: 165 ff.); Brunschwig (2002: 119 ff., 126 ff.); Beever (2004: 43 ff.).

rules out any kind of institutionalized correction of legal rules by their appliers (thus, ruling out *epieikeia* itself). The reason for this is the attempt to draw a political design minimizing the decision-maker's misapplications or errors when she tries to optimize the rule's generalization or formulation by considering it defeated by its justification (instead of entrenched against it). That is, the purpose is to avoid "decision-making errors", and consequently to assume the ordinary "rule-based errors" (i.e., over- and under-inclusiveness) as the lesser of two evils.[22] Now, the Aristotelian legislator would not even admit these latter errors. In fact, his institutional design challenges the very distinction between those two kinds of errors. What is actually an "error" in *practical* rule-making and rule-following? A rule-based error cannot be but a *lawgiver's* error as a practical agent. And this is not a sort of "technical" or "scientific" error, as Schauer assumes when he models rules under the image of *theoretical* universal principles (the principles of game theory here included). It is rather about *evaluative* mistakes when anticipating the future recurrent application of the lawgiver's general statements. Hence the possible errors are not imputable merely to the logical generality of the very statement, but to this latter with each and every one of its applications. Consequently, all of them necessarily depend on what Schauer calls over- and under-inclusive cases, which can only *ex post* be identified as such. It is only from the perspective of each particular case, not from that of the *a priori* generalization itself, that over- and under-inclusiveness arises.

So, from the point of view of the lawmaker, the "perfect rule" in Günther's sense (1989: 156), i.e. one that *a priori* "is notified of all its future cases", does not exist. So there is no such rule that could regulate and perfectly justify its own application as if every problematic situation had been taken into consideration beforehand.[23] According to which reasons, then, is it granted that the *ex ante* judgment of the lawgiver could know and better estimate how many future misapplications of it will be made? It is quite clear that, in any event, this is a kind of practical assessment, that is, a *prudential* one—in the Aristotelian sense. So it cannot be based just on the

[22] "When rule-based decision-making is in place, the most noteworthy error is the failure on some number of occasions to make the best or optimal decision in the particular case. But when particularistic decision-making prevails, the most noteworthy errors will be those in which misguided decision-makers—whether biased, ignorant, incompetent, or simply confused—will make decidedly non optimal decisions. In attempting to design a decision-making procedure, we assess as best we can the expected frequency and consequences of these two types of errors. When the result of that assessment is a preference for rules, there is implicit in this preference a judgment that the errors that might be made by misguided decision-makers are more serious or more likely that the rule-based errors that come from a built-in failure to reach the very best decision in every case" (Schauer 1991: 154).

[23] See Von Leyden (1985): 96. A classic example of such "perfect rules" are those who Kant called "universal laws" (*universale Gesetze*), which are valid in every case (*allgemein gelten*), "as it seems to be required by the concept of a law [*Begriff eines Gesetzes*]." These universal laws sharply oppose "merely general laws" (*bloss generale Gesetze*), which are valid only for the most part (*im allgemeinen*) and where exceptions are added "by groping around particular cases as they come up" (*durch Herumtappen unter vorkommenden Fällen*). Kant (1795 [1968], Anhang II, n.2).

rule as a logical generalization, but rather on its underlying justification, that is, on the *values* the generalization tends to promote, in relation to which the former is just instrumental. This commitment to the substantive values behind the rule absolutely determines the meaning and implications of the "errors" in its application. Indeed Schauer himself acknowledges this when he immediately adds that "at other times or in other things (consider the lesson of the Nuremberg trials) the fears of careless compliance exceed those of careless non-compliance" and that "we cannot divorce the value of entrenchment from the value of what it is that is being entrenched" (1991: 154, 158).

Given that it is not only the generalization but also its axiological backing that sets the practical connection between the legislator and the judge, how can the lawgiver possibly know in advance that there will be *worse* particular applications ("errors") according to the rule's justification than by following its literal formulation? As long as it presupposes the discrimination between better and worse readings of the justifications behind rules, the "preference for rules" cannot be *justifiably* grounded unless it proves to be dependent on the system of material principles that in its turn justifies the axiological content of their generalizations. Otherwise the Schauerian legislator would merely beg the question. For, then, the preference for rules is in fact exclusively based on the preference for the literal meaning of the rule's formulation (i.e. for the generalizations *per se*). Maybe here lies the ultimate assumption of all kinds of rule-formalism.

The "preference for rules" of the Aristotelian legislator, by contrast, assumes the evaluative-laden character of the generalizations in rules right from the beginning. Over and under-inclusive cases are those in which the justification is applicable, cases that the rule *should* have anticipated and thus governed.[24] Now, the cases that the generalization *does* govern and to which it is *prima facie* applicable—those literally designated by the selected properties in which the extension of the generalization consists—are nonetheless equally value-laden (even if they include naturalistic terms). That is the distinctive feature of practical rules and values, which unlike theoretical (such as technical or scientific ones), are essentially *others-related* (*pros heteron*, *EN* V.1.1129b27-1130a8). Justice, particularly, emerges out of the mediating operations of individuals involving the *good* of other individuals.[25] (1130a5) Consequently, *semantic* autonomy of rules is not tantamount to *axiological*

[24] "Factual predicates will therefore in some cases turn on features of the case that do not serve the rule's justification, and in others fail to recognize features of the case whose recognition *would* serve the rule's justification" (Schauer 1991: 33, original emphasis).

[25] So, to take up again Schauer's example, the term "dog" appearing in the rule's formulation "no dogs allowed" is not exclusively a descriptive, but also an evaluative one. It does not take into account only universal-naturalistic properties of dogs (such as biological, physical, etc. ones). In that case any kind of *practical* value would be pushed to the back row, if not disappear entirely. Instead, the relevant properties selected by that formulation have to do with human operations, therefore with goods and evils (annoyance, safety, etc.). It is a value judgment based on them which makes the cases of a guide dog or of disturbances caused by agents or animals other than dogs relevant and which requires reintegrating them to the justified scope of the rule.

autonomy: from the point of view of the common system of substantive values and principles governing their practice, there cannot be a true *asymmetry* between the lawgiver and the law-applier.

There is no doubt that legal rules imply the entrenchment of the literal meaning as the right answer to be adopted by appliers. However this is so *because and as far as* they actually *promote* the underlying value or the justification of the rule, which in such a way becomes socially *generalized*. That is why Aristotle assumes that the legislator must "necessarily" "speak universally" and "take the usual case" (*EN* V.10.1137b15-16), there "where matters hold only for the most part" (*Rhet*.I.13.1374a25). This clearly makes the universality of rules a (necessary) virtue and at the same time a deficit or insufficiency. The virtue lies in the fact that the rule's generalization *will* indeed hold "for the most part." There will be a collection of cases—those positively categorized in the general formulation passed by the lawgiver—in which the literal, regular application of the latter will produce correct decisions. It is the "optimal" area of rules, there where their moral and political authority is justifiably exerted. That is the right way to understand Aristotle's apparently "legalist" sentences above quoted.[26] In the easy cases, the applier's judgment is to be superseded by that of the legislator. This is how the rule of law works as based on a law of rules. Yet that will not be resulting from sheer logical consistence in literal, subsumptive application of those rules, as if it were a "deterministic" process. Rather it derives from the axiological coherence of the practical reasoning of the applier to the principles which underlie the previous practical reasoning by the law-maker. So, here there is no trace of any axiological asymmetry at all. In order to properly grasp this point, we need to reflect again about the mentioned deficit and insufficiency from which the "errors" generated by the logical universality of legal rules stem.

The need to proceed to the "correction of the rule" (*epanorthôma nomou*) is for Aristotle the unavoidable consequence of an "error" generated by its logical universality, which, however, makes it "incomplete" or "defective" (*elleipei dia to katholou*, 1137b26-27) not only logically but *practically* too. Such incompleteness clearly has an epistemological diagnosis and the Aristotelian legislator is aware of it (*hekontôn*). Legislators, in effect, "find themselves unable to define things exactly [*ou dynatai diorisai*], and [nonetheless] are obliged to legislate universally [*anankaion katholou eipein*] where matters hold only for the most part" (1137b29-31). In other words, it is impossible for the legislator to transform his practical determination of the rule's statement, and the generalizations it is composed of, into a *theoretical*, objective kind of definition or reasoning that would hold *always*, invariably (like in *epistēmē*). Here we can see that not the logical generalization incorporated to the rule's formulation but the *moral* framework thereof is responsible for the rule's defectiveness.[27] Yet this presupposes the *necessity* for generalizations, i.e.

[26] See supra n. 6 and infra n. 34.

[27] In the moral, political and legal domains—the realm of *politikē*—"we must be content [...] in speaking about things which are only for the most part true and with premises of the same kind to reach conclusions that are no better." (*EN* I.2.1094b20-23) Unlike in nature, now the standard

logical or somehow universal criteria for the legislative practice to be undertaken as a rational enterprise. For this is the only way for rules not to merely identify and select general, abstract properties[28] but also for them to make possible a regular and therefore *equal application* of them. This will allow those rules to be the recursive instantiation of the same standards to similar cases. That is, universality is required as a necessary condition for equality, the essential component of justice.

Logical universality is necessary; it is not a *sufficient* condition for equality though. A regularity of practices of application (i.e., "entrenchment") where judges act upon citizens assuming the role of mediators who enforce the values there involved is also required.[29] (*EN* V.4.1132a20-25) And of course, logical universality is not a sufficient condition for absolute justice either, since this latter depends on the *material* content of such values, i.e. on how they embody those citizens' moral and political *good*.[30] These values are to be implemented by legislative (distributive justice) and judicial (corrective justice) operations interfering in the practice of individuals, both in public (State) and private law relations. It is then a *normative*, dependent upon moral-political criteria, equality.[31] Political substantive values such as *isonomia* or *eleutheria* are practically constructed and reproduced through this sort of general equality.[32]

Now, the point is that even if we are dealing with the morally and politically *correct* substantive values—the right reasons—, incorporated by the legislator to the formulation of her rules, and even if we assume that these rules should be recurrently applied and instantiated by judges in the future, this does not guarantee that justice is going to be achieved in each and every decision of application of

---

*hôs epi to poly* has a *normative* dimension. As it is known, here lies the difference with the Platonic legislator and his "kingly science." See Brunschwig (2002: 126 ff.) On the relations between Plato and Aristotle on this topic, see Michelakis (1953); Georgiadis (1987: 159 ff.).

[28] Aristotle refers to the good legislator as a competent *technikos* and as someone who has studied his craft theoretically (*theorêtikos*); as Bodéüs (1993: 58) points out, it is "in any case, one who must have attained general knowledge." See *EN* X.9.1180b20ff.

[29] "That is why parties to a dispute resort to a judge, and an appeal to a judge is an appeal to the just; for the judge is intended to be a sort of living embodiment of the just. Moreover, they seek the judge as an intermediary, and in some cities they actually call a judge a "mediator", assuming that if they are awarded an intermediate amount, the award will be just. If, them, the judge is an intermediary, the just is in some way intermediate." (*EN* V.4.1132a20-25; Irwin's translation). It is, then, due to the fact that it consists of the application of pre-existent general rules that judicial justice "treats people as equals." (1132a2-6)

[30] Due to his neglecting this axiological dimension, Schauer (1991: 135 ff.) denies any connection between the generality of rules and justice *simpliciter*: "Thus there is nothing essentially just about rule-based decision-making." (id., 137) This is however inconsistent with his defense of the "asymmetry of authority" and ultimately with his entire vindication of a rule-based "presumptive positivism", according to which rules may result overridden by "particularly severe" substantive reasons. See infra n. 42.

[31] In Leyden's words (1985: 13), equality means "a process of equalisation between people who are different."

[32] *Pol.* IV.4.1291b34-38; see V.1.1301a28-35; VI.2.1317b2-3. So, disagreements about distributive justice are disagreements about equality itself: see *EN* V.3.1131a27-29, *Pol.* III.12.1282b18-23.

those rules. What, then, does produce the "indeterminacy" and hence the "incompleteness" of the rule of rules, ultimately?

The subject matter of the legislator's decision is indeterminate (*aoriston*) and this is so not only due to the fact that social practice "is concerned with the ultimate particular, which is the object not of knowledge [*epistêmê*] but of perception" (*EN* VI.8.1142a26-27). It is above all a consequence of the *second-order* prudential nature of the legislative practice (*nomothetikê*) and thus of the fact that it is continued by the adjudicative practice (*dikastikē*)

Take the example given by Aristotle: a rule on causing injuries by means of weapons. It is a difficult task to formulate such a rule in the light of past practice, for "the endless possible cases presented, such as the kinds and sizes of weapons that may be used to inflict wounds—a lifetime would be too short to make out a complete list of these." Since "a precise statement is impossible and yet legislation is necessary, the law must be expressed in wide terms [*haplôs eipein*]." Once it has been laid down, the general statement by the legislator has positively selected, thus suppressed, practical properties (actions, circumstances, individuals) which in those subsequent particular cases where it is to be applied (since "the decision of the lawgiver is not particular but prospective [*peri mellontôn*] and general [*katholou*]", *Rhet.* I.1.1354b8-9) may prove to be over- and under-inclusive (i.e. they may say too much or too little). "And so, if a man has no more than a finger-ring in his hand when he lifts it to strike or actually strikes another man, he is guilty of a criminal act according to the written words of the law; but he is innocent really, and it is equity that declares him to be so" (*Rhet.* I.13.1374a32ff.). For what equity requires from us all is to be indulgent with "human things" (*anthrôpinois*) and then "to look not at the law but at the legislator, not to the written rule [*logon*] but the purpose the legislator gave to it" (1374b10-13, my translation), because he himself *would not* allow that result to be produced.

The "error" is then a necessary, unavoidable one, resulting from the very structure of the legal *praxis*, or more precisely, from the "legal dynamics", to use Hamburger's words (1971: 93). Actually, its origin is not to be found—as Schauer presupposes—in the very rules considered as *logical* generalizations "automatically" producing over- and under- inclusive effects. The reason rather lies in the fact that rules, being themselves the outcome of a practice, are constantly and inevitably exceeded by those practices they intend prospectively to govern[33]—practices which otherwise have themselves no alternative way of being morally and politically governed than by means of rules. The way in which this happens is graphically described by Schauer when he writes that the rule-based errors—"an error being defined as a result other than that indicated by direct particularistic application of a background justification or theory of justification"—are "produced not by decision-makers but by life, for life, unlike the factual predicate of a rule is probabilistic and not universal, variable and not fixed, fluid and not entrenched" (Schauer 1991: 154).

[33] Compare with Hart (1994:128–129) on the "indeterminacy of purposes" as a nuclear dimension of the "open texture" of rules.

On the one hand, there is the variability of the first-order social *praxis*, which we could compare to Heraclitus river: no rule can ever be applied twice the same way. On the other, there is the correlated variability of the second-order adjudicative *praxis*. Both are anticipated by the legislative one: the former being the substratum that provides the abstract, typified conditions of application of the latter; and this one being the institutional identification of such conditions and the enforcement of the established legal consequences on social practice,[34] but eventually requiring it necessarily being redefined all over again in every particular situation in which that application actually instantiates. Such particular situations, Aristotle says, emerge "at short notice" (*Rhet.* I.2.1354b3), that is, unexpectedly, in the course of the development of social practices. The judicial practice, then, consists essentially of *retrospectively* restating and determining the practical meaning of each case in the light of the general legislative statement. It is the judge, facing *this* particular case, who is the one to consider it as under- or over-determined by the rule's generalization then—that is, by the literal interpretation of the legislative statement, which he has the *prima facie* duty to adopt.

The Aristotelian legislator is perfectly aware of the limited way in which logical principles alone are able to project universality (and hence equality and justice) in the practical domain: "The reason is that all law is universal but about some things it is not possible to make a universal statement which will be correct. In those cases, then, in which it is necessary to speak universally, but not possible to do so correctly, the law takes the usual case, though it is not ignorant of the possibility of error" (*EN* V.10.1137b12-16). But this is not only a logical but also a moral-political defectiveness: there are deeper reasons than epistemological complexity behind it.[35] No legislator can prevent her rules from being defective *in casu*, however correct they may be *a priori*. Consequently, the correctness of a rule can only be affirmed *a posteriori*, in view of the actual course of its particular applications. The lawmaker's rationality cannot reach the absolute certainty that a rule, even if correct (i.e. a correctly deliberated general rule), is going to determine the correct decision for every single case to which it is being applied as a general criterion.

Yet this intrinsic possibility of defectiveness arising in a particular case does not make the generalization nor the legislative practice defective in absolute terms: "It is none the less correct, for the error is neither in the law nor in the legislator but in the nature of the thing [*physei tou pragmatos*], since the matter of practical affairs [*hê tôn praktôn hyle*] is of this kind from the start." It is then about correcting the

[34] "In general, then, the judge should, we say, be allowed to decide as few things as possible. But questions as to whether something has happened or has not happened, will be or will not be, is or is not, must of necessity be left to the judge, since the lawgiver cannot foresee [*proidein*] them", *Rhet.* I.1354b11-16; see also 1354a29-30.

[35] Like in the case of decrees, the epistemological limits ("In fact this is the reason why all things are not determined by law, viz. that about some things it is impossible to lay down a law, so that a decree is needed", *EN* V.10.1137b28-29) come together with the moral-political ones ("no decree can be universal", *Pol.* IV.4.1292a37).

error case by case dialectically. Sound legal decisions are *not* given *ex ante* by the general formulation enacted by the legislator, but rather they are the result of practical reasoning undertaken case by case. Particular correct applications of the law—and therefore the correctness of the rule as a whole—are necessarily to be deliberated *ex post*. It is only the singular deliberation of the judges what guarantees that correctness by safeguarding the indemnity of values and principles behind rules in every single case—thus making those rules somehow behave like the leaden rule used by the Lesbian constructors: "the rule adapts itself to the shape of the stone and is not rigid", for "when the thing is indefinite [*aoriston*] the rule [*kanôn*] also is indefinite" (*EN* V.10.1137b29-32). In order to be *appropriate* for every particular case and not only deliberatively valid or justified from the point of view of the legislator, any legal rule needs to enter a discourse of application which will test whether it could be correctly applied to the given situation; otherwise it must give way to another rule (Günther 1993: 217). In other words, rules must be made so to treat unequally unequal cases: that is the task for the judge, he who "restores equality" (IV.1132a25). And Aristotle establishes how that is to be done: "When the law speaks universally, then, and a case arises on it which is not covered by the universal statement, then it is right, when the legislator fails us and has erred by over-simplicity, to correct the omission—to say what the legislator himself would have said had he been present, and would have put into his law if he had known" (V.10.1137b19-24).

## 11.6 Epieikeia vs. the Asymmetry of Legislator's Authority

This last quotation condenses the negation by the Aristotelian legislator of the "asymmetry of authority" that the Schauerian vindicates. Why, in effect, should those rules' errors be tolerated as a lesser evil instead of being corrected? Moreover, why should "a result other than that indicated by direct particularistic application of a background justification or theory of justification" be *necessarily* considered an "error" at all? Sheer formal-institutional reasons are here claimed for the rigidity of rules, but this neglects the fact that the legal institution necessarily requires dealing with *exceptions* to the rules that are justified by substantive reasons. The Aristotelian argument is that legislative rules should be defeated by their underlying values or by other higher moral-political justifications within the legal system. There is no justification why the literal application of rules should always be entrenched even if the result thereof is suboptimal. It would be fallacious to assume that a judge's re-evaluation against the literal interpretation is necessarily *against* the legislator's underlying reasons too and hence an error that should be *generally* prohibited. Furthermore, it is not at all politically justified in a rule of law system—in which "we do not allow a man to rule, but rational principle", *EN* V.6.1134a35—to limit in that way the decision-making power of judges, and to impose them a strict obedience to rules, claiming that they may commit a significant aggregate number of evaluative errors and thus that they should be, so to speak, turning the perfect into the enemy of the good.[36] The first reason against this line of argumentation is that it attributes a disproportionate importance to the institutional stability of legal rules, whereas for Aristotle that would only be justified at a lower cost in terms of suboptimality.[37] The second reason is quite simple: to impose by all means the submission to rules' formulations and to dissuade judges from resorting to their justifications would imply institutional sanctions and punishments that are eventually *not justifiable*, especially when, as Schauer himself accepts, the authority herself—the legislator—, "were she present", would believe that in this case the rule should *not* be followed.[38] Now, as we have already emphasized, the fundamental reason is this: the literal entrenched application of rules *is* evaluative-dependent too, not merely a logical subsumption of cases under rules that are conceived as general formulae, thus it essentially depends on a congruence of value among judicial decisions.

The assumption that the errors in literal, entrenched interpretation would be aggregately less or less serious than in non literal interpretation—and thus they should be tolerated—is based on the misunderstanding that the appliers' practical reasoning is evaluative-free. Such view, in its turn, depends on a formalist conception on rules that pictures them as a type of objective, mechanical generalizations which "determine" the applications preventing us from any kind of deliberation as if making the way for a series of "infinitely long rails", to recall the famous Wittgenstein's criticism of *Normenplatonismus* (1999: §§218–219). This is how, according to Schauer and an extended opinion (not only in legal positivist circles), they shall grant the limitation of the judiciary's political power within an institutional design under the "rule of law." A pragmatic, post-positivist conception of legal rules emphasizes, to the contrary, that even in literal, allegedly evaluative-neutral application of legal statements, judges are supposed to exert their "best judgment." Not only are they committed, beyond the directive dimension of rules,

---

36 "A decision procedure that aims to optimize in every case may be self-defeating, producing worse results in the aggregate than a decision procedure with more modest ambitions" (Schauer 1991: 101–102).

37 "Hence we infer that sometimes and in certain cases laws should be changed; but when we look at the matter from another point of view, great caution would seem to be required. For the habit of lightly changing the laws is an evil, and, when the advantage is small, some errors both of lawgivers and rulers had better be left; the citizen will not gain so much by making the change as he will lose by the habit of disobedience" (*Pol.* II.8.1269a10-18).

38 Schauer (1991: 131 ff.) "The rule-maker adds sanctions for violating the rule (and thus furnishes prudential reasons for following the rule) even under circumstances in which the subject perceives it to be best, all things considered, to violate it, and even when it *is* best, all things considered, to violate it, for otherwise, there would be little reason for the subject to obey in those cases in which the subject believes, *erroneously*, that this is a case in which violation is justified." (ibid., 132, orig. emph.) See for that criticism Alexander/Sherwin (1994: 1199–1200); Bayón (1996: 156–157).

to those reasons and justifications underlying them, but this is actually the way in which political power is institutionally allocated to judges by the legislative authority, i.e. by means of rules under a "rule of law" system.

We have seen that the point of a rule is the *value* or balance among values making the course of action that is expressed in the formulation not merely (causal-probabilistically) "normal" but "good", that is, a *justified* one. In every single instance of application, that evaluative relationship connecting the formulation of the rule with its underlying justification needs to be renewed by the appliers. A decision from a judge imposing the course of action that is congruent with the rule's formulation and excluding the alternative ones must be taken. For this is the only way for underlying values to prevail in each singular case against eventual counter-values. So the "entrenchment" of rules is based on a *prima facie* "resistance" to re-evaluate in normal or easy cases the individual case directly in the light of the rule's justification (or even of other external justifications), thus a resistance to re-open the original evaluative judgment of the rule-maker. But obviously that does not mean that an applier's commitment to value judgments, and then a deliberative operation, is not present. A fortiori, this is also true when the rule becomes over- and under-inclusive. "Suboptimality" in these cases is not merely a matter of *technically* defective rules. Actually these are rather "hard cases" in the *moral-political* sense which Dworkin (1977) made familiar in legal philosophy. Since no finite legislator can neither anticipate all the *appropriate* applications of her willed rules, nor avoid the emergence of cases which might give rise to conflicts in the light of their underlying reasons and principles, the judge is responsible for a new deliberative practical reasoning which evidently cannot be confined to the alleged "semantic autonomy" of the rule's generalization. This notion intends to rule out any deliberation when applying a rule by presenting it as a mere operation of logical subsumption. However, it is the generalization *itself* which here becomes dramatically challenged by its very grounding values, thus, its own purported universality being challenged. Over- and under-inclusiveness raise indeed the problem of the potential *irrationality* of practical rule-following. The practical rationality of the applier is fully compromised when he is required to carry out a *blind* reasoning in those cases in which not only is the rule said to be "opaque" to its underlying reasons but it also becomes counterproductive or self-frustrating. This is the case when the process of entrenchment of a generalization—that turns it into a rigid one and then allegedly truly a "rule"—requires the applier to take, and not to modify, the course of action as it was encapsulated in its literal extension, also when that means *interrupting* the process of fulfilling and thus of *generalizing* the underlying values. The problem is not merely then about whether a singular case shall be formally qualified or not under a particular rule (as a *logical* property of this latter). It is instead about whether *rules* can exist in practical reasoning at all considering the unavoidable range of cases in which necessary contradictions emerge between their *logical universality* and the *practical universality* of those values underlying them.

So, the generalizations in rules cannot be evaluatively opaque once we have to admit that their suboptimal results in a particular case should be considered "hard cases", that is, pragmatic contradictions between the literal *prima facie* application of the generalization and its underlying justification (which is necessarily "transparent" in the present case), or, in other words, substantive disagreements between the legislator and the rule-applier about whether the rule-formulation is to be considered an appropriate means to fulfill a certain prevailing value. From this it follows that "easy" or "clear" cases, those in which there is "a plain answer on the basis of the materials comprising the rule itself" (Schauer 1991: 209), cannot be the result of an "opaque" reading of the rule either. Not only hard cases ask for an "interpretation", but ordinary, literal reading is essentially an *interpretative* and *evaluative* task too. A case can only appear as an "easy" or "clear" one if its decision not only complies with the rule formulation but also it actually promotes, rather than frustrates, the underlying values given the circumstances of the case. And the crucial point is that this result *cannot* be guaranteed in advance by an institutional structure out of formal rules alone.

Schauer holds that Dworkin's argument[39] according to which "easy cases are only those in which the result indicated by the rule is consistent with the best, all things considered, result" is inconsistent, for it "collapses a rule into its justification" and so "collapses the idea of a rule entirely"—thus allegedly making the rule "superfluous by generating a set of results extensionally equivalent to those that would have been generated without the rule." Accordingly, he writes that Dworkin dissolves the indications of the rule "into the question of what decision should be made on the basis if the full array of justifications", and so "merges the question of what to do with that of how to determine what a rule *means*." Rules, then, would be entirely buried under a "jurisprudence of justifications."[40] Yet it seems quite an *ad hoc* and circular argument that to make the formulation of the rule consistent with the result indicated by its justification implies considering such a rule "extensionally equivalent to the justification being the rule itself" (id., 212). For, from a pragmatic stance, it is clear that this illegitimately conflates the level of the rules with that of values or justifications. The latter can only be turned into rules by being positively specified and categorized in certain actions (and other action-related properties) that will constitute the generalization itself. That is exactly the selective evaluative operation that is carried out by the legislator's practical reasoning and subsequently continued by the recursive literal application of his formulations by the judges. Consequently, the literal meaning is not only about their linguistic meaning *but rather about precisely "what to do" in every single case with those positive formulations*—a pragmatic rather than a semantic matter. Thus, given that the determination of the generalization by the legislator establishes which precise courses of action are those that embody the relevant value, the judge's task is to identify and *impose* them *in casu* and hence to guarantee an evaluative coherence with the purpose of the legislator when the latter promotes those very *same* values or reasons.

[39] See Dworkin (1985: 146 ff.); (1986: 46 ff., 114 ff., 350 ff.).

[40] Schauer (1991: 209–212, orig. emph., 53, n.1); (see also 1987).

Rules (i.e. their linguistic or literal formulations) are neither superfluous nor irrelevant insofar as they are the links which communicate their supporting values or reasons to each present particular case in which they are intended to be in force (from the legislator's perspective). The judicial *praxis* consists then in a constant iterated *materialization* of that link. So, values without rules are empty and rules without values or principles are blind.[41]

And that is the reason why rules, not being superfluous, are not sufficient either when the reasoning to apply them is merely understood as a matter of literal interpretation. If general formulations involve value judgments by the legislators, then their application by the judges cannot be evaluative-detached. Still when the linguistic meaning is clear, it has to be deliberately considered whether the rule is *justifiably applicable* or not in every particular case, all things considered. Such a deliberation does not only involve those values which underlie each individual rule, and hence which limit the universe of possible relevant reasons, but also the second- and further order justifications underlying the whole legal system. So, the "incompleteness" or "omission" (*elleipô*) of legal rules that Aristotle talks about is directly related to the "gaps" in the law[42] as the result of the literal interpretation falling short or surpassing the rationale of the rule (viz. showing itself to be under- or over-inclusive) and then producing an absurd or unacceptable response from the legal system—thus making it speak nonsense. Accordingly, far from being "the systemic analogue of a rule" (Schauer 1991: 199), the law redirects us to those moral and political substantive values and principles grounding it. The possible "errors" arising in the level of rules refer us to legal gaps which are always *axiological*, i.e. only susceptible to being identified from the perspective of the practical reasoning that the applier is supposed to undertake, not from the point of view of the legislative

[41] Schauer himself is otherwise well aware of this in several parts of his book, prior to his vindication of a presumptive, rule-based positivism. Take, for instance, the following passages: rules are more of a "relationship" between "a behaviour within the extension of a rule-formulation and the behaviour that takes place" than logical "entities" or "properties." (1991: 64, 112); rules have to do with "the behaviour the rule seeks to affect", since "we commonly distinguish different rules according *to what they do*, rather than how they say it." (id., 63, emphasis added); "the process of taking a rule to be applicable depends not only on the rule's own designation of applicability, even presupposing internal validity, but of something external to that rule and to the rule system of which it is a part"; "Were all the relevant rule-appliers within a decision-making environment to treat the rules within that environment as overridable, then those rules would be overridable, even though the rules as canonically inscribed incorporated no provisions for override." So concludes Schauer: "something *about* a rule and not the rule itself determines not only what weight the rule will have, but whether it is even a rule at all." (Schauer 1991: 126, 128, or. emph.) The reference of that "something" must clearly be the rule's underlying values—or indeed any other relevant ones. As far as Schauer's "presumptive positivism" eventually recognizes that legal rules can be defeated by "particularly strong" substantive reasons, it departs from a plain rule-formalism. However, there is hardly any distance between this positivism and what he calls "rule-sensitive particularism", as some commentators have pointed out Bayón (1996: 160–161); Alexander/ Sherwin (2001: 68 ff.).

[42] See Shiner (1994: 1254 ff.); Georgiadis (1987: 160ff).

order *in abstracto*.[43] It is according to such reasoning that, after this identification, gaps are corrected and filled in, whether by defeating *prima facie* literal readings of the rules' formulations or formulating new ones.[44] This is how judges, using argumentative means, constantly reconstruct the legal system as such a "system." Material appropriateness, instead of formal rigidity of rules, is the true regulative ideal that must govern the institution of adjudication under the rule of law.

Hence, the unity of the legal system is essentially of a *justificatory* nature and not only of an *authoritative* one. It is the unity imposed by a coherency among principles and values over the simple consistence of the formulations of rules. And that is the only way in which the "rule of law", with its idiosyncratic separation of powers, can be rationally compatible with the submission to the decisions of *another*, as implied in the "law of rules."[45] Aristotle assumes that the legislative rules provide the premises for the deliberations that all those who exercise political activity in all its levels carry out (thus, he calls political prudence *bouleutikê*, for "it is said that

---

[43] See Michelakis (1953: 35). Brunschwig (2002) holds a contrary opinion: "The reparation of apparent gaps in the law could not be entrusted to the judge, at least in a state ruled by law where no one can be convicted except by references to a law enacted and in force at the moment when he was acting. Only the legislator can intervene to complete the gaps of a law, by means of a law making punishable in the future the type of action in question." It is clear, however, that in this case it is not about rules alone since the *principle* of legality in criminal law is here at stake too. This principle precisely implies the exclusion of any "gap" at the level of rules out of moral-political reasons (the protection of individual freedom and autonomy). Brunschwig's conception of the law is positivist, for legal decisions are coextensive to the limits of legal formulations: 'strictly speaking, the law does not manifest "gaps" but "deficiencies" in the etymological sense of the world, i.e. it "falls short" [...] Discrimination of cases where the law speaks and where it is silent as such would depend on neither equity nor justice: it would not be a question of ethical virtue, but only a technical question of reading and knowing the legal code." (Brunschwig 2002: 138–139) See the following note.

[44] Brunschwig finds an "internal duality" in Aristotle's appeal to "the legislator's intention": "The judge says (*a*) "what the legislator would have said if he had been there", and (*b*) "what he would have put in his law, had he known of the case in question." [...] In case (*a*) the judge imagines himself in the shoes of the legislator, who *would say* what must be done in that case, that is, *what he would himself do as judge*; whereas in case (*b*) the judge simply puts himself in the shoes of a legislator who *would write* a supplement to his own law, taking its generality down a notch, but *keeping to his role as legislator*, and leaving it up to the judge to apply the law which he has revised himself." (Brunschwig 2002: 152, or. emph.) In this author's opinion—erroneous in my view—, in case (a) the judge's decision is strictly "singular", "not covered by any general law", and in case (b) the judge would apply an "imaginary law" that does not result "in a real law, integrated into the legislative code." (id., 152–153) As already stated, such a reading is fully dependent on a positivist, rule-formalist conception of the law, which identifies its limits with the limits of its rules. See Zahnd's criticisms on the traditional Anglo-American conception of equity as a gap-filling device in which legal positivism and natural law converge (1996: 268 ff., 291 ff.): "The proper role of equity is not simply to fill gaps in the law. Instead, equity consists primarily in a judge's exercise of practical intelligence to conform universal laws to particular situations" (280).

[45] "When rules are followed, especially in those cases in which the act of rule-following appears to the rule-follower to be within the area of under- or over-inclusiveness, the rule-follower can be characterized as simply deferring to the decision-making capacities of another" (Schauer 1991: 161–162).

those people who deliberate are only the instruments of politics"). Hence political action evolves by way of creating and applying legal rules: politicians "are limited to acting as manual labourers do", that is, they are always executors of general rules by means of new ones.[46] However, to reduce this process of implementation of legal rules—hierarchy relations among political organs here included—to a simple chain of propositional subsumptions of individual cases under generic formulae would in fact imply distorting the very practical rationality, i.e., the moral-political rationality that internally conforms such a process. That is, to dissolve authoritative relations as such, transforming them into non-deliberative connections, "technically" modeled. This formalist image of rules falls far from an adequate reconstruction of evaluative-committed rule-following under the rule of law. To defer to the authoritative decisions of another presupposes, even when punishments are entailed (thus, prudential reasons being involved), the argumentative articulation of practical reasoning in its wider sense. Legal *praxis* is indeed about a complex continuity between practices of interpretation and application of legal formulations, something which necessarily is of a justificatory nature, for it is opened to exceptions and corrections that are not fully foreseeable by the legislator, needing to be determined on every single occasion instead. Rationality of adjudication requires deliberation on the correct applicability and acceptability of legal rules to every particular case in order to make their logical universality coextensive with their axiological universality.

## 11.7 The Self-Correction of the Legal Method and the Impossibility of Particularism

So, the Aristotelian legislator, he who speaks from the perspective of a pragmatic philosophy, clearly recognizes the need for such axiological continuity between legislation and jurisdiction when describing the *epieikeia* as "the sort of justice which goes beyond the written law [*to para tôn gegrammenôn nomôn dikaion*]." To put it in Brunschwig's words, this represents the "legitimate exception to the rule correctly" by a "higher norm of correctness" (2002: 116, 142). This is a norm of reasonableness that does not "suspend" the application of the rule to the case but *rectifies* it (*orthôs*, *EN* V.10.1137b14-16, 22) by way of demanding the relevant value judgments that articulate the judge's deliberation. As Beever (2004: 35) points out, equity consists of those moral judgments that are required to reconcile legal justice with absolute justice, thus also filling the gap between them. That value-dimension is precisely what makes this sort of justice a "better" (*beltion*) and "superior" (*kreittôn*) justice than simply *nomikon dikaion* or rule-based justice (*EN* V.10.1137b28, 10)—that is, better than

[46] *EN* VI.8.1141b28-29 (Bodéüs's translation). See *Pol.* VII.3.1325b21-23.

the type of justice that a Schauerian legislator, he who denies any institutionalized correction of the rules, would foster. Now, for Aristotle it is "not better than absolute justice but better than the error that arises from the absoluteness of the statement" (1137b24-25, 11). Therefore, *epieikeia* is not about a sort of absolute *external* correction of the law, as traditionally understood by the natural law theories.[47] It remains instead *internal* to the law, as an institutional principle that governs its development[48]—even a universal principle, as Aristotle suggests in the *Rhetoric*.[49]

Accordingly, equity is not "a different class of thing [*genos*]" (1137b9) to legal, rule-based justice. It is indeed structurally linked to law, for that kind of justice necessarily consists in the application of rules. The exercise of political justice is not possible at all without rules. Still, the rule of law is not equivalent to the law of rules: against positivism, it must be broadened to incorporate the principles of morality that those rules presuppose as grounds. That is why Aristotle is in no way a "rule-sensitive" particularist, but rather endorses a strict rule-based method (*nomoi* are necessarily *katholou*) together with an internal, corrective practical reasoning in the adjudicative arena. However, now against iusnaturalism, any interpretation of the *epieikeia* claiming that it is exhaustively confined in a moral, pre-legal rationality would dismiss the institutional dimension of the law, in which rules represent the essential element. For rules make of the law an institution that is characteristically authoritative, composed of officials, procedures and coercion, thus irreducible to moral institutions—required by these instead (Miller 1995: 59). To overlook this would mean to dissolve the law in morality. For it would reproduce the legislator's cardinal political—"architectonic"—problem, which is none other than the practical *positivization* of certain moral values after the correspondent balancing and subsequent formulation of rules that aim at *generalizing* the former within the *polis* by way of its continuous "entrenchment." So, the institutionalization of pre-existent moral

[47] Equity as something referring us to "natural justice" or "natural law" was the classical Thomist interpretation: "What is equitable is better than what is legally just but is contained under the naturally just", Th. Aquinas, *In dec. Eth.*, V, 16, n. 1081. Along these lines, see Trude (1955: 124 ff.); Gauthier/Jolif (2002: II-1, 431–432); Hamburger (1971: 100); Leyden (1985: 84); Gadamer (2004: 351 ff.); against, see Brunschwig (2002: 141 ff.); Yack (1990: 227–228).

[48] Along these lines, Zahnd (1996: 280 ff.); Yack (1993: 144 ff., 193 ff.); Shiner (1994: 1250); Georgiadis (1987: 164).

[49] The "common unwritten laws" are to be interpreted as universal values not in an aprioristic ontological sense but rather in the sense of those values "which appear to be universally recognized" [*homologeisthai dokei*], that is, around which there is an *ex post* consensus in every positive legal system (which in essence are always changeable, that is, historical). So, they are universally just: they have "everywhere the same force [*dynamin*]." (*EN* V.7.1134b18) It is only under this light that they appear to be unchangeable, and it is then that Aristotle equates their character with *epieikeia*'s: what is equitable "is ever constant and never changes" (*Rhet.* I.15.1375a30).

norms and values through legislative authorities entails them necessarily undertaking *particular* political deliberations and adopting decisions, thus making a "practical difference."[50] The moral dimension of law (moral criticism here included) is entirely tied to its political dimension: the legal game is a moral-political one. Justice is no longer just a *moral* matter, but also a matter of a deliberative compound by legal authorities of plural and divergent ethical conceptions on virtue in the community. Given this dialectical plurality, a legal decision-making institution must be found to determine the norms that are to govern the society. These norms, Aristotle insists, must be expressed in the form of a general *logos*, that is, in legislation.[51] This may well be called a "politicization" of morality through law and not only the "moralization" of the latter.[52] The controversies concerning the content of virtue and happiness are in fact disagreements as to the content of politics, and thus conflicts over the constitution (*Pol.* III.8.1280a7-22). Therefore the law is neither a mere "specification" or "particular case" of morality, nor is it "merely declarative" in its relation to this latter, as argued by traditional iusnaturalism.[53] It is rather about a second-order relationship, as the Aristotelian architectonic metaphor suggests: political institutions mark a turning point between morality before and after the State, and the law makes the transition from one to another. Legal rules are then not "pure forms" but structures *embodying and moulding* substantial values in order for them to be socially transmitted. Neglecting this is a key error in which both iusnaturalism and legal positivism converge. We find here an important caveat for post-positivism too: for example, in relation to Dworkin's and Alexy's conception of rules as "all-or-nothing" devices, and principles as linked to a "general" morality from which the law would be just a mere "branch" or a "special case."[54]

Aristotelian equity, by contrast, while reintroducing the moral, extra-positive values that necessarily underlie the legislator's deliberation and which are at the basis of the genesis of law as a rule-governed institution, does not dissolve its rationality in morality. Specifically, it neither dissolves judges' reasoning in their *particular* intuition of those values, nor does it make judges substitutes for legislators when correcting them. *Epieikeia* is not about a rectification of the *political* choices of the legislator, of his evaluative "errors." That is, it is not about the rectification of the supposedly correct reading of the constitution by the legislator, but rather the rectification of those errors that concern the

---

[50] Thus the sense of the already mentioned definition of *nomikon dikaion*: "that which is originally indifferent, but when it has been laid down is not indifferent" (*EN* V.7.1134b20-22).

[51] See *Pol.*III.15.1286a15-16; *Rhet.* I.1.1354a22. Beever (2004: 43).

[52] As said before, that is the ultimate reason why the universal, unqualified sense of justice (*haplôs dikaion*) is closely tied to "virtue taken as a whole [*hôlen aretê*]" (see 1130b18-22), for it presupposes the legal institution as the promoting device of all the ethical virtues, now in the political sphere.

[53] See, e.g., Finnis (1980: 281 ff.).

[54] See Dworkin (2011: 405 ff.); Alexy (1989: 212 ff.).

implementation of their political choices in *the form of rules* under the light of an individual situation. The law (and hence the rule of law) is not conceived as a system of self-referred rules, but as an instrument for the fulfillment of the moral-political substantive values which are intended to be incorporated into the political practice. Those values should be the framework from which to examine legal rules and singular decisions hereinafter. Insofar as they fail to capture those values, legal rules must be corrected in order to govern a state in accordance with its constitution. Judges, Aristotle assumes, are no arbitrators[55]—so particularism, even if "rule-sensitive", is not applicable here. In MacCormick's words: "Equity cannot be understood [...] as something particular by contrast to the universality of justice."[56] *Epieikeia* is just the *universalizable* correction of the legal rules. Here exceptions confirm rather than abolish the rule (Von Leyden 1985: 96–97). That is: equity gives rise to new rules that are only legally justifiable as long as they are kept consistently connected to the same values (or balances among them) that the legislator has established, thus they continue and iterate those values in social practices. This is a historical, dialectical process (except for a divine legislator or a herculean judge). And, of course, for Aristotle it is clear that there are many different argumentative ways to project principles and values contained in the standards of one concrete political constitution onto the rest of the legal rules throughout the process of their particular application, especially when there is a plurality of decisors. This is the main source of legal justice's intrinsic variability. The legal development represents an open deliberative process indeed. In Aristotle's proposal of rationalization, thc judge puts himself in the legislator's shoes, as well as vice versa, that is, the legislator puts himself in "the judge"'s shoes, for Aristotle refers to both the hypothetical legislation that the legislator would have introduced (*ei hêdei enomothetêsen*, *Rhet.* 1.13.1137b23-24) had he known the present case requiring correction, and to the *future* potential legislation he may enact, reacting to the judge's current practice (Marcin 1978: 399). The complex continuity of this process will be at all events a matter of the interplay between legislators and judges by means of practical reason—and all this precisely through some sort of rational procedure of coherent universalization in terms of general guidelines of action if justice is to be pursued. The crucial mediating role that legal rules, i.e. the rules of legal justice under the "rule of law", play between moral and politics requires inescapably the unity of practical reasoning. Hence, any axiological asymmetry to be found between legislators and judges is ruled out by incorporating into the legal method the very correction of the method itself.

---

[55] "For an arbitrator goes by the equity of a case, a judge by the law." (*Rhet.* I.13.1374b23) In Beever's words: "Aristotle's claim is only that judges may attempt to realize the intent of the legislator, given that that intent will not and cannot be captured in the form that legislation must take. This, then, is far from the view that equity is justice's rebellion against law" (2004: 45).

[56] MacCormick (2003: 98) "There simply is no such thing as standardless equitable decision-making to Aristotle" (Zahnd 1996: 290).

## Bibliography

Aguiló, J., M. Atienza, and J. Ruiz Manero. 2007. *Fragmentos para una teoría de la constitución.* Madrid: Iustel.

Alexander, L., and E. Sherwin. 1994. The deceptive nature of rules. *University of Pennsylvania Law Review* 142(4): 1191–1225.

Alexander, L., and E. Sherwin. 2001. *The rule of rules. Morality, rules and the dilemmas of law.* Durham: Duke UP.

Alexy, R. 1989. *A theory of legal argumentation. The theory of rational discourse as theory of legal justification* [1978]. Trans. R. Adler and N. MacCormick. Oxford: Clarendon.

Aquinas, Th. 1964³. *In decem libros Ethicorum Aristotelis ad Nicomachum expositio*, ed. R.M. Spiazzi. Taurini: Marietti.

Aristotle. 1995. *The complete works of Aristotle: The revised Oxford translation*, ed. J. Barnes. Princeton: Princeton UP.

Aristotle. 1999. *Nicomachean ethics.* Trans. T. Irwin. Indianapolis: Hackett Publishing Company.

Aristotle. 2002. *L'Éthique à Nicomaque* [1970²], ed. R.A. Gauthier, and J.Y. Jolif. Louvain: Éditions Peeters.

Bayón, J.C. 1996. Sobre la racionalidad de dictar y seguir reglas. *Doxa. Cuadernos de Filosofía del Derecho* 19: 143–162.

Beever, A. 2004. Aristotle on equity, law and justice. *Legal Theory* 10(1): 33–50.

Bodéüs, R. 1993. *The political dimensions of Aristotle's Ethics.* Trans. J.E. Garrett. Albany. New York: Suny Press.

Brunschwig, J. 2002. The rule and the exception: On the Aristotelian theory of equity. In *Rationality in Greek thought*, ed. M. Frede and G. Strike, 115–155. New York: Oxford UP (repr.).

Burns, T. 1998. Aristotle and natural law. *History of Political Thought* XIX(2): 142–166.

Conklin, W.E. 2001. *The invisible origins of legal positivism. A re-reading of a tradition.* Dordrecht: Kluwer.

Dworkin, R. 1977. *Taking rights seriously.* New York: Oxford University Press.

Dworkin, R. 1985. *A matter of principle.* Oxford: Oxford University Press.

Dworkin, R. 1986. *Law's empire.* Cambridge: Harvard University Press.

Dworkin, R. 2011. *Justice for hedgehogs.* Cambridge, MA: Harvard UP.

Finnis, J. 1980. *Natural law and natural rights.* Oxford: Clarendon.

Gadamer, H.G. 2004. *Truth and method* [1975]. Trans J. Weinsheimer, and D.G. Mar. London: Continuum.

Georgiadis, C. 1987. Equitable and equity in Aristotle. In *Justice, law and method in Plato and Aristotle*, ed. S. Panagiotou, 159–172. Edmonton: Academic Printing & Publishing.

Günther, K. 1989. A normative conception of coherence for a discursive theory of legal justification. *Ratio Juris* 2(2): 155–166.

Günther, K. 1993. *The Sense of Appropriateness: Application discourses in morality and law.* Trans. J. Farrell. Albany: SUNY Press.

Hamburger, M. 1971. *Morals and law. The growth of Aristotle's legal theory* [1951]. New York: Biblo & Tannen.

Hart, H.L.A. 1994. *The concept of law* [1961], ed. P.A. Bulloch, and J. Raz. Oxford: Clarendon Press.

Kant, I. 1968. Zum ewigen Frieden [1795]. In *Kants Werke. Akademie Textausgabe*, Berlin: Walter de Gruyter, VIII.

Lockwood, Th.C. 2006. Ethical justice and political justice. *Phronesis. A Journal for Ancient Philosophy* 51(1): 29–48.

MacCormick, N. 2003. *Legal reasoning and legal theory* [1978]. New York: Oxford UP.

MacDowell, D.M. 1993. *The law in classical Athens* [1978]. Ithaca: Cornell UP.

Marcin, R.B. 1978. *Epieikeia*; equitable lawmaking in the construction of statutes. *Connecticut Law Review* 10: 377–400.

Michelakis, E. 1953. *Platons Lehre von der Anwendung des Gesetzes und der Begriff der Billigkeit bei Aristoteles*. München: M. Hueber Verlag.

Miller, F.D. 1995. *Nature, justice, and rights in Aristotle's politics*. Oxford: Clarendon.

Nussbaum, M.C. 1985. *Aristotle's De Motu Animalium. Text with translation, commentary and interpretive essays* [1978]. Princeton: Princeton UP.

Ostwald, M. 1969. *Nomos and the beginnings of the Athenian democracy*. Oxford: Clarendon.

Raz, J. 1990. *Practical reason and norms*, vol. 2. Princeton: Princeton UP.

Schauer, F. 1987. The jurisprudence of reasons. *Michigan Law Review* 85: 847–870.

Schauer, F. 1991. *Playing by the rules. A philosophical examination of rule-based decision-making in law and in life*. New York: Oxford UP.

Schauer, F. 1997. Generality and equality. *Law and Philosophy* 16: 279–297.

Schnoor, C. 1989. *Kants Kategorischer Imperativ als Kriterium der Richtigkeit des Handelns*. Tübingen: Mohr.

Schroeder, D. 2003. Aristotle on law [1981]. In *Aristotle and modern law*, ed. R.O. Brooks, and J.B. Murphy, 38–51. Dartmouth: Ashgate.

Shiner, R.A. 1994. Aristotle's theory of equity. *Loyola of Los Angeles Law Review* 27(4): 1245–1264.

Trude, P.Der. 1955. *Begriff der Gerechtigkeit in der aristotelischen Rechts- und Staatsphilosophie*. Berlin: De Gruyter.

Von Leyden, W. 1985. *Aristotle on equality and justice. His political argument*. New York: St. Martin Press.

Wimmer, R. 1980. *Universalisierung in der Ethik. Analyse, Kritik und Rekonstruktion ethischer Rationalitätanspruche*. Frankfurt/M: Suhrkamp.

Wittgenstein, L. 1999. *Philosophical investigations* [1958[2]]. Trans. G.E.M. Anscombe. Oxford: Blackwell.

Yack, B. 1990. Natural right and Aristotle's understanding of justice. *Political Theory* 18(2): 216–237.

Yack, B. 1993. *The problems of a political animal. Community, justice and conflict in Aristotelian political thought*. Berkeley: University of California Press.

Zahnd, E.G. 1996. The application of universal laws to particular cases: A defense of equity in aristotelianism and Anglo-American law. *Law and Contemporary Problems* 59(1): 263–295.

# Chapter 12
# Legal Vices and Civic Virtue: Vice Crimes, Republicanism and the Corruption of Lawfulness

**Ekow N. Yankah**

## 12.1 Introduction

The appropriate relationship between law and vice is contentious not only because of the implications for any particular law. It is hotly disputed because competing views are ultimately grounded in opposing views of the law and implicate the very boundaries of legal legitimacy.

For staunch liberals, vice crimes represent illegitimate governmental interference. The contention, in very crude terms, is that government power is appropriate to protect people from "moral" wrongs, the violation of one's rights by another. Law is not appropriately concerned with one's ethical failings or the development of personal virtue.[1]

In contrast, virtue jurists and non-liberal moralists of a certain kind believe that communities have an appropriate interest in the development of good character and a virtuous society through the normative guidance of the law. On this view, that gambling or pornography, for example, inculcate vice are reasons that at very least count towards their being legally prohibited.[2] Thus the opposing views lend a window into critical disagreements about the very project of law.

---

Published in 7 Crim. Law and Phil. 61 (2012).

[1] It nearly goes without saying that much of this is contentious. The distinction offered between moral and ethical is not uncontroversial, though I think conventional. Further, many liberals have importantly argued for forms of liberalism which take into account the development of personal virtues. In political theory, a famous example is Raz (1986). In criminal theory, Anthony Duff (2012) leads the way in finding a balance between the demands of liberalism and communitarian and virtue oriented goals.

[2] Just as liberals strive to find a place for the concern of virtue in a liberal system, virtue theorists have worked mightily to balance the drive of virtue theories with the demands of liberal theories. See Yankah (2009: 1192–1197).

E.N. Yankah (✉)
Professor of Law, Cardozo School of Law, Rm. 504, 55 Fifth Avenue, 10003 New York, NY, USA
e-mail: yankah@yu.edu

L. Huppes-Cluysenaer and N.M.M.S. Coelho (eds.), *Aristotle and The Philosophy of Law: Theory, Practice and Justice*, Ius Gentium: Comparative Perspectives on Law and Justice 23, DOI 10.1007/978-94-007-6031-8_12, © Springer Science+Business Media Dordrecht 2013 

Despite being in opposition these two views share a common and misleading premise of the relationship between law and vice. The view is of personal vices out there in the world, with Themis observing with a cool eye. The only question is the extent to which she may unleash her powers to suppress vices – strike at pornography to retard licentiousness or gambling to repress avarice and sloth.

The preoccupation of legal theorists on the way law can suppress private vice is understandable. Private vices stir our visceral imagination; they are the stuff of childhood morality tales and the seven deadly sins. Yet in our fervor to use the law to attack the vices *within* our neighbors, we miss the way misuse of law disrupts the bonds *between* neighbors. Using legal power to repair private vices misses the way legal power can perpetuate vices within the law. These legal defects undermine a form of virtue more intimate to law and legal institutions – our shared civic virtue. That is to say, I wish to explore the ways in which a legal systems exhibit "legal vices" that can undermine civic virtues.

There are several reasons I have for turning our gaze inward on the law. The first is that the presence of legal vices – particularly, the manipulation of law to serve factional interests – is, by definition, within the purview of the law, making their eradication something of special salience for lawyers, academics, judges and others engaged with the law. Secondly, to the extent legal vices are responsible for the undermining of civic virtues, repairing them cuts across liberal and non-liberal lines.

Most importantly, the dangerous way legal vices undermine civic virtue, leading to political cynicism is an underappreciated danger. As citizens become disillusioned and view law as thinly veiled politics, the reasons to contribute to a robust civic society, share the burdens of participatory governance and pledge fidelity to more than their own immediate interests disappear. It is alarming to see how the weakening of civic virtue breaks the power of the law to command obedience, leads to widespread violations of law and eventually results in lawlessness.

I fear this warning reads as the usual academic handwringing. Yet readers would do well to remember that civic virtues are, in some ways, more fragile than personal virtues. The spectacular "Arab Spring", a still ongoing wave of civil revolutions, was spurred as much by a total collapse of faith in the legal and political institutions of African and Middle Eastern nations as by declining economic conditions. Despite the turmoil of revolution, the lives lost and bodies broken, the Arab Spring represents a rebirth of civic virtue where it had once eroded. In America, the "Occupy" movements evidence the outburst of a long submerged hunger to re-engage in civic conversation. In less hopeful cases, recent riots across the United Kingdom come to mind, the breakdown of civic bonds results only in spasms of violence, crime and lawlessness. Much worse is the paralysis that comes when civic virtue is absent without hope of revolutionary restoration. Those who have travelled through such places, African Kleptocracies spring to mind, know well the damage done to the civic soul when citizens regard the law as a tool of power. At a certain point, legal vices can make it clear to citizens that the law is to be manipulated for personal gain and does not take them seriously as agents, draining the legitimacy, the very lifeblood, from a system. In such a situation, there can be no incentive to hew to the law and an otherwise just man may grow contemptuous of law without even the nagging guilt that accompanies personal vice. A breakdown of civic virtue leaves one with no reason to obey the law. Left unchecked, legal vices threaten the very rule of law.

I will argue that the current view of vice crimes as a legal tool to suppress individual vices is mistaken. An inspection of Aristotle's work, foundational for much contemporary virtue theory, reveals that law must first and foremost foster and protect civic virtue. Nor is this a matter of philosophical exegesis. Surveying examples of current legal vices, that is particular defects within the law, shows that Aristotle's concerns about the rupture of civic bonds is if anything more pressing today.

The most basic civic virtue, one which may be thought of as the underpinning of all civic virtues, is law abidingness. By law abidingness I do not mean to focus on the well explored question of whether there is a *prima facia* obligation to obey the law. Rather I mean to describe something rather broader and perhaps more republican in spirit. Law abidingness describes a dispositional civic virtue, a willingness to generally follow roughly just laws and set aside one's personal interests in order to comply with legal requirements. Law abidingness does not mean that one believes themselves obligated to follow every law nor is it defeated by a singular instance of law breaking.[3] Nor can it be captured by any singular analytic phrase or formulation. Rather, law abidingness is shown in a willingness to submit to the sacrifices entailed in the demands of law rather than leap to personal gain. It is captured in Rawls' sense of reasonableness, Aristotle's republicanism and described in his virtue of justice. It is what we mean when we believe a society has internalized the rule of law. It is, after all, a sense of fidelity to the law.

To isolate this virtue requires tying together different philosophical strands, beginning with Aristotle insights into the way in which citizens appropriately internalize legal norms. Throughout his writings, Aristotle highlights the reciprocal nature of defects in law and vice in society; he is keen in reminding us how misshapen laws can undermine civic virtues.

Rather than law simply standing apart from personal vice, capable only of attacking and suppressing it, legal defects can inculcate certain political vices all their own. These defects are of critical interest because they reveal the unique demands of political morality that govern civic institutions and the project of law-making itself.

## 12.2 Part I: Law, Guidance and Virtue

The brief picture of the inculcation of virtue I want to describe is Aristotelian in nature. To start with the plain, Aristotle's position is at odds with the modern liberal contention that the law is not appropriately interested with the development of one's character. The clear *telos* of the state was to support and preserve virtuous citizens and a flourishing society (*Pol.* I. 2. 1252b29-30, Benjamin Jowett, trans.[4]; *see also* Cooper 2005: 72–74).

---

[3] An outstanding explication of this concept can be found in William A. Edmundson (2006).

[4] All future references to Aristotle's *Politics* refer to the same translation.

For Aristotle law plays a critical role in the development of virtue and a flourishing society. Large, if somewhat unexciting, portions of *The Politics* are explications of the ways different political systems are conducive to the formation of different virtues of the citizenry (*Politica*: VII and VIII). Since the education of the young is essential to the continued health of the political constitution, civic education was the business of the state (*ibid.*: VIII.1; VI.6). Legal and political systems, then, play a key role in shaping both excellent personal character and the right relationship between citizens and the law. Aristotle noted, quite sensibly, that a critical component in absorbing moral virtues was habituation to the right values in one's youth:

> It is difficult to get from youth up a right training for virtue if one has not been brought up under right laws; for to live temperately and hardily is not pleasant to most people, especially when they are young. For this reason their nurture and occupations should be fixed by law; for they will not be painful when they have become customary. But it is surely not enough that when they are young they should get the right nurture and attention; since they must, even when they are grown up, practiced and be habituated to them, we shall need laws for this as well, and generally speaking to cover the whole of life; for most people obey necessity rather than argument, and punishments rather than the sense of what is noble (*EN* X. 9.1179b32-1180a4, W.D. Ross, trans.).[5]

This manner of inculcating virtue is perfectly commonplace. Children first learn not to take things from their siblings and playmates because Mom and Dad say so. They are punished when they break the rules, typically accompanied with a scolding with its embedded moral lesson (Burnyeat 1980: 71–73, 78–79). Rules are learned as rules; only later do children truly internalize the values they represent (*see also* Yankah 2009: 1177–1178).

Likewise, legal rules and their associated norms guide us both in learning the right ways to pursue our ends and, on the strongest view, which ends to pursue at all. Individual legal prohibitions can be seen as commanding the right moral acts, training us to behave correctly thus inculcating virtue (*EN* V.1.1129b20-24). So rightfully framed legal prohibitions on assault, adultery, fraud and the like, exist in part to stamp out personal vices and impart moral virtues through habituation.[6]

Of course many moral traits are foremost experienced as moral rather than legal rules (Huigens 2002: 104). But it is also true that legal norms typically inform the boundaries of acceptable behavior, especially where there is no obvious morally correct action. Further, the expression of many character traits are either derived from or supervene upon legal rules (MacIntyre 1985: 150). When I encounter a complex problem of how to treat others justly, I do not typically start with my own conception of justice divorced from legal and communal standards (Kraut 2002: 270–271). The amount of information I need to fairly disclose in a complex financial deal or the amount of waste my factory may produce will often be importantly informed by legal

[5] All future references to Aristotle's *Nicomachean Ethics* refer to the same translation.

[6] For an excellent example of the way criminal prohibitions contribute to and turn on virtue, see Kyron Huigens (1995: 1444–1449). In subsequent work, Huigens has moved away from virtue *qua* virtue and has disavowed his earlier republicanism and focused on an Aristotelian based view of practical reasoning instead.

standards (*ibid.*: 124–125; Edmundson 2006: 37–38). Thus, law provides normative guidance not only in the particular actions but in "setting parameters within which individuals carry on together their social interactions, pursue their individual and common projects, and the means for repairing the relationships when things go wrong" (Postema 2010: 1856). In doing so, law provides the standards by which we evaluate our actions and hold accountable the actions of others in our community (Kraut 2002: 124–128).

Though law informs our view of the right thing to do one rarely races to a law library to research what legal regulations apply to our every situation. Outside of complex endeavors, we relate to legal norms in a rough and ready way; we look around us and often intuit, quite correctly, the applicable legal standards (Edmundson 2006: 13–14). Aristotle observes this feature when he points out that the normative guidance provided by law is not restricted to statutory law. Rather, the virtuous person is shaped by a mixture of the laws in their community and the surrounding legal and social norms which inform them; the broader term for this being the *nomos* (Kraut 2002: 105–106; Solum 2006: 89–90). Thus, the instruction of the positive law, which in turn informs and is informed by social rules and norms, all combine from when we are young to shape our character and habituate us to develop personal virtue.

Central to the development of civic virtue is a particular intellectual virtue, *phronêsis* or practical wisdom (*EN* VI.1). Phronêsis is first the ability to see clearly the manner in which knowledge and action must be combined to reach a desired end. Phronêsis is partly the excellence of means-ends reasoning; it captures the common sense that the cinematic hyper-intellectual scientist lacks and the hero uses to save the day. Phronêsis applies no less in morally sticky situations; it is the virtue that some display when they pick out the important variables in a confusing situation, those who always know the right thing to say or do. Phronêsis, however, requires more than means-ends reasoning. Phronêsis includes a capacity to reflect on the quality of one's ends and to direct one's intelligence towards worthwhile goals (MacIntyre 1985: 149–154, 162).

The intellectual virtues are not independent from law as intellectual and moral values inform each other. One needs a certain amount of intellectual acuity and practical wisdom in order to truly understand the rules, both moral and institutional, which train and guide moral development; that is, to do the right thing for the right reason. Developing insight as to the right reasons to behave virtuously is the difference between internalizing virtue and becoming an unthinking automaton. Likewise, using one's practical wisdom to pursue ends regulated by a well formed character is the difference between showing practical reason and merely a cunning ability to get what one desires.

The picture drawn here has been all too cursory because it is familiar and intuitive. The law inculcates virtue by requiring virtuous actions and prohibiting acts which foster vice. By inculcating legal propositions and their attending social norms (*nomos*) we habituate ourselves to virtuous acts and worthwhile ends. Virtuous character helps us develop excellent traits of thinking, disciplining our thoughts and suppressing the temptation to reason in a self-serving or biased manner. Reciprocally, excellence in our intellectual development helps us develop the theoretical thinking (*sophia*) necessary to determine our legal responsibilities and the practical wisdom (*phronêsis*) to know how and when to pursue and apply them. Taken all together, the idea is that law can guide men to greater personal virtue.

## 12.3 Part II: Relating to Law and Virtue

The deepest modern controversy surrounding the above picture concerns the very legitimacy of political efforts to shape character and its compatibility with a neutral liberal state. Yet even on its own terms, this picture seems to ignore an obvious danger. Even were it a legitimate aim of the state to mold moral people, what assurance have we that the legal system would get it right? It takes no imagination to bring to mind countless despicable legal regimes which have corrupted their citizens; antebellum America decreeing the lack of humanity of enslaved Africans, Nazi Germany aiming at the extinction of Jews and other non-Aryans, Apartheid South Africa, the Jim Crow South and on and on and on.

Aristotle paid careful attention to this when describing the proper relationship between law and virtue. Though the ability of law to instill particular virtues was clearly of importance to Aristotle, more important was whether each citizen related to law and virtue in the right way. One had to access the deeper reasons the laws were virtuous and fostered a flourishing society. To think that the well constituted citizen will be susceptible to any ethical manipulation by law, no matter how vile, is to foist on Aristotle an image of a passive receptacle that he explicitly rejected (*EN* VI.13.1144b14-1145a6, *see also* George 1993: 19–47).

Recall it was earlier noted that a fully capable citizen was not just *nomimos* in that they would internalized the legal and social norms of her community but also *phronimos* in that they possess some measure of the practical wisdom necessary to reflect on the choice-worthiness of those ends. As virtue theorist Lawrence Solum explains

> [I]t is a condition for a norm to count as a *nomos* that the norm must be such that it could be internalized by any fully virtuous human… For short, we can might say that for a norm to be a *nomos* it must be such that it could be embraced by the *phronimoi*—by those humans in full possession of the human excellences. Social norms or positive laws that clearly hinder rather that enable human flourishing could not be internalized by a fully virtuous agent who has grasped the *telos* or proper end that *nomoi* serve (Solum 2006: 104–105).

Aristotle's analysis reveals why the laws of antebellum America and Nazi Germany would be recognized by a virtuous citizen as directly hostile to the flourishing of persons, destructive to a flourishing society and thus incapable of being internalized as a law (*EN* III.1.1110a4-27; III.4.1277a27; *see also* Kraut 2002: 114–118).

Citizens must approach the law with practical wisdom to ensure they are not corrupted with the wrong values. This does not mean that wise citizens simply choose to follow individual laws they like or dislike. After all, the rightly composed citizen is, on this picture, nominos, and thus predisposed to follow the law. The wise citizen knows that some portions of every legal system will be rough, unattractive or conflict with their preferences. Minor defects do not erase the fact that a stable and healthy legal system on the whole secures great goods for the entire community. But the wise citizen does not uncritically swallow whatever values their legal system enshrines.

Aristotle's refinement gives us a perfectly plausible but strangely unsatisfying answer. The best way to discern which of the moral values in your legal system you should absorb is to be wise about which moral values are in fact virtuous. This typically Aristotelian answer is surely true but unhelpful. How can a society be certain that its laws instill the right personal virtues in the population?

Aristotle's answer is short and to the point. We can't. In an ideal world, the law of a political system enshrines the best virtues needed for its citizens flourish in their society and the citizenry to internalize these virtues for the right underlying reasons. One need not be a philosopher to realize that the percentage of people in any group or nation who are ideally virtuous – brave, even-tempered, thoughtful, morally and financially prudent – are a small group (*EN* X.9.1179b4-b19; *see also* Burnyeat 1980: 75–76). Even after we adjust for Aristotle's culturally determined and frankly ugly views on the inability of women, natural slaves and others to be truly virtuous, we can recognize the truth in the claim that only a small portion of any population will exhibit full virtue. Aristotle reveals that our current cynicism is not particularly modern, arguing that most people are simply too impulsive or hedonistic and lack a natural sense of "shame," only "abstain[ing] from bad acts because of their baseness but through fear of punishment" (*EN* X. 9. 1179b11-13). The best of arguments and role models are "not able to encourage the many to nobility and goodness" (ibid. at 1179b10-11). Worse yet, by the time the majority of people have grown to adulthood, it is too late to change their formed character (*EN* X.9.; *see also* Burnyeat 1980: 81). Aristotle presages the truth in the old island saying, "by the time it is grown, the tree will not bend." Those who are not fortunate enough to possess noble loving characters by nature or trained to virtuous actions can only mimic the wise and virtuous. By following the lead of the virtuous, they may be left with some "tincture of virtue."

Despite the fact that much virtue theory is rooted in Aristotelian soil, Aristotle himself rejected the urge embedded in modern vice crime legislation. Law might give people a reason to avoid bad acts because they feared punishment but it could not be counted on to instill true virtue in those who did not already seek it.

Ultimately, I believe little of this is controversial. Even the most ardent modern champions of morals legislation admit that law cannot truly instill virtue (George 1993: 25–47). But while they admit that vice crimes at best protect the background moral culture, they are happy to mobilize the power of the state to chase after that "tincture of virtue." Even if legislating virtue were legitimate, it is unlikely that criminal law can instill moral virtue into the general public. Worse, using the law to instill private virtue distracts from the public virtues that law can successfully instill and the dangers that accompany their erosion. What Aristotle concludes is a profound lesson for modern debates surrounding vice crime and beyond. Instead of focusing on instilling private virtues, Aristotle insists that the primary goal of law is to secure public virtues. It is civic virtue that is most intimate to the way a citizen relates to their political community and to their legal system. Most importantly, it is civic virtue which supplies the ordinary citizen with reasons to obey the law. In arguing for us to preserve civic virtue, Aristotle makes an astounding and powerful claim. All at once he reveals the mistaken focus in the modern debate surrounding vice and crime, the importance of law focusing on civic virtue and the dangers of the erosion of civic virtue to our practical reasons to follow the law. Ultimately, Aristotle argues most convincingly for the priority of civic virtue.

## 12.4 Part III: The Priority of Civic Virtue

If personal virtuous character was too intimate and difficult for law to mold, Aristotle did think the law could secure public or civic virtue. Private virtue demands a level of success in integrating all of the admirable character traits and beyond the reach of law. Civic virtue, on the other hand, requires only that a citizen do their part in supporting the welfare of the community and is more easily attainable (*Pol.* III.4.1276b17-1277a32).

To some, it may seem counterintuitive that civic virtue is easier to attain than private virtue. Many find it easier to exhibit concern towards family and friends than concern for the abstract mass of "fellow citizens." While there is surely some truth to this, the objection miscasts the relationship between personal and civic virtue. Private virtue is not necessarily exhibited because one finds it easier to prefer those near and dear. Rather, private virtue is exhibited in the highest order when one *appropriately* values projects and persons dear to them. The highest degree of virtue, as Aristotle describes it, requires the integration of all the moral virtues in balance. So while a father quite appropriately loves and favors his daughter, his pushy insistence that she is always the captain of the team or the lead in the play does not strike one as virtuous. That his adoration causes him to grant every outlandish request or to manically drive her in chess at all costs equally fails to exhibit the integration of virtue. It is the need for internalization and integration of this moral balance that led Aristotle to be pessimistic of the law's ability to instill virtue.

In contrast good citizens need not be perfect in use of practical reason, they need only exhibit enough civic virtue to contribute to society in their public lives. Civic virtue falls short of the perfect personal constitution of virtue for which legal moralists aim (*EN* X.9.1179b18-20). While civic virtue requires possessing enough practical wisdom that one can be thoughtful about the law and takes turns contributing to governing, it does not require the unity of virtue required to exhibit private virtue. The law cannot and need not make everyone a good person but it must aim to make everyone a good citizen (*Pol.* III.4.1277a1-4; Kraut 2002: 364; *see also* Gararella 2009).

Considering the number of neo-Aristotelians who draw upon him for support of legal moralism, there is some irony in noting Aristotle's explicit prioritization of public virtue as the appropriate aim of legal intervention. Aristotle focused on the preservation of civic virtue not merely because it is more easily affected by legal norms.

Aristotle argued that it is not the person who exhibits internal virtue who is best but the man who exercises the virtue of justice towards others (*EN* V.1.1129b25-1130a14). Failing to exhibit internal virtue is a weighty matter, still that is a matter of falling short of excellence in one's personal life. A failure of appropriate civic concern, in contrast, will oft lead to our wronging another person or harming our community (Kraut 2002: 121). Virtue, Aristotle intones, is most completely found in ensuring that one treats others within your community with due concern.

Indeed, there are surprising moments when Aristotle appears to be almost a modern liberal. Aristotle is keen to point out that though weak-willed persons may corrupt their character, this corruption does not injure the community and thus is a private concern. Aristotle's message is that law should not focus on the elusive ideal of personal virtue but rather law is best able to and ought be most concerned with promoting civic virtue (*EN* V.1.1129b25-1130a14; *see also* Kraut 2002: 265–267, 275, 378–379).

Ultimately, Aristotle prioritization of civic virtue serves as a stark warning in today's debates about the reach of law. Setting aside concerns with legitimacy and efficacy, Aristotle realized that the temptation to focus law on private virtue distracts from the primary political duty of scrupulously guarding our civic virtues. This danger is paramount. Because civic virtue is a certain relationship with one's political community and legal system, giving in to the temptation to misuse law in one area can spread and insidiously weaken our commitment to law in another. Damaging civic virtue threatens the overarching fidelity that citizens reserve for the law and risks having law become a patchwork of fractured interest groups. A lack of civic virtue, Aristotle warns, threatens the great communal goods secured by a stable and orderly legal system. Eventually, the absence of civic virtue undermines the rule of law and leaves citizens with no reason to obey the law at all. One need not scan our modern political debates long to notice the gripping urgency of this warning from the past.

Having set aside the pursuit of personal virtue as the object of law, let us turn to the public excellence described as civic virtue. For Aristotle, civic virtue was not simply a public good, it was a virtue which was good for the individual (*EN* I.7.1097b11; *Pol.* I.2.1253a2-18; *see also* Huigens 1995: 1444–1449). Because human beings are on the whole deeply social creatures, relying on each other for not just material but social and emotional support, it is in our nature to pursue life in common (*Cf.* Cooper 2005: 65–70). Human beings rely on each other for more than the mutual advantage common to today's preoccupation with homo economicus. We are not simply rational agents, we are moral agents with thick and rich ends, both individual and shared (*Cf.* Gararella 2009: 44). Only a completely denuded view of social interaction could ignore that we grow together as communities to pursue a more robust vision of the good life, form friendships and partnerships and care for emotional and social needs (*EN* VIII.12.1162a16-19; *Pol.* I.2.1252b29-30). We are interconnected in complex ways which ground communal (and political) identities. Many of us cheer for our teams during the Olympics. We are saddened and angered when we learn of some injustice done by our government in a way that is different than if the same tragic event was committed in someone else's name (*See*

Cooper 2005: 65–66, 72–73). Of course there are some rare individuals who wish to spend the vast majority of their lives in complete solitude.[7] But these individuals attract our attention precisely because they are rare and when taken to its limits such behavior can be evidence of deep emotional trauma or mental illness. On the whole human beings flourish as part of social and political communities.

If one accepts what I take to be this uncontroversial point, then it is easy to see the next step. In reviewing Aristotle's vision of shared governance, we must once again set aside his disturbing omission of non-citizens; barbarians, natural slaves, women and the working class. What is important is whether his view of what was the good life for those he did consider citizens is compelling. If being a part of human society is a part of the good for persons, then participating fully in society must be part of that good. Part of this participation is to have a voice in the affairs of your society, in its structuring and its governance and even taking turns as public officials (*Pol.* III.4.1276b36-1277b34). This communal good is realized in sharing the joint burdens and sacrifices of participating in that society. Notice on this view there is no tension between civic virtue defined as participating by taking turns in government and civic virtue defined as obedience to the law. Because participating in government and obeying the law are both ways one contributes to the well-being of one's polity and ensures the voice of government is a shared one, both are facets of the civically well-constructed citizen.

Civic virtue finds its foundation in the inherent social nature of people and begins in a sense of shared participation in our political communities. Describing the precise nature of participation captured by civic virtue, however, is somehow trickier. Most of us have a vague sense of what someone means when they trot off to jury duty, mumbling something about doing their civic duty. Nailing down this shared sense though remains difficult. We first noted that Aristotle viewed having a share in the political and governing class of your community as part of the good for every individual. It is easy to understand that a shared voice is constitutive of citizenship, a way of conveying the equal respect and concern for the views and goals of each citizen (Kraut 2002: 362; Hampton 1998). The obvious compliment to sharing in the benefits of holding office is sharing in the burdens of governance; indeed in the overwhelming majority of unglamorous public offices these can be the same thing. It is this sense we understand when someone refers to onerous jury duty as part of civic duty.

Aristotle saw the willingness to share in the burdens of communal governing as a critical feature of civic virtue. He describes the unjust person as grasping, motivated by wanting more than their fair share of benefits and less than their fair share of burdens. The civically motivated citizen is just partly because she views herself as owed fair consideration by her community and as obligated to contribute her fair share to its health. Civic excellence is partly constituted by understanding that the

[7] Here, I avoid for the moment the engaging debate surrounding the tension between Book X of the *Nicomachean Ethics* and much of what comes earlier in the text. See *EN* X.8. What is clear is that in contrast with Plato, who viewed engagement in politics as sullying the philosopher, Aristotle views political engagement as part of the good for man. See Hitz (2011); Kraut (2002: 132–133).

well-being of other members of your community is of equal public importance as your own (*see* Cooper 2005: 74–75, 78). While this obviously applies to the material benefits, it is clear that Aristotle does not limit our civic bonds to only economic goods. Civic excellence means understanding that one has both the right to contribute to lawmaking and governance but is also duty bound by legal norms and sometimes (perhaps often) must submerge individual interests to that of the whole (*see* Kraut 2002: 102, 265–267, 275).

Ultimately civic excellence, in people and institutions, is evidenced in caring and contributing to the public good ahead of the interests of the individual or their own small interest group. Civic virtue is the realization that kin, tribe, race, political party and lobby group are owed a share of political concern but no more than a share. So understood, laws and legal institutions which display civic excellence will be those that aim at the shared health and common good of the entire political community (*Pol.* III.6.1279a17-21, 1284b4-19). "[T]rue forms of government [ ] are those in which the one, or the few, or the many, govern with a view to the common interest; but governments which rule with a view to the private interest, whether of the one, or of the few, or of the many are perversions." (*Pol.* III.7.1279b28-32)

Because civic excellence is exhibited in citizens, laws and legal institutions which aim at the common good, it is naturally when laws aim at securing private advantage that civic virtue is undermined. When citizens try to game the system to hoard outsized shares of social benefits and elevate their own good above the common good, they disrupt the civic bonds that hold us together. This is particularly clear in legal and political settings, where office holders seek personal gain for themselves, their family or special interest group. Of particular concern, no less in Aristotle's day as in ours, is when laws are used as a tool of class warfare, particularly where the wealthy and powerful use the law to dominate the poor and advance their exclusive interests. Indeed Aristotle reminds us in stark language that the civic bonds that hold us together are under greatest strain where systemic and institutionalized gaps exist between the rich and the poor and the middle class is diluted (*Pol.* IV.1.1295b2-1296b2). Still, it is important to note that the civic concern owed each citizen was not limited to economic wellbeing. Economic wellbeing is only one aspect of the large project of law, to ensure that each citizen had the opportunity to flourish within society and to contribute to the project of joint governance. Aristotle cautions that legal practices and institutions ought to be infused with a spirit of equal care for the interests of all and constructed to avoid entrenched power and the naked pursuit of the interests of a few factions (*Pol.* V. 8.1307b37-1308a24).

Though I cannot undertake a full defense of republicanism, two points deserve attention. For some, civic virtue raises the specter of communal concern that is insufficiently shared; recall that while Aristotle believed the bonds of civic friendship bound all citizens, his view of who could be counted as a citizen was shamefully impoverished, excluding foreigners and "barbarians," the working class, natural slaves and women. Likewise, there are those today who speak of "the undeserving poor," "those type of people," etc., as a way of denoting those imagined to be outside our civic concern. Further, one might worry that a theory of civic virtue too closely relies on the kind of bonds prevalent in a small homogenous society hostile to outsiders. Even at its most expansive, civic virtue will leave some outside its concern; obligations of global justice, for example, cannot be accounted for by civic virtue. These concerns lead to believe it is better to focus on some broader values such as human rights.

It is important to grant that civic virtue cannot capture all of our important political duties. Civic bonds have boundaries, indeed, that is part of their intrinsic value. Those outside our political community possess human dignity and we have duties of beneficence and charity towards them. Nonetheless, it is true these duties are not grounded in civic virtue. Yet it would be a mistake to conclude that civic virtue is an exclusionary concept. Just as modernity has expanded our conception of what can constitute the good life, so too we have expanded our conception of the fellow citizen. The very point of civic virtue was that unlike the rarified standards of personal excellence or the intimacy of personal bonds, a sufficient level of civic virtue is attainable by and owed to all. Rather than stand as the exclusionary ideal represented by personal virtue, civic virtue is inclusive enough to be shared by all citizens.[8]

At the same time, civic friendship offers a richer conception than that of rigorous liberalism, for civic virtue highlights duties that run both from society to the individual as well as from the individual to society. Whereas liberalism can be satisfied when individuals respect their duty of non-interference, a republican is aware certain civic harms call for redress by any and all members of society and civic duties apply to all in virtue of our being bond together in a community. Republicanism highlights the missing element in the liberal sentiment, "Why should I fix this? It wasn't my fault!" and rejects the liberal sentiment that all political duties must be ground in something akin to the harm principle. Classic republicanism provides an answer for why one ought to participate in the Parent Teacher Association even when it cannot be justified by cost/benefit analysis. The republican idea that we live together, not merely beside each other, provides a distinct political justification for claims of distributive justice that do not rely on tortured liberal rationales that one must somehow be causally responsible for social needs (Lacey 1988: 171–176; Duff 2007: 44–56). The well-constructed citizen realizes that over pronounced stinginess in matters of distributive justice is yet another facet of parochial factionalism, a trait that undermines civic virtue across legal domains – from tax policy to criminal law.

On the other hand, because it is thicker than many forms of liberalism, some may fear that republicanism is too thick. The philosophy 101 version of this fear goes something like, "aren't you a communist?!?" More subtly, how can one ensure the republican ties that bind do not choke out liberty?

---

[8] It cannot be denied that even on this picture civic virtue is exclusionary because it extends only weakly to those outside our polity. It is important to repeat, however, that those outside our civic community retain a broad array of human rights. Further, even Aristotle's conception of civic rights maintained a minimum amount of regard for the well-being of those who fall outside of citizenship, a concept I hope to develop further in the future (*Pol.* I.6.; III.6.). I am grateful to the referee for pressing this point.

This concern has been at the root of modern civic republicanism which takes as its primary value the citizens' freedom from interference or domination (Skinner 1978, 1984; Pettit 1997a, 2002). While I have adopted the more classically influenced "civic humanism" which rejects non-domination as a singularly important political value there is clearly much overlap between these variants of republicanism. Remember that one of the critical distinctions between personal and civic virtue is precisely that civic virtue demands far less of the citizen. On any reasonable conception, civic duties must leave space that exhibits concern for each citizens' individual projects, be they selfish, silly or wrongheaded. On Aristotle's picture, it was important in a healthy polity to maintain personal room for citizens to pursue their own projects and to view personal projects or those of others near and dear as more important or intimate than the community's needs (*EN* VIII.9.). The picture Aristotle invoked is that of many different sailors on a ship; each is able to pursue their own project and their own task so long as he contributes his fair share to the common duties. It is important to note that classical republicanism does not view the state as a unity in which individual interests must be drowned out but rather as a cooperation where joint interests are pursued vigorously so long as self-serving does not subvert one's duties to the common good. This is why Aristotle rejects Plato's arguments that citizens merge their personal projects, including their wealth and turning over their very children for communal parenting, in order to erase the boundaries between them. Aristotle understood that the ability to favor those dear to you and to give time and energy to those projects most meaningful to each individual are important parts of civic flourishing. Thus we need not surrender to G.A. Cohen's worry that republicanism requires one to "level down" the entirety of one's excess wealth.[9] Civic virtue only requires that each does their fair share and refrain from warping laws, legal institutions and politics into weapons to pursue their selfish or parochial interests.[10]

Because civic virtue is first and foremost care for the common good and a willingness to share the benefits and burdens of society, it is destroyed by citizens, laws and institutions which seek parochial advantage above all (Dagger 2009: 150). A society where factions use the law to institutionalize discrimination or cynically take advantage of other groups cannot long hope to keep alive the bonds of civic friendship. This, one might think, is a grave enough danger. But to leave this picture here would be to still underestimate the threat. Where civic virtue erodes, citizens realize that their wellbeing is no longer of equal concern. Without equal concern for

[9] Though I do think that Cohen's intuition that there is republican virtue in feeling compelled to contribute more personal wealth to civic society when one understands that the polity's distributive justice scheme is plainly unjust is entirely correct (Cohen 2008).

[10] The exact requirements of civic duty will obviously differ from community to community, country to country. At various times in countries such as Israel and Germany, all or some citizens have been required to serve in national or military service for a year. In New Zealand, citizens are required to vote by law. While I would not import any particular practice wholesale into the United States, it is critical to note that both these examples show how different communities require civic duty while ensuring that citizens maintain ample space to pursue their life plans.

the common good citizens have no reason to submit to the law and bear their share of the sacrifices of governing. The erosion of civic virtue threatens the very rule of law leaving citizens no reason to obey the law.

## 12.5 Part IV: Civic Virtue and the Obligation to Obey the Law

The same concern that leads a citizen to understand that he must sometimes temper the naked pursuit of his own interest in deference to the common good results in a willingness to abide by the community's legal norms in regulating that pursuit. This is because the law is one of the focal ways in society produces stable norms by which we can all arrange our affairs and organize our conflicts. In so doing the rule of law, even when imperfect, allows us to secure many great social goods (*EN* V.1.1129b24-25). The very fact that a legal system exists most often promotes the good of a community, thus, citizens with civic virtue have reason to adopt a supportive and deferential attitude towards much of the law. In the normal thoughtful case, attitudes of law abidingness will be embedded in the same civic virtue which fosters and promote the common good and each citizen's part within it (*Cf.* Kraut 2002: 117–118, 271; *see also* Edmundson 2006: 26–28).

Conversely, where law and legal institutions have become implements of raw factional politics, one's reasons to obey the law disappear. Indeed, Aristotle saw the use of law to pursue advantage for oneself, family or interest group as tyrannical (*Pol.* IV.4.1292a4-37; IV.6.1293a30-34). In such cases, the law can go so badly corrupt as to issue orders that no good person could possibly obey (*EN* III.1.1110a4-27; *see also* Edmundson 2006: 22). Again, the claim here is not that any disagreement or minor bias within the law allows one to pick and choose which laws to obey (*EN*VIII.9.1159b26-27). But when powerful factions shape the law without even regard to the well-being of those outside their group, according only as much as needed to prevent outright revolt, the project of law giving which commands respect has failed (*see also* Kraut 2002: 116–118; Waldron 1994: 259–264).

This is why Aristotle so feared deep-rooted factions. When factionalism grows too deeply entrenched, competing groups cease to think of themselves as committed to governing for the common good and the bonds of civic trust cannot remain (*Pol.* IV.1.1295b2-1296b2; V.9.1309b18-35). Aristotle warned against allowing a permanent underclass which would feel it could only advance by committing crimes or manipulating the system to siphon off from the rich. More likely, he warned against allowing the wealthy and powerful to suppress and exploit the poor and bend the law to their own advantage. Further, Aristotle understood that where the happiness and flourishing of some parts of the population was purchased by exploiting and limiting the success of other portions, civic virtue was doomed. He was concerned that some citizens could grow so wealthy and powerful that they were, for all practical purposes, no longer governed as ordinary. It takes little imagination to think of today's ultra-wealthy who exercise great political influence and all too often seem above the law.

In an environment rife with factions seeking partisan interest "rulers grow so contemptuous of others that they do not provide the city with a general and stable legal framework of legitimate expectations...[In such a system,] the law has no authority, and it is the individuals who rule" (Kraut 2002: 106, 276). Law becomes merely a tool for sustaining the power of the most powerful faction. We intuitively understand how factionalism and its attendant lack of respect for other groups leads to the corruption of the law. In Fuller's classic, The Morality of Law, he reminds us that while the law can go wrong in many ways, its defects have a common trait. The important thread throughout these defects, Fuller reminds, is that they each in some way show insufficient respect for the citizen who is attempting to plan her life. When laws command that something be done and at the same time orders the punishing of those who do that act, the law fails to play its primary role, providing normative guidance for citizens who are trying to plan productive and successful lives (Fuller 1964: 36–40, 65–68, 82–91; Waldron 1994: 266–267). Rulers who have no basic respect for the planning needed to live autonomously have no reason guard the law against internal defects.

A political community that no longer considers one's interests and uses law as simply an instrument of power is owed no fealty (Waldron 1994: 278, 280–281). Legal requirements are often controversial and costly. They benefit some more than others and rarely gain universal accord. As Jeremy Waldron points out, in the pursuit of any particular goal, the government will surely discard the preferred goals or methods of some. Legal goals will require sacrifice from many, only some of who are independently motivated to do so. For those who have misgivings about any legal goal, the fact that the government has taken their interests into account, that laws make space for their life projects and that their interests may win out in the next case may be the only thing which secures allegiance to the law. Law which is used capriciously or does not pay one's group interests due consideration cannot ground their fidelity.

While this conclusion is perfectly intuitive, it is a republican perspective which best highlights the connection between civic virtue and being bounded by law. Anthony Duff, whose important work in criminal law straddles republican and communitarian foundations, underlines this point, noting that where groups of individuals are systematically excluded from participation in political life, "the law sounds to them as an alien voice... [and] the claim that they, as citizens, [are] bound by the laws and answerable to their community becomes a hollow one" (Duff 2001: 195–196; 2007: 49–56). Where the law serves factional rather than common interests, it is suffers from a particular sort of vice which leads to legal alienation and ultimately undermines the citizens' understanding that they are bound by the law (Gararella 2009: 179–180). The realization that legal institutions are instrumental tools of power undermines law globally; the belief that law is to be used to consolidate power and advantage is one that necessarily casts doubt on the very point of legal obedience.

Weakening civic virtue and the accompanying fidelity to law is a dangerous and infectious disease. Aristotle reminds us that fidelity to law and a willingness to sacrifice for shared legal goals is a political body's most precious and fragile resource, "In all well-tempered constitutions, there is nothing which should be more jealously guarded than the spirit of obedience to the law, especially in small matters; for transgression creeps in unperceived and at last ruins the state" (*Pol.*

V.8.1307b30-34). Thus the eroding of civic virtue drains the willingness of citizens to obey laws they dislike specifically and the law generally. The point here is simple. As we undermine civic virtue, law becomes simply a tool for the pursuit of crass interests. But citizens cannot be fooled forever and will eventually come to realize that the common good has been pushed aside and their interests are not being considered (*ibid.*: IV.1297a13-1297b28; V.8.1307b30-34). Without being grounded in a pursuit of the common good, law increasingly becomes little more than an instrument to be used to serve one's own purposes, whatever they may be (*See* Fuller 1964: 92). With such an instrumental view of the law, citizens eventually discard the civic virtue needed to be restrained by law at all.

Some readers may find such warnings either overly abstract or melodramatic. Yet examples of widespread breakdowns in civic virtue leading to gross legal defects and the eventual collapse of fidelity to law are sadly not distant. Take the odious *Plessy v. Ferguson*, the Supreme Court holding that a Louisiana law that established "separate but equal" accommodations for African-Americans as opposed to White Americans did not violate the equal protection clause of the Constitution.[11]

It should be obvious that a legal regime which institutionalized separate statuses is the archetypal danger earlier explored. It is entirely predictable that this purposeful destruction of civic bonds to exclude some citizens and preserve the power and advantages of others would lead to an immoral warping of the legal system. There is no serious argument that the separate but equal doctrine of the Jim Crow South represented anything equal or lived up to the legal principles enshrined in the 13th and 14th Amendments (Lyons 2008: 36–42). The extent to which judges could convince themselves that their legal analysis was anything less than a sham only reveals the intellectual corruption Aristotle warned of when one is driven by vicious emotion and prejudice (*Pol.* III.9.1280a14-16; *see also* Kraut 2002: 144).

The glaringly legal hypocrisy of the Jim Crow South shows the emergence of vice within the law as a result of eroding civic virtue. What is striking is the way this warping of civic virtue led to the collapse of the obligation to follow the law. Even if one could imagine the impossible, that judges were genuinely confused about whether the dingy restrooms, dilapidated hotels and crumbling schools reserved for Southern blacks could be described as separate but equal, what was clear was that countless systematic acts of violence and degradation aimed at blacks were clearly illegal even by the lights of Southern laws (Lyons 2008: 29–33). Blacks were beaten, stolen from, kidnapped, raped and lynched with tragic regulatory; such actions occurred with the full knowledge and often the active assistance of legal officials. Police officers were all too often the (un)official arms of terror protecting the social fabric by terrorizing and humiliating black citizens. Legal officials not just corrupted the law to serve their own purposes but ignored it all together when it interfered with using power and violence to enforce racial superiority. I suppose some could seek cold, cold comfort in the conclusion that such lawlessness was localized to the treatment of African-Americans. Even were this true, it is startling to realize how quickly hypocritical law leads to the absence of lawfulness at all.

---

[11] 163 U.S. 537, 540 (1896).

While I cannot pretend to have an exact causal story, social scientists and legal scholars are increasingly focused on disentangling the complex reasons people have for not complying with the law. The picture is far from clear. For example, there is no evidence that African-Americans flout drug laws, a source of wide racial disparity in the criminal justice system, more than Caucasian-Americans (Yankah 2011). One might question, however, if these finding obscure the damage done precisely because so many across racial lines already view the drug laws, particularly the marijuana laws, with such suspicion that they are widely disregarded. Further, there is disturbing evidence that as citizens increasingly view the ways laws are enforced as tools of factional power, they comply less with the laws (Jackson et al. 2012). Just as importantly, lost respect for law is evidenced not simply in a decrease in compliance, after all one may not want to use drugs or steal, but in a decrease in cooperation with police and legal officials (Tyler and Fagan 2008). It is startling that in many communities respect for the law is so low that cooperating with the police marks one as a sucker, a snitch and a traitor to the community. Lastly, it strikes me as a dangerous and corrupt hope that widespread disregard for law can be forever cabined. Experiences with other regimes which have shown such failures in civic virtue – Apartheid South Africa and Nazi Germany, for example – give no reason to "hope" that lawlessness can be so easily contained.

Law which secures the advantage of small factions within society while systematically ignoring the interests of others undermines civic virtue and eventually lacks the ability to command obedience at all. The restraint of powerful individuals and factions in society comes not simply from particular laws which check their power but from a wider culture of law abidingness and legal accountability (Postema 2010: 1852–1853). If Brian Tierney's observation that "in the last resort, if a people become corrupt, its laws will be corrupted," is moving, the total disregard for law in Jim Crow South reminds us that the opposite is equally true (Tierney 1963: 395). As Gerald Postema reminds, "When law is a convenient instrumentality of the ruling, but only that, it imposes no constraint reflexively on those who wield the instrument" (Postema 2010: 1857). Simply put, the failure of civic virtue eliminates the obligation to obey the law and destroys the rule of law.

## 12.6 Conclusion: Civic Virtue, Legal Vices and Lawfulness – Legal Obligation and the Criminal Law

I fear I have gone the long way about to reach a rather simple conclusion. One would hardly think it worthwhile if the debates surrounding a number of topics in the legal academy – vice crimes, state neutrality and the limits of liberalism – had not largely ignored these conclusions. Despite the fact that modern virtue jurisprudence has often used Aristotle's views as philosophical foundation, those virtue theorists have long ignored Aristotle's persuasive and tempered conclusions.

Aristotle cautioned that using law to widely instill private virtues was a doubtful project. Instead, he urged that the law be primarily concerned with civic virtue. Unlike the rarity of extraordinary virtue, the achievement of civic virtue, an appropriate level of concern for your community and willingness to sacrifice a fair share in supporting it, was not only attainable by the ordinary citizen but was essential to the health of a functioning political society. Most importantly, Aristotle recognized the mutually intertwined nature between civic virtue and biased and corrupt law. Because civic virtue is grounded in seeking out the common good and sharing in common sacrifice, the erosion of civic virtue results in increasingly disparate political factions willing to use the law to secure their own advantage. As the law becomes merely an instrument of power and rule, civic virtue is further eroded and the obligation to obey the law withers and dies. There is no simple causal connection. This, of course, is a feature of many complex relationships. Much like the person who allows their health to erode and thus finds it ever harder to get back into shape, the failing of one feeds into the failing of the other.

Examining such complex and interwoven relationships can be frustrating precisely because there is no easy causal switch. This, however, does not leave us powerless to attend to and repair the erosion of civic virtue and its symptoms where we find it in our modern landscape. In conclusion, I would like to mention an area in the modern legal landscape where civic virtue is in all too short supply. It is startling how clearly we can see the symptoms of eroding civic virtue, laws which turn away from a shared common good, entrench factions and disproportionately burden others leads to increasing instrumentalization and finally collapse of fidelity to law.

Despite the amount of attention it receives, I remain convinced the American "War on Drugs" remains underappreciated. Particularly difficult has been the intersection of race, crime and policing, from the notorious disparity between punishment for crack and cocaine to the mass incarceration of Black and Hispanic men. Ideal reforms in this area of criminology are difficult to determine; it is often the same neighborhoods which are victims of drug violence which are gutted by mass incarceration. Yet one facet of the war which is impossible to ignore is the racial component of the most widespread illicit drug, marijuana. Though Blacks make up roughly an eighth of the population, one-third of those arrested and one half of those imprisoned for drug offenses are black. Thus Black men are more than 12 times as likely to be imprisoned for drug offenses than White men. The prohibition on marijuana is a tremendous driver of this disparity, as four in ten drug arrests were for marijuana possession. The cruelty of this racial disparity is brought home by the fact that both races use illegal drugs at roughly the same rates.

Our current drugs current policing system results in wealthy whites (and wealthy blacks) comfortably flouting the law at home while the police stop, harass, pat down and arrest poor people of color, often using the marijuana laws as a standing license to police a segment of the population. This breeds frustration and resentment among minorities, drug offenders and law abiding citizens alike. Black and Hispanic citizens realize that any serious attempt to control marijuana would have to focus on leafy suburban colleges campuses as much as on urban street corners. The realization that those of a different race and class are immune from police attention invites the dark suspicion that these laws do not genuinely aim at marijuana control but are rather tools of social control and factional power to allow the policing of certain communities. The sheer scale of law breaking and the continuous shrinking of marijuana punishment illustrates the failure of marijuana laws to command fidelity.

Let me conclude by refocusing on the theme. What Aristotle made clear so long ago has been lost in our modern conversation about vice crimes. Aristotle recognized naturally enough that the normative guidance of law was important in guiding the development of our character. But he cautioned all not to be too zealous in using law to pursue private virtue; it was, he concluded, an unlikely and distracting project. Instead, Aristotle consistently prioritized public virtues and warned that the first goal of law and legal institutions was to guard against the vices of factionalization and the seeking of private benefits. Aristotle saw that it was the vices within the law that most threatened our communal well-being. Insufficient attention to civic virtue allows law to be warped into an instrument of power and exploitation. Reciprocally, allowing the law to become an instrument of exploitation further damaged our shared civic virtue. Ultimately, civic virtue can be so badly damaged that citizens have no reason to follow the law at all. What the rampant lawlessness of the Jim Crow South and the increasing disregard for the criminal sanctions surrounding marijuana prohibition reveals is that when we seek to use the law crassly for our own advantage, it is finally as bad for us as it is for our victims.

## Bibliography


Aristotle. 1941a. Politica. In *The basic works of Aristotle*, ed. Richard McKeon, 1113–1316. New York: Random House.

Aristotle. 1941b. Nicomachean ethics. In *The basic works of Aristotle*, ed. Richard McKeon, 927–1112. New York: Random House.

Braithwaite, John, and Philip Pettit. 1990. *Not just deserts: A republican theory of criminal justice*. Oxford: Clarendon.

Burnyeat, M. 1980. Aristotle on learning to be good. In *Essays on Aristotle's ethics*, ed. A. Rorty, 69–92. Berkeley: University of California Press.

Chishti, Mazaffar A. 2006. Immigration reform: Balancing enforcement and integration: Enforcing immigration rules: Making the right choices. *New York University Journal of Legislation and Public Policy* 10: 451, 452–455.

Cohen, G.A. 2008. *Rescuing justice and equality*. Cambridge: Harvard University Press.

Cooper, John. 2005. Political animals and civic friendship. In *Aristotle's politics: Critical essays*, ed. Richard Kraut and Steven Skultety, 65–89. Lanham: Rowman & Littlefield Publishers, Inc.

Dagger, Richard. 2009. Republicanism and crime. In *Legal republicanism: National and international perspectives*, vol. 184, ed. Samantha Besson and José Luis Martí, 147–166. Oxford: Oxford University Press.

Duff, Anthony. 2001. *Punishment, communication, and community*. Oxford: Oxford University Press.

Duff, Anthony. 2007. *Answering for crime*. Portland: Hart Publishing.

Duff, Anthony. 2013. Virtue, vice and criminal law: A response to Huigens and Yankah. In *Law, virtue and justice*, ed. Amalia Amaya, and Hock Lai Ho. Oxford: Hart Publishing.

Edmundson, William A. 2006. The virtue of law-abidance. *Philosopher's Imprint* 6: 1.

Fuller, Lon. 1964. *The morality of law*. New Haven: Yale University Press.

Gararella, Robert. 2009. Tough on punishment: Criminal justice, deliberation and legal alienation. In *Legal republicanism: National and international perspectives*, vol. 184, ed. Samantha Besson and José Luis Martí, 167–184. Oxford: Oxford University Press.

Garrett, Aaron. 2010. Rights, equality, and justice: A conference inspired by the moral and legal theory of David Lyons: Courage, political resistance, and self-deceit political law's ethos: Reflections on the public practice of illegality. *Boston University Law Review* 90: 1771.

George, R.P. 1993. *Making men moral: Civil liberties and public morality*. New York/Oxford: Clarendon Press/Oxford University Press.

Hampton, Jean. 1998. Punishment, feminism, and political identity: A case study in the expressive meaning of the law. *Canadian Journal of Law and Jurisprudence* 11: 23.
Hitz, Zena. 2011. Aristotle on self-knowledge and friendship. *Philosopher's Imprint* 11: 1.
Huigens, K. 1995. Virtue and inculpation. *Harvard Law Review* 108: 1423, 1444–1449
Huigens, Kyron. 2002. Homicide in aretaic terms. *Buffalo Criminial Law Review* 6: 97, 104.
Jackson, J., B. Bradford, M. Hough, A. Myhill, P. Quinton, and T.R. Tyler. 2012. *Why do people comply with the law? Legitimacy and the influences of legal institutions*. Electronic copy available at: http:// ssrn.com/abstract=1994490.
Kraut, Richard. 2002. *Aristotle*. Oxford: Oxford University Press.
Lacey, N. 1998. *State punishment*. London: Routledge.
Lyons, David. 2008. The legal entrenchment of illegality. In *The legacy of H.L.A. Hart*, ed. M. Kramer, C. Grant, B. Colburn, and A. Harzistavrou, 29–43. Oxford: Oxford Scholarship Online.
MacIntyre, Alasdair. 1985. *After virtue*. London: Duckworth.
Marshall, S.E., and R.A. Duff. 1998. Criminalization and sharing wrongs. *Canadian Journal of Law and Jurisprudence* 11: 7.
Martí, José Luis. 2009. The republican democratization of criminal law and justice. In *Legal republicanism: National and international perspectives*, vol. 184, ed. Samantha Besson and José Luis Martí, 123–146. Oxford: Oxford University Press.
Murphy, Jeffrie. 1973. Marxism and retribution. *Philosophy & Public Affairs* 2: 217.
Pettit, Philip. 1997a. *Republicanism: A theory of freedom and government*. Oxford: Oxford University Press.
Pettit, Philip. 1997b. Republican theory and criminal punishment. *Utilitas* 9: 57.
Pettit, Philip. 2002. Keeping republican freedom simple: On a difference with Quentin Skinner. *Political Theory* 30: 339.
Postema, Gerald J. 2010. Rights, equality, and justice: A conference inspired by the moral and legal theory of David Lyons: Law's ethos: Reflections on the public practice of illegality. *Boston University Law Review* 90: 1847, 1856.
Raz, Joseph. 1986. *The morality of freedom*. Oxford: Clarendon.
Skinner, Quentin. 1978. *The foundations of modern political thought*, vol. 1. Cambridge: Cambridge University Press.
Skinner, Quentin. 1984. The idea of negative liberty: Philosophical and historical perspectives. In *Philosophy in history*, ed. R. Rorty, J.B. Schneewind, and Q. Skinner, 193–221. Cambridge: Cambridge University Press.
Solum, Lawrence B. 2006. Natural justice. *American Journal of Jurisprudence* 51: 65, 89–90.
Tierney, Brian. 1963. The prince is not bounded by the laws: Accursius and the origins of the modern state. *Comparative Studies in Society and History* 5: 378.
Tyler, T.R., and J. Fagan. 2008. Legitimacy and cooperation: Why do people help the police fight crime in their communities? *Ohio State Journal of Criminal Law* 6: 231.
Waldron, Jeremy. 1994. Why law – Efficacy, freedom or fidelity. *Law and Philosophy* 13: 259.
Wishnie, Michael J. 2008. The low wage worker: Legal rights—Legal realities: Labor law after legalization. *Minnesota Law Review* 92: 1446.
Yankah, Ekow N. 2004. Good guys and bad guys: Punishing character, equality and the irrelevance of moral character to criminal punishment. *Cardozo Law Review* 25: 1019.
Yankah, Ekow N. 2009. Virtue's domain. *University of Illinois Law Review*: 1167.
Yankah, Ekow N. 2011. A paradox in overcriminalization. *New Criminial Law Review* 14: 1.
Yankah, Ekow N. 2012. Crime, freedom and civic bonds: Arthur Ripstein's force and freedom: Kant's legal and political philosophy. *Criminal Law and Philosophy* 6: 255.
Yankah, Ekow N. 2013. Liberal virtue law. In *Virtue and justice*, ed. Amalia Amaya, and Hock Lai Ho. Oxford: Hart Publishing, Forthcoming 2013.

# Chapter 13
# A Neo-Aristotelian Notion of Reciprocity: About Civic Friendship and (the Troublesome Character of) Right Judicial Decisions

**Iris van Domselaar**

## 13.1 Introduction

It is generally held that in modern liberal societies it is a matter of fairness that the exercise of political power is justified by referring to reasons that in principle are acceptable for each and every citizen. If political institutions are to be legitimate, they should work to the advantage of all citizens. We do not allow citizens to be sacrificed or their fundamental interests to be violated just because it serves the common good or the good of a fellow citizen. This requirement of fairness follows from a liberal interpretation of reciprocity as it was succinctly articulated by John Rawls: "our exercise of political power is proper only when we sincerely believe that the reasons for our political action may reasonably be accepted by other citizens as a justification of those actions" (Rawls 2005, xliv).

Obviously, this principle of liberal reciprocity or fairness directly bears on the institution of adjudication. One does not need to highlight dramatic examples of judges imposing the death penalty, ordering the custody of a child or the expropriation of a house to understand that adjudication may have extremely burdensome consequences. Adjudication genuinely and directly affects the lives of concrete citizens. More formally put: adjudication has the exclusive right to some rather extreme forms of coercion. Adjudication can be distinguished from other (political) institutions by its comprehensive scope and its regulative powers with respect to other associations. In Western constitutional democracies adjudication is generally considered the final authoritative institution for defining the limits of all other activities that take place.

For these reasons it is not surprising that liberals, committed as they are to the protection of the individual citizen, want adjudication to fulfil the principle of

---

I. van Domselaar (✉)
Amsterdam Law School, University of Amsterdam, Amsterdam, The Netherlands
e-mail: I.vanDomselaar@uva.nl

L. Huppes-Cluysenaer and N.M.M.S. Coelho (eds.), *Aristotle and The Philosophy of Law: Theory, Practice and Justice*, Ius Gentium: Comparative Perspectives on Law and Justice 23,
DOI 10.1007/978-94-007-6031-8_13, © Springer Science+Business Media Dordrecht 2013

reciprocity. Therefore the justification of adjudication is a major concern for liberal legal scholars. As David Lyons puts it: "Not just anything a court might think up as a way of deciding a case will do. For not just anything is capable of *justifying* a decision" (Lyons 1984b, 184).

In this chapter, I will (critically) discuss 'adjudication as applied moral theory' as a way of accounting for reciprocity (par. 2). This is the kind of justification most liberal legal scholars and philosophers of law come up with. It justifies judicial decisions by conceptualizing them as the institutional implications of principles of political morality that reasonable citizens can accept. This approach aims to meet the requirement of reciprocity in a rather direct way: if reasonable citizens can accept a certain set of principles of political morality, they can also accept the implications of these principles when applied to concrete legal cases. In this approach the practice of adjudication and hence of reciprocity has a certain clarity, stability and safety to it.

I will argue that 'adjudication as applied moral theory' does not suffice as an account of reciprocity (par. 3). Instead, I will propose and examine a neo-Aristotelian approach to adjudication, which does not rely on moral theory and its principles, but rather on judicial virtues to do the justificatory work regarding adjudication, judicial decisions and the burdens they bring (par. 4). Moreover, I shall consider to what extent a neo-Aristotelian approach to adjudication can indeed satisfy the requirement of reciprocity (par. 5) and I will propose Aristotle's concept of civic friendship as a necessary amendment (par. 6). We shall see that in this neo-Aristotelian approach to adjudication reciprocity comes with some messiness, limited intelligibility and transparency, and also with moral loss. Reciprocity cannot always be fully realized through judicial decision-making and this cannot be explained in terms of negligible incidents.

Finally, I will tentatively discuss (par. 7) whether the concept of civic friendship makes sense for all kinds of legal cases, that is, unabridged.

## 13.2 Reciprocity in Adjudication as Applied Moral Theory

'Adjudication as applied moral theory' is a relatively straightforward and top-down account of the requirement of reciprocity for the institution of adjudication. According to this approach the requirement of reciprocity is presumed to be satisfied if the corpus of settled law sufficiently complies with the moral background principles of the system and if judges generally apply the law and these principles adequately. Judicial decisions are held to be justified by foundational moral principles that are generated through a theory of political morality.[1]

[1] See for a discussion of foundationalism in moral philosophy: Timmons (1987). For a critical discussion of foundationalism in moral and political thought see: Williams (1985) and Hawthorn (2005).

Adjudication is considered a practice that is primarily about drawing the right conclusions from moral premises in a discursive, argumentative activity for which the judge primarily needs to have analytical skills and legal intelligence. The relevant considerations that are decisive for the concrete judicial decisions are given a priori. Whether a legal decision is right will be determined by a standard or set of standards external to the judge and which facts are relevant is exhaustively determined by the principles of political morality (Soper 1977, 504). This fits well with the widely shared assumption in Western constitutional democracies that judges should be bound by the law, i.e. that they have "a duty to decide in a particular way, for the express reason that the law requires that decision" (Dworkin 1985, 49). 'Adjudication as applied moral theory' explains why judges often "speak and write as if they were discovering what the law requires and allows for, even in hard cases" (Lyons 1984a, 93).

In contemporary legal theory Dworkin is among the prominent defenders of 'adjudication as applied moral theory'.[2] He holds that a decision is right if it is supported by arguments stemming from the best moral background theory of law. "A proposition of law, like the proposition that Tom's contract is valid, is true if the best justification that can be provided for the body of propositions of law taken to be settled provides a better case for that proposition than for the contrary proposition […]" (Dworkin 1985, 142). If decisions cannot in one way or another be reduced to these moral background principles, these are to be met with suspicion because they may be the result of tradition, prejudice or mere preferences of the judge, rather than of moral reasons (Lyons 1984a, b, 104).

Surely not every moral theory can function as the final justificatory ground for judicial decisions and not every moral theory can account for reciprocity. The theory that can best fulfil these functions, is the theory that scores highest on two dimensions, compared to other theories. The first is the dimension of fit between moral theory and the legal system "in the sense that it requires less of the material to be 'mistakes'" (Dworkin 2005 (1977), 340). The second is the dimension of justification in the sense that the moral theory offers "a morally more compelling justification" for the legal order as a whole (Dworkin 2005 (1977), 340). That is, it must come closer to "capturing sound political morality" together with the minimum requirement that it offers an adequate interpretation of reciprocity (Dworkin 2005 (1977), 340).

It must be noted that the liberal principle of reciprocity itself is a formal principle, which does not prescribe which particular foundational principles of political morality can perform said function, other than their being an expression of an adequate equilibrium between the interests of all citizens conceived as free and equal.

[2] See for instance: Dworkin (2005 (1977)), Dworkin (1985), and Dworkin (2011). Robert Alexy also defends 'adjudication as applied moral theory' where he argues that because the law has a claim to correctness, this by itself provides an argument for the idea that "morality is necessarily included in the law" and that "[m]oral reasons can and must participate in the justification of legal decisions when authoritative reasons run out." Cf. Alexy (2004, 165).

It should be possible to understand these principles and their application as benefitting each and every citizen.

For Western constitutional democracies several theories of political morality or of justice come to mind as prima facie candidates for offering the justificatory ground for judicial decisions and the understanding of reciprocity for the institution of adjudication. Rawls' Justice as Fairness and Nussbaum's Capabilities Approach are among them, if only because both theories explicitly claim the function of moral background theory for adjudication that has both fit and justificatory force.[3] For instance, Nussbaum presents her Capabilities Approach as a theory that offers "the foundations of political entitlements and constitutional law" in liberal constitutional democracies (Nussbaum 2007, 7). In a similar vein, Rawls states that his principles of justice apply to the "judiciary and above all to a supreme court in a constitutional democracy with judicial review" (Rawls 2005, 216). Moreover, both Nussbaum and Rawls discuss concrete case law from the viewpoint of the principles of political morality that their theories offer. These are positive indications that they consider their theories of political morality as adequate background theories of law.

For our understanding of how 'adjudication as applied moral theory' accounts for the liberal requirement of reciprocity, Rawls' theory offers recourse. Via the device of the "original position" he famously gives an account of why reasonable citizens have reasons to accept the abstract and general principles of political morality that underlie their society (Rawls 1971 (1999), 102–160). In addition, through his notion of the "four stage sequence" he explains how this requirement can also be satisfied for the central political institutions, including that of law and adjudication, and for the relations between citizens (Rawls 1971 (1999), 102–160; 171–176). As to law and the judiciary Rawls asserts that the requirement of reciprocity is fulfilled if the law more or less complies with the background principles of political morality and if the judge applies this law fairly to the particular case. Citizens would only have a legitimate complaint if the judge fails "to apply the appropriate rule or to interpret it correctly" (Rawls 1971 (1999), 207).

This is also expressed by the interpretation of the rule of law that Rawls offers, which he considers an elaboration of formal justice for the legal domain and which he sees as an implication of the principle of liberty that his version of political morality aims to protect. In his account of the rule of law, the law constitutes the "grounds upon which persons can rely on one another and rightly object when their expectations are not fulfilled. If the bases of these claims are unsure, so are the boundaries of men's liberties", Rawls says (Rawls 1971 (1999), 207).

Again, hence in 'adjudication as applied moral theory' the idea is that if a legal order complies with the principles of justice and is rightly administered by the judiciary, than the requirement of reciprocity is exhaustively fulfilled for judicial decisions. These decisions, however painful they may be, can be taken as expressions of

[3] Cf. Rawls (1971 (1999)); Rawls (2005). See for the most recent version of Nussbaum's Capabilities Approach: Nussbaum (2011).

a genuine equilibrium between 'self and others', between the good of society or of concrete other citizens and that of the affected citizen. Once the law is rightly applied there is no need to be bothered by these sometimes painful outcomes. These are assumed to be already accommodated for in the principles of political morality which citizens have reason to endorse and which exhaustively determine their legitimate expectations. "We cannot at the end count them a second time, because we do not like the result", Rawls says (Rawls 1971 (1999), 71).

As to the moral bearing of the painful consequences of judicial decisions in 'adjudication as applied moral theory' Dworkin makes a telling distinction between "bare harm" and "moral harm". He considers bare harm a loss that is merely subjectively experienced by the affected citizen. It is "the suffering or frustration or pain of dissatisfaction of desires that he suffers" through a judicial decision that by itself is not problematic because of its being fair (Dworkin 1985, 80). Moral harm, by contrast, is an objective loss which "assumes that someone suffers a special injury when treated unjustly" and which stems from the injustice factor (Dworkin 1985, 80).

So, in Dworkin's terms, in 'adjudication as applied moral theory' losing citizens will only have a legitimate complaint about having suffered a moral harm as a result of a particular judicial decision, if settled law including the background principles of political morality is violated by this decision. The losing litigant is then likely to be conceived as a "wounded left on the battlefield" in the crusade for the good of society or of a fellow citizen (Dworkin 1986, 213). In all other cases the burdens stemming from judicial decisions can be considered to potentially constitute bare harm or a rather inconvenient side effect of living in a principled, just society.

Having discussed the central features of 'adjudication as applied moral theory', we can now see that in this approach the fulfilment of reciprocity in adjudication has a certain pureness, stability and safety to it. For one thing, for judges to honour the principle of reciprocity primarily amounts to exercising their cognitive qualities that enable them to determine the legally right decision on the basis of the corpus of settled law and the moral background principles. These qualities will be indispensible for 'finding' the true legal proposition. In the words of Rawls: "Judicial virtues such as impartiality and considerateness are the excellences of intellect and sensibility that enable us to do these things well" (Rawls 1971 (1999), 453). Their person will not be at stake in answering the question whether the requirement of reciprocity has been fulfilled and whether the burdens their decisions impose on concrete citizens are legitimate. Judges can consider themselves as cognitive tools for the realisation of fairness; the fulfilment of reciprocity safeguards judges from having to bear personal responsibility for the burdensome consequences of their decisions.

Moreover, as he observes this principles of reciprocity the judge will not need to "stain" his robe and consequently the legal order as a whole also stays morally immaculate. This because 'adjudication as applied moral theory' holds that if a judicial decision is right, reciprocity is exhaustively satisfied. Both the judge and the legal order can rest assured that the affected citizen has a good reason to accept the burdens of his decision. Also, the reasons as to why citizens should accept these burdens are held to be fully explicably; in the end these are moral principles provided by moral theory. In the ideal of 'adjudication as applied moral theory' neither the judge nor the affected citizens will have to deal with unintelligibility, insecurity or moral loss.

## 13.3 Objections Against Adjudication as Applied Moral Theory

We have briefly and therefore rather crudely sketched 'adjudication as applied moral theory' and its way of accounting for reciprocity. In this section two objections against this approach will be raised in a tentative way in order to give more depth to the exposition of the neo-Aristotelian approach to adjudication that follows. The merits of the latter approach stand out better against the background of the weaknesses of 'adjudication as applied moral theory'. The objections can be taken as an internal critique on 'adjudication as applied moral theory', for the upshot of both objections is that this approach cannot genuinely account for its own understanding of reciprocity.

For these objections we not only draw on the notorious sceptic criticism of Legal Realists and the Critical Legal Studies Movement on main stream theories of adjudication, but also on the criticism of political realists like Bernard Williams and Raymond Geuss on theories of political morality in general (Williams 1985; Hawthorn 2005; Geuss 2008).

As to the first, notwithstanding the wide variety of their views, the sceptics belonging to the Legal Realists and the Critical Legal Studies Movement[4] agree in their more or less categorical rejection of the idea that judges are in any meaningful sense 'bound by law', among others because of the inherent indeterminacy of the law. They hold that because of this indeterminacy judges have strong discretion in determining the legal correct answer in every case.[5] The law is the sum of the "prophecies of what the courts will do in fact, and nothing more pretentious" (Holmes 1897).

By this 'strong discretion thesis' the sceptics do not mean to deny that judicial decisions can (sometimes) be foreseen, that is, predicted. However, for them this predictability is based on empirical regularities in the behaviour of a particular judge or a group of judges, rather than on the normative force of reasons drawn from

[4] Because the Legal Realists and proponents of the Critical Legal Studies Movement produced a vast literature and because their mutual relations and differences are deeply complex, for this context the sceptic approach will be presented by primarily drawing on secondary literature, in particular on: Altman (1986) and Kramer (2007).

[5] Obviously, the background assumption is that "[t]he extent to which there are determinately correct answers to legal questions is inversely proportional to the extent of the leeway left to legal officials in arriving at concrete decisions". Cf. Kramer (2007, 14).

the legal system. If it exists at all, the predictability of judicial decisions is due to "extra- legal factors such as shared psychological inclinations, rather than to the terms of legal requirements and entitlements" (Kramer 2007, 26).

To be sure, if it is right, the sceptic view obviously has serious implications for the extent to which judicial decisions can be seen as an expression of reciprocity among citizens. If judges are not (to a sufficient degree) constrained by the law, judicial decisions cannot be conceived of as realizing the principle of reciprocity. Being the outcome of the subjectivity of the judge, judicial decisions will bring burdens for particulars citizens without there being a good reason for it.[6]

I will briefly delve into the arguments the sceptics offer for strong discretion and connected to this for their indeterminacy of law thesis. The sceptics hold that the law is indeterminate firstly because there is never just one single and clear rule solely relevant to a case – it is always a group of (vague and contradictory, at least competing) rules (Altman 1986, 187). There are always "multiple potential points of indeterminacy" so that in each and every case the judge is free to choose which rule to apply (Altman 1986, 186). Also, where legal precedents are concerned, the judge according to the sceptics can more or less randomly decide what rule a certain precedent stands for; every precedent can be interpreted in several ways so as to make it stand for conflicting or competing rules and therefore the judge has an unrestricted leeway to choose the outcome of any case (Altman 1988, 186; Hart 1961, 134–135).

Secondly, according to the sceptics the judge has wide elbowroom in selecting and establishing the relevant facts of the case. The law itself does not prescribe how to select the relevant facts and neither do these facts present themselves with labels with the applicable rule on it. Consequently, as the judge knows the law he can select and establish the facts such, that his legal argument arrives at the decision that he intuitively favoured from the start (his "hunch") and for whatever reason that pleases him. Jerome Frank has eloquently stressed this point where he states "[t]he judge, in arriving at this hunch, does not nicely separate his belief as to the 'facts' from his conclusion as to the 'law'; his general hunch is more integral and composite, and affects his report – both to himself and to the public – concerning the facts" (Frank 1931, 116). Due to these inherent features of the law, the sceptics hold that the law by no means guides or constrains judges. The law, or so it is held, does not channel the decisions of the judge, but other legally arbitrary factors do. Adjudication therefore is to be taken as a matter of contingent fact, rather than as the embodiment of law or reason.

Clearly, we cannot simply accept the aforementioned critiques without also addressing the main replies. One line of argument put afore against the 'strong discretion thesis' is that besides explicit rules and precedents the law is also

[6] Here we see that legal scholars and political philosophers generally share the assumption that legal indeterminacy matters, for it directly concerns the question of the legitimacy of the judiciary as a crucial political institution of society. In order for the judiciary to be legitimate in a normative sense, the law at least to a certain extent must be able to constrain the judge.

constituted by a range of legal conventions, i.e. of social norms that exist independently of the judge and that do constrain him. Judges, precisely because of the fact that they fulfil the role of a judge, will experience social constraints that are definitive for what it means to be a judge (Soper 1977, 477). Hart famously argued that if we want to understand the way in which judges are bound by law and hence fulfil their professional responsibilities, we should take the relevant secondary rules, rules of recognition and rules of adjudication into account. By accepting his role the judge subscribes to these rules that belong, as Hart puts it, to the "common public standards of official behaviour" (Hart 1961, 116). The whole edifice of (primary and secondary) rules is based on an ultimate rule of recognition. This ultimate rule is not really a rule, but rather a complex of historical, social and institutional facts that imply that the system of valid law as a whole and its adjudication exist and are accepted (Hart 1961, 108–110, 116–117). In spite of the discretion allowed for by settled law, due to this rule the judge will decide in conformity with the legal system; his decisions will be based on his knowledge of the system, of social morality and other non-subjective determining factors such as the judicial virtues of impartiality and neutrality (Hart 1961, 205).

Another line of argument does not so much focus on the determinacy of legal conventions, but rather stresses the relative determinacy of language in support of the claim that the judge is to a sufficient degree constrained by the law. Briefly, the argument is that language can and frequently does speak with a sufficiently clear voice so that the law, being constituted by language, does constrain the judge "from overstepping what are admittedly pre-theoretical and almost intuitive linguistic bounds [...]" (Schauer 1985, 431).

So, both lines of argument boil down to the idea that although the law does not give an account of determinate correctness, it does not offer one correct answer for each decision, it does exclude a wide range of reasons as non-legal so that the judge at least to some extent can be said to be constrained (Kramer 2007, 4). The idea is that "though multiple contrary answers will each be correct, many answers are incorrect" (Kramer 2007, 15).

However, from the viewpoint of reciprocity, at least as it figures or is meant to figure in 'adjudication as applied moral theory', these answers do not suffice. Neither legal conventions nor language can secure that the law at least to a minimum degree has a moral content due to which citizens have a good reason to accept painful or detrimental judicial decisions. Although both conventions and language do exclude certain reasons, they do not necessarily exclude heinous or immoral norms (Kramer 2007, 13). In order for the law not only to be sufficiently determinate in a legal sense, but also sufficiently determinate in a moral sense so as to account for reciprocity in adjudication, more is needed.

It is at this point that the principles of political morality enter the stage. As indeed Dworkin and others have argued and as is suggested in 'adjudication as applied moral theory', it could possibly be the case that legal discretion can be held "within the sway of general principles of political morality" (Kramer 2007, 25). However, according to the sceptics moral background principles cannot make the law's moral content more determinate. This because these principles are themselves indeterminate and legal systems are generally characterized by contradictory principles (Altman 1986, 189). In the words of Altman: "the jurisprudential invocation of principles only serves to push back to another stage the point at which legal indeterminacy enters and judicial choice takes place" (Altman 1986, 189).

This sceptical critique finds support in the work of both Bernard Williams and Geuss who both deny that theories of political morality can be action-guiding. They hold that principles of political morality, being theoretical constructions and hence being abstract and general, are too indeterminate a viewpoint to actually offer concrete guidance to the institutions in a concrete society. For Williams principles of political morality are typically constituted by 'thin concepts': concepts that lack sufficient descriptive content, due to which their application is not 'world-guided'; the conditions of their application are not constituted by facts about the world that can be easily perceived (Williams 1985, 140–152; Hawthorn 2005, 48–50). As a consequence the 'application' cannot be reduced to these principles themselves, but this 'application' is tantamount to the choices actually made by the agents who have authority to 'apply' them. How the principles of political morality will work out in a concrete case (either directly or indirectly through legislation and policy) will therefore depend on an "opaque aggregation of actions and forces" such as the power-relations in parliament or the preferences of the judge.[7] Thus, by implication the critique of Williams and Geuss supports the sceptic approach to adjudication, which entails that principles of political morality cannot provide for the moral constraint that is necessary to account for reciprocity in adjudication.

But, before indeed coming to this conclusion, let me expound Rawls' and Nussbaum's potential replies to this charge. As we have seen, their theories of justice both claim to have sufficient practical force so as to account for the moral bearing of institutional decision-making in society. For his argument that justice is sufficiently determinate to be action guiding on a local level, Rawls for instance introduces the device of 'the four stage sequence' (Rawls 1971 (1999), 171–176). It serves "to simplify the application of the two principles of justice" for constitutional democracies (Rawls 1971 (1999), 171). It is meant to accommodate the institutional data of constitutional democracies and to make justice a "workable political conception" and not merely a theoretical exercise (Rawls 1971 (1999), 171–176).

This 'four stage sequence' distinguishes the level of the philosophical principles, the level of the constitution, the level of legislation and policy and that of the application of rules to concrete cases "by judges, administrators, and the following of rules by citizens generally" (Rawls 1971 (1999), 175).

Because on each level there will be a considerable flow of relevant information available, Rawls thinks there will in fact be little leeway for public officials or citizens in determining what justice requires (Rawls 2005, 340). Justice will also become more substantial through the guidelines to be used for its application, or so Rawls states. These in any case include "accepted general beliefs and forms of reasoning found in common sense and the methods and conclusions of science when these are not controversial" (Rawls 2005, 224).

---

[7] By accepting Williams' point one could even question whether the concept of "application" in Rawls' *Theory of Justice* is appropriate.

Where the viewpoints specified by the 'four-stage sequence' or these guidelines for application do not give a determinate answer, Rawls holds that "justice is to that extent likewise indeterminate" (Rawls 1971 (1999), 176). For these situations a well-ordered society can legitimately "fall back upon a notion of quasi-pure procedural justice: laws and policies are just provided that they lay within the allowed range, and the legislature in ways authorized by a just constitution has in fact enacted them" (Rawls 1971 (1999), 176).

Likewise, Martha Nussbaum offers an elaboration of how the principles of political morality that she proposes, the Central Human Capabilities, can be conceived as sufficiently determinate to have practical force in decision-making processes in society. As to legal reasoning Nussbaum states: "What judges must interpret is the list validated by their own nation, together with the tradition of precedent that interprets it" (Nussbaum 2007, 15). Moreover, public reasoning about these principles should according to Nussbaum in general be a matter of "wise practical reasoning", the discretion assigned to the authoritative institutions to determine the concrete basic claims of citizens in society is curtailed by "parameters" (Nussbaum 2006, 79). Differences between different societies may exist only "at the margins" (Nussbaum 2006, 180).

Yet, neither of these accounts suffices to secure for the normative determinacy of the law, and a fortiori not for reciprocity in adjudication. This because the sources that are allowed to play a role in the specification of the principles of political morality seem to be so diverse that they can hardly be conceptualized as speaking with one homogenous voice. In case these sources offer conflicting arguments, it is not clear what (comparative) weight should be given to different kinds of knowledge, to common sense ideas, or to different conceptions of reasoning. These conflicts cannot be solved in terms of the principles of political morality itself – because precisely these very principles are to be made more determinate through these sources. Moreover, neither Rawls nor Nussbaum explains why the different kinds of sources that are used to make political morality more determinate, do indeed bring about results that can be traced back to the values that the principles of political morality aim to protect. It is not clear what these sources have in common so that they can be said to genuinely lead to specifications of political morality, rather than to just contingent states of affairs. Everything considered we do not yet have an answer why citizens will have a reason to bear the burdens of particular judicial decisions.

Another problem that arises if principles of political morality are to secure the moral determinacy of the law is that these principles can genuinely conflict with one another. They may prove incompossible, meaning that the realization of one value comes with the violation of another.[8] If such a situation is at hand in a legal proceeding, a background theory of political morality does not offer a decisive

[8] Brett Scharffs offers an instructive description of incompossibility: "The term 'incompossibility' usually refers to states of affairs, either choices or acts. For example, your being in both New York and Paris right now is an incompossibility; they are jointly impossible states of affairs. But values might also be incompossible if the realization of one necessarily makes it impossible to realize the other [...]." Cf. Scharffs (2000, 1395).

reason for the judge to decide one way or the other; the viewpoint it offers is internally conflicted and thence may point in two opposite directions in the same respect at the same time. In such situations political morality does not constrain the judge. It may then well be that the outcome will be determined, not by good reasons, but by the "judge's own moral, political, psychological, Oedipal, or intestinal predilections" (Schauer 1985, 410). The judge may still offer legal reasons and he may refer to moral principles, but his reasoning would be a "*post hoc* legal justification for the non-legally derived result in order not to affront the accepted myths of society, including the myth of the rule of law" (Schauer 1985, 410–411).

Particularly when fundamental interests are at stake, this kind of judicial discretion is at odds with the liberal notion of reciprocity indicating that the exercise of political power "should in principle be capable of explaining itself at the tribunal of each person's understanding" (Waldron 1987, 149). In case of such conflicts, the reason why the particular citizen should bear the burden of the decision remains completely in the dark. To simply assume that these outcomes fall within the domain of political morality is to beg the question, for it is not clear what these outcomes have to do with morality at all.

To be sure, it must be noted that both Rawls' and Nussbaum's theories do also provide for arguments as to why genuine conflicts will not occur. They both hold what can be qualified as a specificationist's view on conflicts between political values, which relies on the power of reflection to specify political values in such a way that they fit together "like pieces in a jigsaw puzzle" (Wenar 2011).[9] The idea is that reflection can make these values 'fluid', as it were, so as to avoid genuine conflicts.[10] Exceptional circumstances aside, both Nussbaum and Rawls assert that if constitutional principles, policy and legislation are in place, it is a matter of practical fact that in a fully just society genuine conflicts between political values need not occur.

Clearly, the force of such an approach to potential conflicts of political morality is that it can function as a strong antidote against "stupidity", "laziness" and "malice" in public reasoning regarding citizens' basic claims (Nussbaum 2000, 1016). However, stupidity, laziness and malice are not necessary conditions for all practical conflicts between political values. Because of its variability the practical world is such that even the best reasoning available at a particular time and place cannot forestall that values of political morality will prove to be incompossible. Unless we hollow out the substantive scope of political morality, it seems problematic to a priori push the issue of conflicts to the periphery of public decision-making. It is more fitting to conceive of these conflicts as genuinely part and parcel of what characterizes society.[11] This, at pains of relying on an image of the world that is difficult to maintain, namely that it can be so organized that no conflicts will arise

---

[9] See also Wellman (1995) and Richardson (1997).

[10] See for the use of the term "fluid" in this context: Richardson (1997, 139).

[11] For critical comments on Nussbaum's attempt to toning down the prevalence of genuine practical conflicts between the Central Human Capabilities see among others: Hackett (2001) and Claassen and Düwell (2012).

when realizing plural values.[12] The ethical reality of everyday adjudication does not support the claims of a 'specificationist' conflict resolving strategy. Take for instance the continuing conflicts between the freedoms that orthodox Christians claim for the meaningful exercise of their religion and the rights of other (groups of) citizens who want to be treated as equals. These do not 'stop' once a more abstract and general viewpoint – as for instance embodied by the judiciary – has tried to resolve them. It seems more realistic to accept that the values that political morality is committed to may prove incompossible so that in adjudication genuine decisions need to be made about these conflicts. Again, for these cases it remains to be seen to what extent the decision of the judge complies with the requirement of reciprocity.

## 13.4 A Neo-Aristotelian Approach to Adjudication

As said, a neo-Aristotelian approach to adjudication can possibly solve the problems that an applied moral theory approach faces when accounting for reciprocity. We have seen that 'adjudication as applied moral theory' cannot account for the rightness of judicial decisions as they can be the outcome of the mere preferences of the judge or other contingent facts, thereby violating the principle of reciprocity.

A neo-Aristotelian approach sees adjudication as an institution that is primarily concerned with practice and hence demands that theories of adjudication accommodate the characteristics of practice: one must "not look for precision in the same way in everything, but in accordance with the underlying material in each sphere, and to the extent that is appropriate" (*EN* 1098a26-28). According to Aristotle the practical world is characterized by indeterminacy. Due to this characteristic principles, rules and other kinds of general precepts can never fully and in advance capture all that is relevant from a practical viewpoint (*EN* 1140b3-5; Nussbaum 1990, 66–75; Wiggins 1980, 232).

In a neo-Aristotelian approach to adjudication the attempt to accommodate for the central characteristics of practice does not lead to scepticism about adjudication, but rather leads to the claim to offer an adequate justification for adjudication in Western constitutional democracies.[13] Although, as we shall see, some messiness, limited intelligibility, transparency, and moral losses are the consequences of

[12] To be sure, practical reasoning as displayed in clever constitutional arrangements, adequate social policies and wise judicial reasoning obviously can and will reduce the likelihood and scope of conflicts between citizens' legitimate claims. On all these levels of generality reflection may point out that many apparent conflicts of justice in fact can be 'genuinely' solved, i.e. without moral loss.

[13] By 'neo-Aristotelian' is meant that this approach is based on some of Aristotle's fundamental ideas; it is neo-Aristotelian because some of his conclusions about women, slavery and manual labour, as well as some of his metaphysics are rejected. Cf. Simpson (1992).

relinquishing moral theory as the final justificatory criterion for a judicial decision, a neo-Aristotelian approach is not tantamount to leaving adjudication to the workings of contingent facts, rather the contrary. How this can be, will become clear once we have expounded the central features of this neo-Aristotelian approach.

First of all, as already suggested, this approach is committed to the 'priority of the particular'; it takes the particularities of a legal case as its central focus. These particulars are the "ultimate authorities against which the correctness of particular choices is assessed" (Nussbaum 2001, 299), because it holds that in the end all actions relate to the particular or ultimate (*EN* 1143a25). Whether a judicial decision is right will therefore in the first place depend on the extent to which these particularities have been rightly addressed.

Obviously, the particularities of a case do not impose themselves on the mind or "jump to the eye" of the judge (Wiggins 1980, 232). They are a matter of perception and interpretation and thence in order for the judge to see the 'truth' about these particularities his capacity of perception is crucial. Particularities are not the "object of systematic knowledge, but of perception", Aristotle says (*EN* 1142a27-28).

A neo-Aristotelian approach to adjudication thus stresses the importance of a capacity that has been qualified as "perceptual capacity" (McDowell 1998, 51), "situation sense" (Llewellyn 1960, 59–61, 121–57, 206–208), "situational appreciation" and "moral vision" (Wiggins 1980, 233). This capacity that Aristotle names practical wisdom (*phronêsis*) should enable a judge to come to know the truth of the particularities of a case and to assess what the concrete situation requires of him. It points to an excellence of the character of the judge, a "reliable sensitivity" which enables him to see rightly and to be properly affected as he confronts a case (McDowell 1998, 51). To phrase it in Nussbaum's terms: "Practical insight is like perceiving in the sense that it is non-inferential, non-deductive; it is, centrally, the ability to recognize, acknowledge, respond to, pick out salient features of complex situations" (Nussbaum 2001, 305). As a consequence, a second feature of a neo-Aristotelian approach to adjudication is that it sees the final criterion for the rightness of a judicial decision in the virtuous person of the judge.

To be sure, the virtue of practical wisdom obviously does not suffice for a theory of adjudication, because it does not explain how the perspective of a judge differs from the perspective of a citizen or how they can be distinguished. Thereto a neo-Aristotelian approach offers an account of the judicial virtues. In two illuminating articles Solum designates a wide range of virtues that are crucial for a judge and thus offers a theory of neo-Aristotelian adjudication, drawing on the general virtues that Aristotle singled out in his *Nicomachean Ethics* (Solum 2003, 2012). The virtues that Solum designates are judicial wisdom,[14] judicial temperance,[15]

[14] For Solum this virtue indeed refers to "a judge's possession of the virtue of *sophia*, or practical wisdom (*phronêsis)* in her selection of the proper legal ends and means. Practical wisdom is the virtue that enables one to make good choices in particular circumstances" Solum (2003).

[15] Solum sees this virtue primarily as "that one's desires be in order." Cf. Ibid.

judicial courage,[16] judicial temperament,[17] judicial intelligence[18] and the virtue of justice. The latter he deems cardinal for adjudication: "If we know anything about judges it is that they ought to be just. If judges should possess any virtue, then surely they should possess the virtue of justice" (Solum 2003, 194). These judicial virtues are inextricably linked to each other. For Aristotle it is characteristic that, as John McDowell has put it, "no one virtue can be fully possessed except by a possessor of all of them, that is, a possessor of virtue in general. Thus the particular virtues are not a batch of independent sensitivities" (McDowell 1998, 70).

So, in a neo-Aristotelian approach the judicial virtues enable judges to know how to decide in a concrete case and they are central in explaining why a judicial decision is right and crucial for the justification of judicial decisions. As Solum has put this: "[…] a *lawful* decision is a decision that would characteristically be made by a virtuous judge in the circumstances that are relevant to the decision" (Solum 2003, 198).

This focus on the virtues is not to deny that principles and precedents do have an important place. They can function as "summaries or rules of thumb", they can "speed up the working through of complex material that could not be surveyed by perception in the available time" and they can "guard against corruption in situations where bias could easily distort judgments, and in general to provide a context of choice for those whose reasoning we do not really trust" (Nussbaum 1990, 99). Solum particularly underlines this in his discussion of the virtue of justice. A judge who is just "cares about the law and norms of her community. She is disposed to do that which is lawful, because she respects and internalized the *nomoi* of her community" (Solum 2012, 29).

At the same time it must be noted that within a neo-Aristotelian approach adjudication has a highly personal character, as the virtues are about the character of the judge, about his dispositions and him being affected properly when confronted with a particular situation. This, not in the least because the priority of the particular implies that the judge will always have to make a choice and decide

---

[16] According to Solum this is the virtue that "corresponds to the vice of civic cowardice. Courage is a mean with respect to the morally neutral emotion of fear. Judicial courage is a form of 'civic courage,' distinguishing this quality of character from courage with respect to physical danger. The courageous judge is willing to risk his career and reputation for the ends of justice." Ibid.

[17] As to judicial temperance Solum states: "Good temper is a mean between a disposition to excessive and deficient dispositions to anger. The virtue of good temper requires that judges feel outrage on the right occasions for the right reasons and that they demonstrate their anger in an appropriate manner." Ibid.

[18] The virtue of judicial intelligence Solum describes as follows: "The corrective for the vices of judicial stupidity and ignorance is a form of *sophia* or theoretical wisdom. I shall use the phrase "judicial intelligence" to refer to excellence in understanding and theorizing about the law. A good judge must be learned in the law; she must have the ability to engage in sophisticated legal reasoning. Moreover, judges need the ability to grasp the facts of disputes that may involve particular disciplines such as accounting, finance, engineering, or chemistry." Cf. Ibid.

how to give effect to the truths he perceives and knows. The priority of the particular demands that the judge is always open to an "indefinite or infinite range of contingencies" and realizes that he does not have an a priori answer to them (Wiggins 1980, 234).

A final important feature of a neo-Aristotelian approach to adjudication is that its logic -the logic of practice and not of theory- allows legal remainders. Such a framework acknowledges that other than as negligible incidents there will be situations in which the practical world genuinely forms an obstacle for the realization of (all of) the values that the legal order aims to protect. In such cases, even though virtuous judges can 'perceive' and 'know' what is the right thing to do, 'all things considered', they cannot get around the moral bearing of the losing interest. Sometimes judicial decisions that are right will nonetheless come with a moral loss. In spite of its commitment to moral ideals a neo-Aristotelian approach allows room for the acknowledgment of that in practice these ideals are fallible; it proposes the virtues as final justificatory ground, but at the same time it opens the door for a critical perspective by accommodating for the possibility that an excellent judge will sometimes do wrong (Barbour 1983).

On the basis of this examination we can now see that – as said – within a neo-Aristotelian approach to adjudication the practice of adjudication comes with a kind of messiness, intransparency and possibly with moral loss. In this approach judicial decisions are not justified by a fully explicit reason that supports a specific decision. The notion of the 'virtuous judge' does not provide a fully articulate set of reasons as to why citizens should accept the particular burdens they face due to adjudication. The explicit reasons that do figure are offered in terms of the law and the judge's perception of the facts. But the judge cannot offer a justificatory reason for why he sees the case as he does.

Also, as this approach takes the virtuousness of the judge as the final justificatory ground for a legal decision, it allows that two opposing judicial decisions can both be right in the same case in the same respect. That is, virtuous judges can come to conflicting decisions in similar cases.

Finally, giving primacy to the particularities of the case, these particularities may be such that a judicial decision will sometimes forego a value that in itself qualifies for legal protection.

## 13.5 Objections Against a Neo-Aristotelian Approach from the Viewpoint of Reciprocity

We have seen that the liberal principle of reciprocity demands that the burdens of the workings of political institutions (adjudication among these) can be defended in front of each and every citizen. These burdens must in some way be backed up by reasons that all citizens can endorse as part of their own good.

Below I shall assert that a neo-Aristotelian approach to adjudication as described above and defended in Solum's virtue jurisprudence has difficulty to comply with this requirement. Even if we grant that reciprocity does not require a completely transparent, intelligible, discursive answer to the question why citizens should accept the burdens brought by adjudication, a neo-Aristotelian approach to adjudication has a hard time to account for the idea of reciprocity.

We have seen that this approach takes the judicial virtuous judge as the central justificatory ground for judicial decisions. Judicial wisdom enables the judge to know what the particularities of the case legally require from him, so that he can take a 'right' decision. Obviously, this kind of justification does some work to live up to the requirement of reciprocity. The notion of the virtuous judge implies that citizens can be sure that judicial decisions are not made by incompetent judges. Also, the virtue of justice guarantees that the virtuous judge has a genuine commitment to the *nomoi*, to settled law and widely accepted legal principles. The concept of the judicially wise judge suffices to rebut the sceptical view of adjudication, namely that legal decisions are merely the enforcement of the subjective insights of the judge, a view that certainly flies in the face of reciprocity.

But it is doubtful whether the notion of judicial virtues is a sufficient condition for full compliance with the liberal requirement of reciprocity. As the 'reason' why the affected citizen must bear the burden of a particular decision it may be too elitist from the viewpoint of the liberal idea of reciprocity, or one could say: too 'mysterious' a reason. Simpson brought this point in more general terms against Aristotle's ethics: "Aristotle's court of appeal is not reason or argument but opinion – and not the opinion of all, but only of a few. These few turn out to be generally identifiable with a particular social class, the class of gentlemen. It is prejudice, not philosophy" (Simpson 1992, 513). In the end the judge can only say to the loser: "that's the way I see it, and I am a competent judge. I cannot say more than that" (Solum 2003, 201).

This is problematic, not in the least because according to this approach two virtuous judges may perfectly well come to different but nevertheless both 'right' decisions in similar legal cases. To ask citizens to accept the specific perception of the presiding judge, knowing that the outcome could have been completely different if another judge had presided, is like asking him to have some kind of awe for the judge or a deep but blind faith and trust in him. From the viewpoint of their good losing citizens have little reason to accept the judicial decision. In any case, merely referring to the 'excellence' of the judge will not suffice for liberal reciprocity.

From the viewpoint of liberal reciprocity there is another weak point in a neo-Aristotelian approach to adjudication. In this approach the actual moral losses that 'right' judicial decisions may cause, will possibly be "silenced" and thus stay unacknowledged.[19] This approach logically allows for right legal decisions coming

[19] For this notion of 'silencing' see: McDowell (1998). This critique is of course a matter for debate. Martha Nussbaum for instance contrary to McDowell suggests that it is precisely part of what it means to be virtuous that one also attends to the "intrinsic ethical character of the claim that on balance is not preferred." Cf. Nussbaum (1990, 65).

with a moral loss, but it does not secure that these situations will indeed be identified, and hence be acknowledged to the losing litigant. Precisely because he is virtuous and has to be focused on taking right decisions, the judge may decide with a kind of professional indifference. The judicial virtues do not prevent that he regards the remainders of his decision with that selfsame professional indifference.

From the viewpoint of reciprocity this is troublesome, because as a minimum it requires that genuine losses should at least be identified and recognized as such. An approach to adjudication that simply 'silences' the values that are not honoured by a particular decision, risks to not take seriously the fact that the good of citizens has indeed been sacrificed. It does not give expression to the fact that we really do not want citizens to bear such losses. In the words of Williams, to ignore such losses would be at odds with society's "decency, citizenry and respect" Williams (2001, 101). Moreover, the acknowledgement of these morally troublesome remainders may stimulate moral progress and lead to better law or more reflection that may help to avoid such losses in the future.[20]

So, from the viewpoint of reciprocity the concept of the virtuous judge does not (yet) sufficiently account for the actual acknowledgment of the possibly "troublesome" character of the moral losses it logically allows.[21]

## 13.6 The Judge as a Civic Friend

I will argue that the objections from the viewpoint of the liberal principle of reciprocity that can be raised against a neo-Aristotelian approach to adjudication can at least to some extent be resolved by the Aristotelian concept of a *civic friend* (*philia*)[22] This concept can make intelligible how judicial decisions can indeed be understood as at their best embodying an equilibrium between the interests of the losing party on the one hand and the interest of his fellow citizens and the good of society on the other. Let us therefore examine this concept more closely.

For Aristotle friendship is a relation in which the parties involved have a genuine and effective concern for one another. Friendship is "good will between reciprocating parties", Aristotle says (*EN* 1155b33-34).[23] Friendship consists of a reciprocal wanting what one thinks good. Also, it suggests the inclination to do good for the other. In the words of Cooper, friendship means "that the fact that the other person needs or wants, or would be benefited by, something is taken by the agent as by itself a reason for doing or procuring that something [...]" (Cooper 1999, 314).

---

[20] See for this point also: Nussbaum (2000).

[21] See for a discussion of the tragic character of judicial decisions: van Domselaar (2010).

[22] There is little literature on the value of the concept of friendship for law and adjudication. See for an important article on this topic: Leib (2006).

[23] Unless otherwise indicated, all the translations are made by Sarah Broadie and Christoffer Rowe. Cf. Aristotle (2002).

What is more, friends are also aware of each other's well wishing (Cooper 1999, 332). "If there is to be friendship, the parties must have good will towards each other, i.e. wish good things for each other, and be aware of the other's doing so [...]", Aristotle says (*EN* 1156a11). This mutual awareness is necessary for developing the sense of mutual trust that characterizes friendships.

At the same time a friendship relation also serves the self-interest of each friend separately because of the good that is realized through the relation. Aristotle's concept of friendship thus points to a complex and delicate combination of self-seeking and selfless concern for the good of the other (Cooper 1999, 317). Cooper illustrates this subtle conjunction by the relation that a businessman can have with a regular customer. "Such a businessman looks first and foremost for mutual profit from his friendship, but that does not mean that he always calculates his services otherwise than as a means to his own profit. So long as the general context of profitability remains, the well-wishing can proceed unchecked; the profitability to the well-wisher that is assumed in the well-wishing is not that of the particular service rendered (the particular action done in the other person's interest) but that of the overall fabric of the relationship" (Cooper 1999, 327).

Aristotle distinguishes three species of friendship: advantage friendship, pleasure friendship and character friendship (*EN* 1156a7-8).[24] As the name indicates, advantage friendship is a friendship that works to the personal advantage of the involved friends. In pleasure friendships the relationship is based on the pleasure it gives the friends. In both cases the persons involved "[...] do not love by reference to the way the person loved is, but to his being useful or pleasant" (*EN* 1156a11-17). These friendships are conditional, "for the one loved is not loved by reference to the person he is but to the fact that in the one case he provides some good and in the other some pleasure" (*EN* 1156a11-17). Character friendships, by contrast, are relations that exist exclusively because of (elements of) goodness in the character of the other. Aristotle sees this kind of friendship as the most complete because it is unconditional and hence does not depend on someone giving pleasure or being useful (*EN* 1156b7-b12). What these forms have in common is that the relations they express all involve reciprocal and effective well-wishing (*eunoia*) of the persons involved.

Now, Aristotle would probably characterize civic friendship or the kind of friendship that is relevant for a neo-Aristotelian approach to adjudication as a species of advantage-friendship (Aristotle 2002, 1160a9-a14; Cooper 1999, 333).[25] In a political community that is "animated by civic friendship, each citizen has a certain measure of interest in and concern for the well-being of each other citizen

[24] Note that for Aristotle friendship is by no means limited to the "intimate relationships between persons not bound together by near family ties". It also applies to relations between parents and children, siblings, espouses, to relations between business partners, common membership in religious and social clubs and political parties. Cf. Cooper (1999, 312).

[25] See for the assertion that civic friends cannot be genuine friends because they lack intimacy and are not genuinely living together: Annas (1987).

just because the other *is* a fellow-citizen" (Cooper 1999, 371). Civic friendship is a relation that is based on the mutual expectation of citizens that the political community, the *nomoi*, works for the sake of their good. "Civic friendship makes fellow-citizens' well-being matter to one another, simply as such" (Cooper 1999, 371). So, civic friendship, like other forms of advantage-friendship, is really a friendship. In a society animated by civic friendship citizens are willing to bear the burdens of the central political institutions because they expect them to work for their good, although this need not be the case for every single political decision. Civic friends are willing to sacrifice their own particular interests for one another, lest this does not cost too much. The sacrifice for instance must not put into peril the overall profitability of being an active participant in a political community (Cooper 1999, 328).

Aristotle thus makes the psychological claim that once citizens have experienced the benefits of living in a political community, as a result they are willing to sometimes sacrifice their own immediate interests and to act in the interest of their fellow citizens (Cooper 1999, 323). In the words of Cooper, as civic friends "[t]hey are accommodating rather than suspicious, anxious to yield a point rather than insisting on the full letter of their rights whenever some dispute arises" (Cooper 1999, 333).

This accommodating attitude is also due to the fact that (civic) friends see one another as 'another self' (*heteros autos*), as people with whom, depending on the character of their friendship, they share a fundamental aspect of their identity. In a society constituted by civic friendship fellow citizens approach one another as people with similar powers and vulnerabilities, with a similar overall interest in leading a good life.

So, if we understand adjudication also as an expression of civic friendship this would yield that citizens who are detrimentally affected by judicial decisions can be confident that these decisions are also the result of an effective attempt to serve his concrete good. The judicial decision not only is right, but can also be taken as an expression of well-wishing by the judge. It is this well-wishing together with the judge being virtuous that qualifies the discretion that the judge inherently has. Despite the painful consequence, the losing litigant can accept this decision as the result of a genuine attempt to balance the common good – including the good of other citizens – on the one hand and his concrete good that the judge as a civic friend aims to take into account and serve on the other.

The judge and the legal order as a whole in turn can rely on the citizens to accept the decision as being a part of fulfilling his duty as good-willing fellow citizen who because of his accommodating attitude will be willing to make a sacrifice. Moreover, by accepting the judicial decision, the losing litigant is assumed to express his commitment to the political order and the values it aims to protect.

Again, the 'application' of the *nomoi*, of settled law and the underlying principles of political morality to the particular case is not enough to come to this accommodating attitude. As we have seen, it leaves too much discretion to the concrete judge so as to prevent arbitrary outcomes. In addition, this 'application' is not sufficient for the trust needed on the part of the affected citizen for him have a reason to accept the painful decision. Aristotle articulates this point clearly where he says that "[…] there is no need for rules of justice between people who are friends, whereas if they are just they still need friendship […]" (*EN* 1155a23-25).

To be sure, reciprocity in adjudication cannot be realized through the mere 'well-wishing' by the judge and the litigants concerned. Whether the principle of reciprocity is satisfied, will obviously also depend on the background institutions and the workings of the main political institutions in society. It is doubtful whether citizens have reason to accept the burdens of judicial decisions if they do not also rely on the 'well-wishing' of all political institutions, such as the constitution, legislation and policy.

If we understand reciprocity to be secured through the civic friendship between the judge and the affected citizen, we can also make intelligible why it is that a judge will sometime consider himself smeared or stained by a judicial decision that seriously harms the chances of a concrete citizen to lead a dignified life: as a civic friend he genuinely cares, he can feel the loss himself, experience a sense of tragic remorse, and give expression to the principle of reciprocity by acknowledging that a genuine good is foregone.[26]

Civic friendship allows that a judicial decision is right and made by a virtuous judge, but nonetheless comes with a moral remainder due to the contingencies of the particular case. That is, the losing citizen may be asked to sacrifice a value or an interest that the political community truly sees as it tasks to protect.

In the previous section we have seen that a neo-Aristotelian approach to adjudication by itself by no means guarantees that the judge will perceive this wrongdoing, this moral loss as troublesome. Being virtuous and having fulfilled his professional role as judge well, does not guarantee that the judge will give a 'moral-remainder-responsive reaction'. Civic friendship makes explicit acknowledgement of the actual moral losses that are sometimes produced by judicial decisions more likely. If the judge is both virtuous and a civic friend – albeit qualified by his viewpoint as a judge – than the good of the litigants is part of his own good.

## 13.7 Does Civic Friendship Apply to All Areas of Law?

We have seen that civic friendship is a useful concept for coming to grips with the requirement of reciprocity in adjudication. However, the question then arises whether the concept makes sense for all (kinds of) legal cases that judges may have to decide.[27] Perhaps it does not suit some kinds of cases or requires further elaboration or a particular caveat for others.

For reasons of scope, I will briefly discuss only one category of legal cases for which the concept of civic friendship arguably is not appropriate. This is the category of cases about citizens having seriously violated or being accused of having

[26] See for a discussion of tragic remorse: de Wijze (2004).

[27] I am indebted to an anonymous referee for this point.

seriously violated the basic interests of other citizens and thence also the good of society.[28] For instance, it could be argued that serious penal cases are characterized by disrupted or broken relations, rather than by relations of civic friendship.

A few remarks suffice as a provisional answer to this point. First Aristotle's notion of advantage friendship and the interpretation of reciprocity implied therein does not entail that each and every activity in the context of the relation must be advantageous to the parties involved. Nor must these activities be in the interests of the parties involved as they conceive them at that particular occasion. This is also expressed in the fact that Aristotle sees failures and wrongdoings as part of the practices that constitute friendship. Some of the relations that Aristotle qualifies as forms friendships, e.g. the relation between parent and child, or between teacher and pupil are clear examples of relations that cannot be characterized without pointing to experiences of failure and wrong-doing (Cooper 1999, 312).

As we have seen, what matters is whether the overall fabric of the relationship is advantageous (Cooper 1999, 327). In order to maintain a friendship with a citizen and hence to wish this citizen well he must be able and willing to provide for the kind of good "in that respect in which they are friends" (Cooper 1999, 326). What is important when a citizen fails or does wrong is that he remains the person who in one way or another can be thought of as advantageous to the political community. This means that the advantageous character of a relation obviously does not directly 'evaporate' when a citizen commits a serious crime or other kinds of serious unlawful wrongdoings. Having committed serious wrong does not indicate that the profitability of the relation and therefore the friendship has ended. Even if this behaviour is structural, the citizen in question may still maintain the properties that enable him to contribute to a mutually advantageous relation.

One could say that civic friendship comes with the acknowledgment that a fellow citizen is never "intrinsically evil, but a human being like us, with diverse frailties and weaknesses, who has encountered circumstances – whether personal or social – that bring out those weaknesses in the worst possible way" (Nussbaum 1998). It is precisely because the wrongdoer is understood as 'another self', as someone with in principle similar potentials and frailties as all others, that in these kinds of situations one can hold him responsible and at the same time stay committed to the friendship.

Also, upholding a relation of civic friendship in case of wrongdoing may further the likelihood of a future cooperative attitude of the 'failing' citizen. Maintaining a relation of civic friendship with a wrongdoer can still be profitable for the common good, for instance through the several social roles that he will (continue to) perform in society.

---

[28] These cases may fall in the domain of criminal law, but may also come up in other areas of law, like tort cases in civil law or negligence and nuisance cases in environmental law. Here I simply accept the legitimacy of the criminal law. It remains to be seen whether civic friendship can also lead to a fundamental critique on the criminal law system as it is now in Western constitutional democracies.

In addition, to approach citizens who have committed crimes as civic friends also has an epistemological value for society as a whole. This sympathizing attitude may enhance our understanding of what citizens in general need in order to be able to become good citizens because it "creates incentives [...] to think hard about those circumstances, so that we do not put people under pressures that many normal agents cannot stand" (Nussbaum 1998).

Hence, the provisional answer to the question whether the concept of civic friendship applies to legal cases of serious wrongdoing is affirmative. In these cases the judge can confront him with the conditions of civic friendship and hold him responsible according to these conditions. He can highlight the fact that the wrongdoer is part of a practice that aims to serve the good of society, including his good and the good of his fellow citizens.

Of course, a judge may find it difficult to have a sympathizing view of criminals or citizens who otherwise deeply violate the interest of others. He may have a hard time trying to 'see' them as "other selves". This not in the least because such a sympathizing view may confront the judge with serious moral dilemmas, because the commitments of a civic friend on the one hand and those of a virtuous judge on the other may not always harmonize.

Obviously, in extreme cases the relation of civic friendship does come to an end, simply because reciprocity is lacking. This may for instance be the case if the wrongdoer lacks the cognitive and moral capacities to be effectively committed to the good of society and of others. It may also end if a citizen who has violated the interests of others has structurally and genuinely expressed that he does not wish his fellow citizens well and that he has no concern for the good of society. Then there will be no way of seeing the relation as (potentially) advantageous for society. In these extreme cases the judge still has to comply with the basic values and rules of society on how to treat the wrongdoer in question, but he need not act as a civic friend.

## 13.8 Conclusion

In this chapter I have discussed the question how we can best account for reciprocity in adjudication in constitutional democracies. Judicial decisions seriously affect citizens' life and we want these decisions to be fair. We want to offer a "losing" litigant a reason why he should accept the particular burden stemming from that decision.

For this purpose I have examined two approaches of adjudication, i.e. 'adjudication as applied moral theory' and a neo-Aristotelian approach to adjudication. Characteristic for 'adjudication as applied moral theory' is that it accounts for reciprocity in a rather straightforward way and that it conceives of adjudication as a stable, transparent and safe practice. I have argued that this approach cannot account for reciprocity for concrete judicial decisions, because moral theory cannot prevent arbitrary factors to determine concrete judicial outcomes. Therefore, I introduced a neo-Aristotelian approach to adjudication. This approach does not rely on moral theory for the justification of judicial decisions, but rather on these decisions being made by a virtuous judge. In this approach the practice of adjudication has a certain messiness, intransparency and painfulness to it.

I have argued that from the viewpoint of reciprocity Aristotle's concept of civic friendship is a necessary amendment to a neo-Aristotelian approach to adjudication. Civic friendship can indeed offer affected citizens a reason to accept the burdens of judicial decisions, namely that the judge uses his leeway from a disposition of well-wishing, with a keen eye for the concrete good of the citizens involved. Also, to see judicial decision-making as an expression of civic friendship makes it more likely that the judge will acknowledge the moral losses that judicial decisions possibly produce. Being a civic friend, the judge as 'another self' will experience the limits of realizing reciprocity himself as painful and will feel duty-bound to acknowledge the genuine loss that flows from his decision and the burden that the losing citizen is asked to bear.

Obviously, this rather tentative proposal for this specific neo-Aristotelian approach to adjudication and hence interpretation of reciprocity in adjudication needs further elaboration. Its (practical) implications must be fleshed out. For instance, it remains to be seen what the implications for legal education are if the judge's professional task is to be both judicially virtuous and a civic friend. Next, Aristotle's concept of civic friendship may be too 'thick' a concept to be a prominent part of our understanding of adjudication in liberal societies. Another question that should be addressed is what consequences a neo-Aristotelian approach to adjudication has for the 'art' of talking to the loser. To be sure, we do not want adjudication to be a practice that fosters sentimentality and pathos or categorically denies that its decisions sometimes produce genuine moral loss. Where to establish the equilibrium remains to be seen.

**Acknowledgement** The author wishes to thank her colleagues Liesbeth Huppes-Cluysenaer, Frans Jacobs, Dorien Pessers and Joep van der Vliet for discussions and their generous help in preparing this chapter. In particular she owes thanks to an anonymous referee for his constructive and insightful comments.

## Bibliography

Alexy, Robert. 2004. The nature of legal philosophy. *Ratio Juris* 17(2): 156–167.

Altman, Andrew. 1986. Legal realism, critical legal studies, and Dworkin. *Philosophy and Public Affairs* 15(3): 205–236.

Annas, Julia. 1987. Comments on John M. Cooper's "Political animals and civic friendship". Paper read at XI. Symposium Aristotelicum, at Friedrichshafen/Bodensee.

Aristotle. 2002. *Nichomachean ethics*: *Translation, introduction, and commentary.* ed. Sarah Broadie, and Christoffer Rowe. Oxford: Oxford University Press.

Barbour, John D. 1983. Tragedy and ethical reflection. *The Journal of Religion* 63(1): 1.

Claassen, Rutger, and Marcus Düwell. 2012. The foundations of capability theory: Comparing Nussbaum and Gewirth. *Ethical Theory and Moral Practice* 1–18. http://link.springer.com/article/10.1007/s10677-012-9361-8

Cooper, John M. 1999. *Reason and emotion: Essays on ancient moral psychology and ethical theory*. Princeton: Princeton University Press.
de Wijze, Stephen. 2004. Tragic remorse: The anguish of dirty hands. *Ethical Theory and Moral Practice* 7: 453–471.
Dworkin, Ronald. 1985. *A matter of principle*. Cambridge, MA: Harvard University Press.
Dworkin, Ronald. 1986. *Law's empire*. Belknap: Cambridge, MA.
Dworkin, Ronald. 2005. *Taking rights seriously*. London: Duckworth (1977).
Dworkin, Ronald. 2011. *Justice for hedgehogs*. Cambridge, MA: Harvard University Press.
Frank, Jerome. 1931. *Law and the modern mind*. New York: Brentano's Publisher.
Geuss, Raymond. 2008. *Philosophy and real politics*. Princeton: Princeton University Press.
Hackett, Rosalind I.J. 2001. Is religion good news or bad news for women? Martha Nussbaum's creative solution to conflicting rights. *Soundings* 83(3–4): 615–25.
Hart, H.L.A. 1961. *Clarendon law series*. Oxford: Oxford University Press.
Hawthorn, Geoffrey (ed.). 2005. Bernard Williams. In *In the beginning was the deed: Realism and moralism in political argument*. Princeton: Princeton University Press.
Holmes, Oliver Wendell. 1897. The path of the law. *Harvard Law Review* 10: 457.
Kramer, Matthew H. 2007. *Objectivity and the rule of law, Cambridge introductions to philosophy and law*. Cambridge: Cambridge University Press.
Leib, Ethan J. (2006). Friendship and the law. *University of California Law Review* 54: 631–707.
Llewellyn, Karl. 1960. *The common law tradition: Deciding appeals*. Boston, MA/Toronto: Little, Brown and Company.
Lyons, David. 1984a. *Ethics and the rule of law*. Cambridge: Cambridge University Press.
Lyons, David. 1984b. Justification and judicial responsibility. *California Law Review* 72(2): 178–199.
McDowell, John. 1998. *Mind, value and reality*. Cambridge, MA: Harvard University Press.
Nussbaum, Martha. 1990. *Love's knowledge: Essays on philosophy and literature*. Oxford: Oxford University Press.
Nussbaum, Martha. 1998. Victims and agents: What Greek tragedies can teach us about sympathy and responsibility. *Boston Review*.
Nussbaum, Martha. 2000. The costs of tragedy: Some moral limits of cost–benefit analysis. *Journal of Legal Studies* XXIX: 1005–1036.
Nussbaum, Martha. 2001. *The fragility of goodness: Luck and ethics in Greek tragedy and philosophy*, Rev ed. Cambridge: Cambridge University Press
Nussbaum, Martha. 2006. *Frontiers of justice: Disability, nationality, species membership*. Cambridge, MA: Harvard University Press.
Nussbaum, Martha. 2007. The Supreme Court, 2006 term. Foreword: Constitutions and capabilities: 'perception' against lofty formalism. *Harvard Law Review* 121(1): 4–97.
Nussbaum, Martha. 2011. *Creating capabilities: The human development approach*. Cambridge, MA: Harvard University Press.
Rawls, John. 1971. *A theory of justice*. Cambridge: Harvard University Press (1999).
Rawls, John. 2005. *Political liberalism*. New York: Columbia University Press.
Richardson, Henry S. 1997. *Practical reasoning about final ends, Cambridge studies in philosophy*. Cambridge: Cambridge University Press.
Scharffs, Brett G. 2000. Adjudication and the problems of incommensurability. *William and Mary Law Review* 42: 1369–1435.
Schauer, Frederick. 1985. Easy cases. *California Law Review* 58: 399–440.
Simpson, Peter. 1992. Contemporary virtue ethics and Aristotle. *Review of Metaphysics* 45(3): 503–524.
Solum, Lawrence B. 2003. Virtue jurisprudence: A virtue-centered theory of judging. *Metaphilosophy* 34(1/2): 178–213.
Soper, Philip. 1977. Legal theory and the obligation of a judge: The Hart/Dworkin dispute. *Michigan Law Review* 75(3): 473–519.
Timmons, Mark. 1987. Foundationalism and the structure of ethical justification. *Ethics* 97(3): 595–609.

Van Domselaar, I. 2010. Tragic choice as a legal concept. *Archiv für Rechts- und Sozialphilosophie* 119: 105–116.
Waldron, Jeremy. 1987. Theoretical foundations of liberalism. *The Philosophical Quarterly* 37(147): 127–150.
Wellman, Christopher Heath. 1995. On conflicts between rights. *Law and philosophy* 14(3/4): 271–295.
Wenar, Leif. 2011. Rights. In *The Stanford encyclopedia of philosophy*, ed. Edward N. Zalta. Stanford: Stanford University.
Wiggins, David. 1980. Deliberation and practical reason. In *Essays on Aristotle's ethics*, ed. Amélie O. Rorty. Berkeley: University of California Press.
Williams, Bernard. 1985. *Ethics and the limits of philosophy*. London: Fontana Press.
Williams, Bernard. 2001. Liberalism and loss. In *The legacy of Isaiah Berlin*, ed. Mark Lilla, Ronald Dworkin, and Kelly Aileen. New York: New York Review Book.

# Chapter 14
# Synallagma as a Paradigm of Exchange: Reciprocity of Contract in Aristotle and Game Theory

**Mariusz Jerzy Golecki**

## 14.1 The Problem of Equivalence in Contracts

The problem of reciprocity in contracts is closely related with the notion of commutative justice.[1] It is Aristotle who distinguished the difference between distributive and commutative justice in the V-th book of *Nicomachean Ethics*.[2] Distributive justice is based on geometric proportion, depending on the hierarchy of public goods. Generally it may be said, that the norms of distributive justice allocate the goods to persons according to the criterion of distribution established by the public authority. Corrective and rectificatory justice is based on arithmetic proportion between the gain and loss, especially in case of torts or unjust enrichment. According to the contemporary interpretations, corrective justice encapsulates certain notion of equality (Benson 1992, 535). This equality is often called "formal equality of treatment". There is, however, the third kind of justice, namely commutative justice. Commutative justice concerns equality in exchange. Within the literature commutative justice is very often identified with corrective justice. This is

[1] For discussion about the Aristotelian concept of justice in contracts cf: Gordley (1981, 1587), Gordley (2001, 297–326), Murphy (2002, 85), Atiyah (1985, 1), Golecki (2008, 26–111 and 266–288).

[2] The significance of this distinction for contract law has been emphasized in Gordley (1981, 1588–1592).

M.J. Golecki (✉)
Department of Legal Theory and Philosophy of Law, Faculty of Law and Administration, University of Łódź, Łódź, Poland, ul. Kopcińskiego 8/12, 90–232
e-mail: mjgolecki76@gmail.com

L. Huppes-Cluysenaer and N.M.M.S. Coelho (eds.), *Aristotle and The Philosophy of Law: Theory, Practice and Justice*, Ius Gentium: Comparative Perspectives on Law and Justice 23, DOI 10.1007/978-94-007-6031-8_14, © Springer Science+Business Media Dordrecht 2013

partly due to the confusion in the text of *Nicomachean Ethics*. Aristotle in one place opposes commutative justice to distributive justice. He explains, however, the difference between voluntary legal relations, namely transactions, and involuntary relations, such as wrongdoing. Voluntary transactions are not assessed from the perspective of equality between the loss and the gain. They are based on commutative justice identified with terms of contract itself. The confusion about the character of commutative justice is in my opinion the reason for many inadequate theories of contract. The question arises whether contract may be objectively just and if so, whether just contract is in fact a contract according to which the gains of two parties are equal.

All these issues seem to pertain to the problem of interdependence between two parties of contract and their respective obligations. Such interdependence has so far been analysed in moral philosophy and could eventually be modelled by game theory. Historically speaking the tension between the free-bargaining principle and the requirement of equity or fairness in exchange may be indicated. On the one hand, the majority of legal systems do not require the existence of equivalence between obligations, especially in case of bilateral (synallagmatic) or reciprocal contracts. According to this doctrine of genetic synallagma, the inner structure of such a contract consists of reciprocal obligations i.e. obligations of two parties. The interdependence between the obligations of both parties means than both of them are potentially liable, not that the economic values of obligations are equal. The system of common law does not require equivalence of consideration, the French law regards reciprocity as subjective as well as BGB. At the same time all legal systems contain some mechanisms to correct harsh bargains or grossly unfair transactions.

The problem of equivalence in contracts is encapsulated in the doctrine of synallagma, and thus it belongs to the wider theory of commutative justice. The term *synallagma* refers to any interaction between the parties and an existing interdependence of their respective legal and economic positions. Therefore, if it is considered as a strategic interdependence, synallagma may be analysed from the perspective of ethics and game theory. Nevertheless, before implementation of the game theory, it is necessary to present briefly the legal meaning and content of the term of synallagma as a part of Aristotelian theory of justice in exchange.

## 14.2 Aristotelian Reciprocity Based Theory of Commutative Justice

The analysis of the concept of justice in reciprocity proposed by Aristotle originally aimed at generalising the rules of distribution of goods within the public and private context. The notion of distributive justice explicitly refers to the rule on allocation of a part of the given whole based upon the principle of geometrical proportionality. Additionally Aristotle applied the concept of corrective (rectificatory) justice pertaining to restoration of initial balance between parties. The compensation for wrongful acts was thus to be based on the principle of arithmetic proportionality.[3] According to Aristotle, the damage which is being committed by the wrong-doer and the damage which is suffered by the victim, are not necessarily identical, therefore the wide scope of indeterminacy exists between the subjective nature of an act of injustice and the actual consequences of this act (*EE* V.5.1133 a).

Concurringly the exchange of various goods could at least to some extend be connected with loss to one party and the profit to the other one.[4] These observations led Aristotle to distinguishing a third kind of justice, namely the commutative justice based on the so called mixed proportion.[5] Commutative justice pertains to voluntary exchange and reflects the proportion between the valuations of the exchanged goods and services referred to relative needs of both parties.[6] Aristotle used the notion of proportionality and natural justice in order to define the "natural" limits of exchange,[7] and he linked these notions with the Greek concept of harmony (Anscomb and Geach 1961, 73–74). He also emphasized that commutative justice as the requirement of morality should not be based the rules of distributive or corrective (rectificatory) justice (*EE* V.5.1132 b 20–25 and *EE* V.5.1132 b 30–34).

The crucial issue concerning the application of commutative justice within the social sphere refers to the fact that in social relations the sum of subjective values is not necessarily equal to the same values evaluated in reference to some objective benchmark. As a response to this problem Aristotle developed the notion of social aggregation as the sum of subjective utilities greater than the whole sum evaluated according to some objective measures (Lee 1937, 129–131; Joachim 1951, 130). The concept of utility as endorsed by Aristotle had been referred to the subjective measure of need (gr. *chreia*) ascribed to the parties.

The process of evaluating the just proportion within the exchange between two parties whose aim is to get the goods from each other heavily depends upon the will of party A to possess the goods of his/her counterparty B more than possessing one's one goods, and vice versa. Accordingly let A and B denote two respective parties to the contract whereas C denotes A's obligation and D denotes B's obligation. The bargain could thus be characterised by the relation between A's valuation of D and B's obligation of C.

[3] An example may serve the rule that two thirds of the profit should belong to this partner who had given two thirds of the original capital for a given enterprise. See Lowry (1969, 47–49).

[4] Within this context Aristotle used the example of exchange between a house-builder and a shoe-maker. (*EN* V. 5.1133 a 10–15 and *EN* V.5.1133 b 1).

[5] Cf. Aristotle plainly distinguishes between corrective and commutative justice, admitting that: "Some hold that the reciprocity is just without qualification. This was the claim of the Pythagoreans, since they defined, without qualification, what is just as reciprocity with another. Reciprocity however fits neither distributive nor rectificatory justice, since often they conflict. (...) Again voluntariness and involuntariness make great difference" (*EN* V.5. 1133 a 1–10, trans. R. Crisp).

[6] In classical comment on the function of commutative justice in *Nicomachean Ethics*, D. G. Ritchie endorses that: "It seems to me quite certain that Corrective Justice is intended to apply to voluntary contracts, only when the terms of the contract have not been fulfilled." Ritchie (1894, 188).

[7] These reflections were later applied within the scholastic debates concerning the notion of a "fair price". See Lowry (1969, 49).

The ratio A/C : B/D refers to the utility based valuation of two respective, mutual obligations from the perspective of A, whereas the other ratio A/D : B/C pertains to the utility based valuation of two respective, mutual obligations from the perspective of B (Wesoły 1989, 218). Eventually the just exchange should then be founded on following proportion:

$$A / C : B / D :: A / D : B / C$$

These relations form initial conditions for the existence of any commercial transaction (Meikle 1990, 161–162). It may be assumed therefore, that the motive encouraging people to conclude agreements and to perform respective obligations arising from contracts is not duress or relation of dependence but rather the existence of mutual benefits arising from exchange. One then may ask a question whether it is possible to make an exchange which would be unjust in relation to the party engaged in this exchange.

The answer offered by Aristotle is based on the assumption according to which in case of lack of mutual benefits there is no exchange, and hence the sole fact of exchange contains also its explanatory factor. This assumption had been an integral part of the Greek law of sale, in which the contract was not binding until the proper exchange was done. On condition that it was performed in a voluntary way, this exchange was regarded as a just one (Pringsheim 1950, 130–137). The notion of a subjective utility within exchange seems to refer to the analysis of reciprocity on the basis of a mathematical proportion. The fundamental statement of Aristotle concerning the exchange among two parties on the basis of the commutative justice is made in regard to the example with a house-builder and a shoe-maker[8] (*EN*.V.5.1133 a 5–15) (Fig. 14.1).

Aristotle treats the subjective utility as an introductory condition of exchange. It seems that Aristotle used the "natural" basis of a voluntary exchange in which both of the parties compare one's own goods with the goods of the other party in a subjective way, and in the same time defining the sphere of mutual benefits conducive to the act of exchange.

The issue concerning a legal justice within exchange was regarded by Aristotle as aiming at defining the middle of the sphere of a voluntary choice by both of the parties (NE V, 5, 1133 b 30). He assumed that there exists the "best" set of laws or the "fairest" price of exchange within the limits of choice and mutuality. The relationship of competition among the parties is evident even before the conclusion of a contract. The factual price is precisely defined either by the parties themselves or by the court as a result of a judicial process.

Thus the social relationship of exchange reflected by the term *synallagma* may be regarded as a meta-legal structure of contractual transaction. This transaction is supposed to have a pre-legal character and it may be pointed out that game theory

---

[8] It is not certain whether it was Aristotle himself or the work of the later commentators who had attached a diagram of the "figure of exchange" to the text. See: Lowry (1969, 57–58).

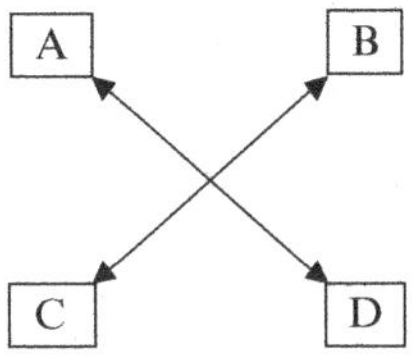


**Fig. 14.1** The relation between exchanged commodities and exchanging parties

and the economic analysis of contracts stress the non-legal essence of contract. Thus *synallagma* has been placed in a different normative system, based on the model of allocations, which is performed accordingly to Pareto-optimum. The model of such exchange seems to have been purported by political economy. Here it is worth to concentrate on the relation between *synallagma* perceived as a legal "matter" and the model of Pareto-optimum regarded as a meta-legal "form". Consecutively the ethical doctrine of Aristotle has been compared to the neo-classical school of economics and perceived as a background for economics concept of value formation.[9] J. Schumpeter explained that the conceptions precisely defined by Aristotle had been applied as tools for analysing the mechanisms and economic institutions existing in ancient Greek society. Aristotle's scientific interests led him to a broader analysis of the market and a rudimentary price theory. According to Schumpeter, his concepts were used later on in neo-classical welfare economics and it seems that Aristotle himself would have used the prices defined within a context of the competition process as the criteria of commutative justice (Schumpeter 1954, 58–61).

The theory of marginal utility presented by Jevons may also be characterised as containing the solutions "in spirit and form, assumptions and conclusions" analogous to the theory of Aristotle.[10] K. Polanyi criticised the analysis of Schumpeter, making a claim that he did not understand the Greek economic conditions in a proper way. According to Polanyi, the aim of Aristotle was not to create a unique economic theory but rather to elaborate on a social theory enabling one to analyse phenomena of a serious meaning and a profound moral dimension, i.e. aiming at reaching "unlimited profit" within commercial activities.[11]

M. Finley stresses that the lack of economic analysis within the theory of Aristotle does not stem out from his lack of knowledge about free-market mechanisms but rather from his steadfast conviction that the morally justifiable degree of richness, which is necessary in order to lead an ethical life is very much limited (Finley 1970, 3–25). Moreover, Aristotle regarded economic activities as rooted in social practices and this conviction resulted in his lack of interest in analysing the mechanisms of the functioning of the market itself.

[9] The connection between Aristotelian theory of justice and development of economic theory of value has widely been discussed in literature. See: Worland (1984, 107–112), Langholm (1979, 164), Tieben (2009, 3–11).

[10] See Soudek (1952, 66–72).

[11] Moreover, K. Polanyi points out that the mechanisms of supply and demand, which are acting within the sphere of free-market, and which are the subject of analysis of the neoclassical theory of economics, were not known to Aristotle; see Polanyi (1957, 64–94).

It seems that the linkage between the theory of Aristotle and the economic theory may be determined also within the normative perspective. If the moral choice contains the making of selection among competitive aims, then the moral agent should seek the maximisation of the highest "meta-aim", pursuing the other aims only until the moment when the benefit resulting from the indirect aim is equal to the costs included in the non-undertaken alternatives. Within the economic sciences the notion of the highest aim concerns utility, and the theory of Aristotle points out in this regard to happiness. This principle seems to be the fundamental moral imperative concerning the general market equilibrium theory (Arrow 1951). The above-mentioned principle is linked to the rule of Aristotle, which does not have solely the character of a moral norm or a technical rule connected to the moral choice.[12] If the aim of economic activity is defined as supplying in a greatest possible degree the material tools for the full cultural development of men, then society should elaborate on a kind of procedure for achieving efficiency while using the existing resources, similar to that of a free-market economy, based on the principle of justice.[13] The above interpretation leads to distinguishing two kinds of relationship between the moral theory of Aristotle and the neo-classical theory of economics. The theory of Aristotle may be regarded as the ethical foundation for the economic theory, whereas the economic theory of contract determines necessary conditions for a proper moral choice.

The other subject is connected with the formalisation of the concept of commutative justice within the field of contracts. This line of reflection aims at specification and further development of the model of mixed proportion, which was formulated by Architast of Trent and subsequently presented by Aristotle in the *Nicomachean Ethics*, into the direction of possible application within the contemporary economic and social conditions of the so-called market society.[14] It seems that it is possible to implement game theory in a way which enables one to find out the relationship between the concept of reciprocity and the formal expression of this relation in a form of mathematical model corresponding to the Aristotelian notion of arithmetical proportion. Therefore, it is a method of creation of social and economic conditions, which make it possible to implement the postulate of equality in relations of exchange.

---

[12] W. D. Ross in his classical work on Aristotle states however, that: "*There is no moral virtue in commercial justice as described by Aristotle. "Justice" here is not a virtue, but a sort of "governor" in the economic machine which keeps exchange prices from swinging far from the actual value, for human needs, of the goods exchanged*". Ross (1953, 213).

[13] Within this context one can underline an interesting remark of J.N. Keynes. While dealing with the scope of methodology of economic sciences. He stressed the existence of an intermediate sphere between the science of political economics and the area of applied economics, which subject is situated not only within the economic perspective but also the moral dimension of economic activity of the society. Cf. Keynes (1955, 61–63).

[14] As far as the notion of "market society" is concerned, see: Polanyi (1944, 23–45).

## 14.3 Synallagma as a Paradigm of Reciprocity Based Contracts

*Synallagma* is an important concept within the contemporary legal science. It is particularly significant within a context of the structure of bipolar obligatory legal acts. Modern theory of contract law concentrated on systemic reflection upon the characteristics of *synallagma*, finally having distinguished different forms of *synallagma*, referring to different character of relations between parties to the contract. According to this proposition the principle of reciprocity in contracts may thus be expressed in three forms. Firstly, it concerns the genetic *synallagma* when the existence of obligation of one party is dependent on the corresponding obligation of the other party. Secondly, the conditional *synallagma* refers to the execution of obligation and, in particular, the situation when the execution of task by one party creates the obligation to execute the task by the other party of a contract. Traditionally, this type of *synallagma* is connected with the possibility of raising the exemption of *non adimpleati contractus*. Thirdly, the functional *synallagma* concerns the internal aspect of contract and is connected with the notion of mutuality and subjective equivalency of tasks. The essence of the relationship among obligations in this case is that the obligation of one party is the equivalent of the other party's obligation or that the obligation of one party is being regarded by the other party as an equivalent one. As far as the contemporary civil law systems are concerned, generally they adopt the subjective version of equivalency.

Moreover, the true meaning of classification of the various kinds of *synallagma* depends on the type of construction of the obligatory contract, adopted within the legal system. Thus, the issue of genetic *synallagma*, or the state of dependence of the creation of one obligation on the creation of the other, within the German legal system is linked to the general rule of causality of obligatory legal acts and the system of making the obligatory contracts based upon §§ 107, 134, 138, 139, 154, 155, 306 BGB. Within the French legal system genetic *synallagma* has a close link with the rule of causality of obligatory legal acts, which is expressed in art. 1131 c.c.fr. and within common law it is combined with the doctrine of consideration. Similarly, Polish Civil Code in art. 487 § 2 defines the mutual contract as a contract where both parties agree to create corresponding mutual obligations. The notion of genetic *synallagma* is also linked to a group of theories developed within the dogmatic studies of civil law about the mechanism of completion of obligatory contract, such as the theory of aim exchange, theory of the basis of legal act, conditional theory and the theory of legal act's.

While the reflection upon the meaning and construction of *synallagma* within legal theory and philosophy of private law concerns a group of particular legal acts such as obligatory contracts and is linked primarily to relationships among the obligations made by both parties to the contract, the meaning of the notion of *synallagma* within philosophy of law seems to be much wider.[15]

[15] The relationship between *synallagma* and legal regulations of contract and tort in Ancient Greece has been widely discussed. See: Lee (1937, 131), Hamburger (1951, 39–65), Maffi (1980, 17).

The notion of *synallagma* may be traced back to the philosophy of Aristotle, and in particular to his definition of commutative justice based on arithmetic proportion (Despotopoulos 1968, 115; Lowry 1969, 47; Harrison 1957, 44). It seems worthy to explain in this moment some terminological misunderstandings. In stricter terms, justice is defined as the practice of justice, and hence human acts may be analysed as just or unjust. Aristotle distinguishes two categories of justice: distributive justice and commutative (corrective) justice (Finnis 1980, 177–184; Golecki 2008, 26–46). As in the latter case there exists two names, it is useful to explain that commutative justice is a more suitable term to indicate voluntary social relations (*synallagmata hecousia*) while corrective justice should indicate involuntary social relations (*synallagmata acousia*).[16]

In fact, the reflection upon *synallagma* aims at the reconstruction and operationalization of the conception of corrective and commutative justice. Both notions are linked to *synallagma* understood as an obligation-raising relationship of a social (inter-human) character or as a private relationship. Therefore, reconstruction of the theory of justice seems to be a necessary condition for the reconstruction of the notion of *synallagma.*[17]

As far as the reconstruction of the theory of commutative justice is concerned, the main issue is linked to establishing the relations between *synallagma* and arithmetic proportion, which is a form of commutative justice. In a broader sense, however, this problem indicates a general economic analysis presented by Aristotle in *Nicomachean Ethics* when he used mathematical theory in order to conceptualize moral and political theory concerning the normative conditions of social relationships or such theory that would provide these conditions.

The other problem is linked to the question whether *synallagmata* are created as a kind of remedy for the principles of commutative justice, i.e. as a reaction for the existing inequality between the parties and the necessity to restore the previous state through the execution of obligation to give back all the gains by the party which in fact had received unjustified benefits. Another situation is linked to the principle of equality among the parties which is expressed in arithmetical proportion as a normative postulate resulting from Aristotle's ethics concerning such formation of obligation-raising relationships of a social character which would in fact be the execution of this very postulate.

There is also a question on the ontological status of *synallagma* understood as an obligation-raising relationship of a social character. In particular one should confirm whether such relationship has a legal or pre-legal character or whether the obligations resulting from the existing social relationships are legal or moral in nature. It seems that the answer to the former issue indicates that Aristotle's theory

---

[16] *EN* 1131 a 2–3. The distinction has been widely discussed in literature. See: Jackson (1879, 76), Gauthier and Jolif (1958, 391).

[17] D. G. Ritchie states that: "It seems to me quite certain that Corrective Justice is intended to apply to voluntary contracts, only when the terms of the contract have not been fulfilled" Ritchie (1894, p. 188).

of justice is linked to legal positivism while the latter demands further explanation if these relationships have only an empirical character being an example of human activity on communal level or whether they belong to the other normative system than positive legal order, having moral or natural legal character (Winthrop 1978, 1201–1216; Yack 1990, 216–237).

One should also bear in mind the differences between the conception of *synallagma* in the philosophy of Aristotle and the later interpretations identifying *synallagma* as well as the arithmetical proportion being a foundation of commutative justice, indicating a necessary element of an objective character, namely the equality among the obligations of both parties of a given relation or the damage and compensation, in case of corrective justice and involuntary *synallagmata.*[18] These differences are evident in comparison of Aristotelian theory of commutative justice and the theory of St. Thomas Aquinas.[19]

The next question concerns the analysis of the notion of equality among the subjects of social relationships. This notion lies at the basis of commutative justice. The assumption that the subjects of social relationships are equal among themselves in regard to the area of commutative and corrective justice may be treated as the explanation of putting these relationships on arithmetical proportion, similar to those social relations existing within the political sphere, results in the use of geometrical proportion as concerning the distributive justice (Manthe 1996, 7–26). The other important issue is the question concerning the character of *synallagmata hecousia*, i.e. voluntary social relationships resulting in the rise of obligations. One should ask about the relationship between the element of will and the external factors indicating this will, as well as about the character of this voluntary notion: whether it concerns the reciprocal will of the parties or whether it is an independent factor, and if it concerns the consent, what is the basis among the voluntary element and the rise of obligation?

The above issue concerns also whether Aristotle had formulated the principle of the will autonomy and the possible scope and meaning of such a principle: is it merely a postulate arising from moral and political philosophy of Aristotle or is it also directed towards the positive law? This question is situated within the sphere of relationships between Aristotle theory and the ancient Greek law as well as between this theory and the contemporary private law. One may conclude that the problem may be formulated as the need for a reconstruction of the theory of commutative justice within the philosophy of Aristotle.

It seems, however, that the mere reconstruction is valid solely on a historical basis and may be regarded as the possible commencement of research on the meaning of this theory and its impact upon the contemporary philosophy of law and private law. As this task is far beyond the scope of this paper which aims at presenting some issues pertaining to the fundament of the contemporary philosophy of law, having its roots in Aristotelian theory of commutative justice and his concept of *synallagma*.

---

[18] *EN* 1135 b 5–10.

[19] This point has been raised by many authors, including: del Vecchio (1956, 60–65), Finnis (1980, 177–184), Gordley (1981, 1604–1606).

The contemporary researchers tend to explain the omissions and inadequacies of the theory of Aristotle.[20] In this respect the critique of his theory developed within the philosophy of law by H. Kelsen plays an important role for the understanding the expectations generated by legal systems and addressed to any concept of justice. Kelsen suggested that the Aristotelian theory of justice had been founded on the error of *petitio principii* as far as the tautological character of the relationship between the proportion and the rule of justice was concerned (Kelsen 1957, 135–136). Additionally he had observed that the difference between the commutative and distributive justice was purely elusive.

It seems that this critique demands a more coherent reconstruction of the Aristotelian theory of commutative justice, the reconstruction or reinterpretation which would give a new sense to the Aristotle's notion of *synallagma* and would therefore aim to its application in contract law. As a response to the inadequacies of the Aristotelian theory of commutative justice three potential approaches towards the explanation of the foundations of contract and contractual relations seem to be interesting and plausible. Firstly, a potential development of the Aristotelian concept of reciprocity-based relations within a wider framework of Aristotelian tradition, especially moral and legal philosophy, seems to offer a space for further elaboration of the objective theory of contract (Gordley 2001, 297–326; Murphy 2002, 85). The application of these concepts seems however to be strictly limited by the fact, that the majority of legal systems define reciprocity in contracts in purely subjective terms, referring to the exchange of promises and the underlying consent of parties rather than to the objective equivalence of obligations as a moral and legal reason for protecting legitimate expectations of parties. Traditional approach may than produce important reasons for legal reform rather than the explanation of actually existing rules, principles and doctrines of contract law. It should take the form of redefinition of some legal concepts such as obligation, agreement, or equivalence, which potentially might take form of the fundamental concepts of moral and political philosophy of Aristotle. The opposite direction is however also available, namely a redefinition of some basic principles of Aristotelian theory of social exchange in order to use them as a foundation for a renewed neoaristotelian moral philosophy and philosophy of private law. This line of argument may take two different forms.

It could concentrate on the procedural aspect of *synallagma* as a basic framework of the theory of autonomy of private law, constructed on the basis of broadening Aristotle's conception of commutative justice with the notion of subjective right and the autonomy of will of the parties. This attitude has already been presented by some authors inclined to combine Aristotelian theory of justice and some other aspects of his moral philosophy with the concept of subjective rights and freedoms

[20] Cf. a critique of Aristotelian concept of commutative justice presented by A. W. R. Harrison who states that: "*The precise interpretation of Aristotle's mathematical scheme here is notoriously difficult and I do not pretend to understand it fully*." Harrison (1957, 45); M. Wesoły endorses that: "(…) *his conception of rectificatory justice seems to be somewhat strange and unsatisfactory*"; cf. Wesoły (1989, 217–218). Similarly Hardie (1968, 191–192), del Vecchio (1956, 52–56) and Kelsen (1957, 135–136).

reflecting the voluntary nature of contractual agreements as postulated by such philosophers as I. Kant and G. W. Hegel (Weinrib 1994, 277–280; Benson 1992, 535). The advantages of this line of thinking about philosophical foundations of contracts are however balanced by cost of syncretism and the lack of precision when different concepts and meanings are forged in a new, synthetic justification for the enforcement of some promises as in case of contracts and the refusal to enforce some others, as in case of donations. Moreover, it seems that the contemporary highly formalised legal concepts are somehow resistant towards any philosophical traditions, regardless whether Aristotelian, Kantian or Hegelian.

The shortcomings of existing legal doctrines consist in lack of explanation of many different factors affecting the ultimate outcome of negotiation process or the formation of contracts. This problem becomes obvious after the examination of contemporary theories of the principle of *synallagma* in contract law. There are generally four alternative legal and doctrinal explanations of the interdependence between the obligations of the parties to the contract. The first theory, characterised as the theory of the aim of exchange (*Austauschzwecktheorie*), is based on the construction of obligatory contract as an exchange of obligations. Such a notion of exchange is connected with the bipolar obligatory legal act rather and thus it is an exchange of a legal rather than economic character. The relationship between the obligations or tasks of the parties is linked to the aim of contract understood as the exchange of obligations. The aim of exchange is to ensure the obligation of one party, which is the response for the obligation of the other party. The less direct aim is situated not only on the legal sphere but also within the economic one and it demands the execution of obligation by the other party of obligatory contract.

According to the second theory on the legal basis of obligation (*Geschäftsgrundlagetheorie*) the notion of genetic synallagma as a basis for relationships between the obligations of the parties to a contract refers to the value of obligations and the equivalency of obligations in strictly economic sense.

The third theory, namely the theory of conditional obligations (*Bedingungstheorie*) primarily concerns the construction of functional and conditional *synallagma*. It seems that some influence on such a formulation of relationship between the conditional character of the mutual obligations of the parties and the notion of *synallagma* may be placed on art. 1184 c.c. fr., which is a rule concerning the conditional character of obligations arising from the creation of a mutual contract while the condition on claiming the execution of obligation by one party is in itself the execution of obligation by the other party of a mutual contract.

Last but not least, the theory of legal reason (*Causatheorie*) is highly relevant mainly within the French legal system. Nevertheless, also in the German legal system there were undertaken some attempts at identifying the construction of genetic synallagma with the notion of causality of obligatory contracts. The starting point for such theories is the statement than the structure of a mutual contract has a causal character, i.e. it concerns the aim of a party and this aim is a real reason for the creation of obligation.

It seems, however, that the genetic *synallagma* understood in this way, as a construction of mutual obligations and related to the aim of a party having a subjective character, has no essential influence within the German legal system. In contradiction to the French legal system, the creation of any contract does not require the existence of a legal reason for its validity. Nevertheless, the notion of causality may have a meta-legal meaning in regard to *causa* understood as a legal reason for the completion of a contract.

The critics of these theories underline the fact that the underlying legal reasoning explains neither the character of relationships existing between the parties nor the reason for their respective obligations. Therefore there is no sufficient theoretical explanation for the existing regulations on the creation of contracts in majority of the European legal systems. Additionally it does not seem that the interpretation of any of these theories in according to the philosophical background of any particular system or tradition would have offered a better explanation of current legal rules. The problem of reciprocity in contracts remains imbued against any attempt of systemic reasoning also due to the fact that different legal systems adopt different theories or refer to different intellectual and cultural traditions. Meanwhile law strives to become neutral from any of these particular explanations. Thus it seems that the functional approach may offer a better chance to explain the nature of the interdependence between parties, the character of reciprocity and the meaning of equivalence of obligations. The functional approach would not promote any particular philosophical tradition. However it would adopt Aristotelian strategy, namely the application best analytical tools in order to express and reveal a profound philosophical problem, and to solve it if it is a matter of practical reasonability rather than purely speculative endeavor. This approach focuses on the possible application of mathematics to the problem of exchange and distribution of goods. This was proposed, among others, by R.A. Posner and may be described as an attempt to formalise the problem of reciprocity in contracts and to give a new meaning to the idea of commutative justice (Posner 1983, 73). It seems therefore that the postulate of the commutative justice has the aim of shaping the legal regulations in such a way as if to enforcing the contracts reaching the optimal allocation of resources. This could only be achieved however by the application of game theoretic models to the problem of reciprocity in contracts.

## 14.4 Game Theoretical Interpretation of Synallagma as a Solution to Bargaining Problem

Game theory does offer a basis for the principle of commutative justice regarded as an exchange ratio established through the bargaining process. The solution to the bargaining problem depends on a disagreement point and includes the attitude toward risk. At the same time it provides a Pareto-efficient outcome (this being one of the axioms formulated by J. Nash). In those circumstances it seems that the basis of the legitimacy of such solution is endogenous, thus the solution predicts the result of bargaining process.

A bargaining problem may be presented as a non-cooperative game, namely Rubinstein's bargaining game (Rubinstein 1983, 97–109), which resembles the situation presented by Aristotle in *Nicomachean Ethics*.[21] In this game both players are making proposals (offers and counter offers) until one of the offers is accepted.[22] The factor of time in which the agreement is reached is taken into account, so that $\delta$ represents the amount of decrease for a party for each period of time.

If A offers x, he retains the share $1-x$. Additionally the discount of time should be taken into account. In these circumstances the counteroffer from B is more attractive for A than his next offer if it gives $(1-x)\,\delta$. The game illustrates the thesis that the outcome of the bargaining process is diminishing in time ("the cake is shrinking"), so that the sooner one offer is accepted, the better.

This game has a unique subgame-perfect equilibrium: A offers B: $\delta/(1+\delta)$ and does not accept any counteroffer from B. B accepts any offer equal or greater than $\delta/(1+\delta)$ or makes a counteroffer of $(1-x)\delta$. A receives $1-x$ or $1/(1+\delta)$. The strategy of A is never accepting a counteroffer, taking into account that the B's counteroffer is not larger than $(1-x)\delta$. The best strategy of B is to take the initial offer. Thus A makes the offer large enough so that B is not able to make a counteroffer preventing repetition of the same offer. The model assumes that in special case $(\tau \rightarrow 0)$ where there is no time interval between the rejection of proposal and a new proposal there is virtually the advantage of the party who makes the offer first.

There are no incentives to cheat in this game and no mechanism for sustaining commitments is required. Within time the game converges to Nash bargaining solution. Additionally the possible asymmetries between the parties result from the different attitudes to the passage of time. In fact the interpretation of Rubinstein's bargaining game stresses that the more patient party has more bargaining power. The difference does not lie in the bargaining skill because both parties are rational optimisers.

The result of the game corresponds to the Nash solution to the bargaining problem (Nash 1950, 155–162). Bargaining solution is a function $f\,(P,c)$, where $P$ denotes payoff space, and $c$ denotes conflict point, such as,

$$(P,c) \rightarrow f\,(P,c) = \arg\max\,(u_A - c_A)(u_B - c_B)$$

The bargaining outcome is represented in utilities $(u_A, u_B)$. The Nash bargaining solution satisfies four axioms: independence of equivalent utility representation, symmetry, independence of irrelevant alternatives and weak efficiency (Pareto-optimality).

[21] For the application of this game to the solution of bargaining problem with reference to the Aristotelian theory of justice cf. Binmore (2005, 105).

[22] The same model of Rubinstein bargaining game has been implemented to solve the bargaining problem in *Peenvyhouse v. Garland*. Cf. Baird et al. (1995, 224–241).

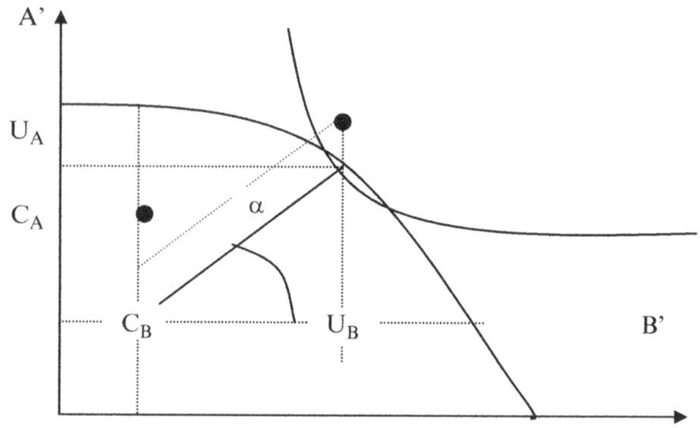


**Fig. 14.2** Nash bargaining solution (Based on Nash 1950)

According to the Fig. 14.2 point *Q* coincides with the Nash bargaining solution. At the same time tan $\alpha$ expresses the ratio of exchange $u'/v'$ and tan $\alpha=(u_A-c_A)/(u_B-c_B)$.

The result of the game corresponds to the Nash solution to the bargaining problem. This solution has been characterized by J. Nash in following way:

> The economic situations of monopoly versus monopsony (...) and of negotiation between employer and labour union may be regarded as bargaining problem. A 'solution' here means a determination of the amount of satisfaction each individual should expect to get from the situation, or, rather, a determination of how much it should be worth to each of these individuals to have this opportunity to bargain (Nash 1950, 155).

## 14.5 Conclusion

The notion of a subjective utility within the performed exchange seems to point out to the possibility of using the analysis on the basis of a mathematical proportion. The fundamental statement of Aristotle concerning the exchange among two parties on the basis of the commutative justice is made in regard to the example with a house-builder and a shoe-maker. This fragment treats the subjective utility as an introductory condition of exchange but it is directly related neither to the emergence of the factor of mutual benefit arising from the exchange nor to the fairness of exchange within the sphere of a voluntary choice. It seems that Aristotle used the "natural" basis of a voluntary exchange in which both of the parties compare one's own goods with the goods of the other party in a subjective way, and in the same time defining the sphere of mutual benefits conducive to the act of exchange. The principle of commutative does not have the character of a moral norm or a technical rule connected to the moral choice exclusively. If the aim of economic activity is defined as supplying in a greatest possible degree the material tools for the full cultural development of men, then the society should elaborate on a kind of procedure for achieving effectiveness while using the existing resources, similar to that of a free-market economy based on the principle of justice.

The concept of the solution of bargaining problem seems to supplement the requirements of the Aristotelian commutative justice with mathematical rigour. Commutative justice refers to the result of face-to-face bargain between the parties in single transactions. The problems of relations between commutative justice and fairness refer partly to the relation between commutative justice seen as a single contract and norms of distributive justice based on social contract. On the other hand both contracts have the same legitimacy, and so does commutative and distributive justice. Both formulations of justice may not satisfy the requirement of fairness. This only means that there is no conceptual priority of distributive justice over commutative justice, as both of them result from the contractual process (individual or social respectively).

## Bibliography

Aristotle. 2006. *Nicomachean Ethics*. Trans. and ed. R. Crisp. Cambridge: Cambridge University Press.

Arrow, K.J. 1951. *Social choice and individual values*. New York: Wiley.

Atiyah, P.S. 1985. Contract and fair exchange. *University of Toronto Law Journal* 35: 1–24.

Anscomb, G.E.M., and P.T. Geach. 1961. *Three philosophers*. Oxford: Basil Blackwell.

Baird, D.G., G.H. Gertner, and R.C. Picker. 1995. *Game theory and the law*. Cambridge, MA: Harvard University Press.

Benson, P. 1992. The basis of corrective justice and its relation to distributive justice. *Iowa Law Review* 77: 515–624.

Binmore, K. 2005. *Natural justice*. Oxford: Oxford University Press.

del Vecchio, G. 1956. *Justice. A historical and philosophical essay*. Ed. A.H. Campbell and Trans. Lady Guthrie. Edinburgh: Edinburgh University Press.

Despotopoulos, C. 1968. La notion de σύνάλλάγμά chez Aristotle. *Archives de philosophie du droit* 13: 115–127.

Finley, M.I. 1970. Aristotle and economic analysis. *Past and Present* 5: 3–25.

Finnis, J. 1980. *Natural law and natural rights*. Oxford: Clarendon.

Gauthier, R.A., and J.Y. Jolif. 1958. *L'Etique a Nicomacque*, vol. II. Paris: Louvain.

Golecki, M.J. 2008. *Synallagma. Filozoficzne podstawy odpowiedzialności kontraktowej* [Synallagma. Philosophical foundations of contract]. Toruń: Wydawnictwo Adam Marszałek.

Gordley, J. 1981. Equality in exchange. *California Law Review* 69: 1587–1656.

Gordley, J. 2001. Contract law in the Aristotelian tradition. In *The theory of contract law: New essays*, ed. P. Benson, 297–326. Cambridge: Cambridge University Press.

Hamburger, M. 1951. *Morals and law: The growth of Aristotle's legal theory*. New Haven: Yale University Press.

Hardie, W.F.R. 1968. *Aristotle's ethical theory*. Oxford: Clarendon.

Harrison, R.W. 1957. Aristotle's Nicomachean Ethics, book V and the law of Athens. *Journal of Historical Studies* 23: 42–47.

Jackson, H. 1879. *The fifth book of the Nicomachean Ethics*. Cambridge: The University Press.

Joachim, H.H. 1951. *Aristotle. The Nicomachean Ethics, a commentary*. Oxford: Clarendon.

Kelsen, H. 1957. *What is justice?* Berkeley: University of California Press.

Keynes, J.N. 1955. *The scope and method of political economy*, 4th ed. New York: Kelly & Millman.

Langholm, O. 1979. *Price and value theory in the Aristotelian tradition*. Bergen: Universitetsforlaget.

Lee, H.D.P. 1937. The legal background of two passages in the Nichomachean Ethics. *The Classical Quarterly* 31: 129–140.

Lowry, S.T. 1969. Aristotle's mathematical analysis of exchange. *History of Political Economy* 1: 44–60.
Maffi, A. 1980. *Synallagma*' e obbligazioni in Aristotele: spunti critici. *Atti II. Seminaria Romani Gardesano,* 13–35. Milano.
Manthe, U. 1996. Beiträge zur Entwicklung des atiken Gerechtgkeitsbegriffes I: Die Mathematisierungdurch Pythagoras und Aristoteles. *Zeitschrift der Savigny–Stiftung für Rechtsgeschichte Romanistische Abteilung* 113: 1–31.
Meikle, S. 1990. Aristotle and exchange value. In *A companion to Aristotle's politics*, ed. F. Miller and D. Keyt, 156–169. Cambridge: Cambridge University Press.
Meikle, S. 1995. *Aristotle's economic thought.* Oxford: Oxford University Press.
Murphy, J.B. 2002. Equality in exchange. *American Journal of Jurisprudence* 47: 85–121.
Nash, J.F. 1950. The bargaining problem. *Econometrica* 50: 155–162.
Polanyi, K. 1944. *The great transformation.* New York: Rinehart.
Polanyi, K. 1957. Aristotle discovers the economy. In *Trade and market in the early empires: Economies in history and theory*, ed. K. Polanyi, C.M. Arensberg, and H.W. Pearson, 64–94. New York: The Free Press.
Posner, R. 1983. *The economics of justice.* Cambridge, MA: Harvard University Press.
Pringsheim, F. 1950. *The Greek law of sale.* Weimar: H. Böhlaus Nachfolger.
Ritchie, D.G. 1894. Aristotle's subdivision of particular justice. *The Classical Review* 8: 185–192.
Ross, W.D. 1953. *Aristotle*, 5th ed. London: Methuen.
Rubinstein, A. 1983. Perfect equilibrium in a bargaining model. *Econometrica* 50: 97–109.
Schumpeter, J.A. 1954. *History of economic analysis.* New York: Oxford University Press.
Soudek, J. 1952. Aristotle's theory of exchange: An enquiry into the origins of economic analysis. *Proceedings of the American Philosophical Society* 96: 45–59.
Tieben, B. 2009. *The concept of equilibrium in different economic traditions. A historical investigation.* Amsterdam: Rozenberg Publishers.
Weinrib, E. 1994. The gains and losses of corrective justice. *Duke Law Journal* 23: 277–297.
Wesoły, M. 1989. Aristotle's conception of justice as equality. *Eos* 77: 211–220.
Winthrop, D. 1978. Aristotle and theories of justice. *The American Political Science Review* 72: 1201–1216.
Worland, S.T. 1984. Aristotle and the neoclassical tradition: The shifting ground of complementarity. *History of Political Economy* 16: 107–135.
Yack, B. 1990. Natural right and Aristotle's understanding of justice. *Political Theory* 18: 216–237.

# Chapter 15
# The General Principle of Proportionality and Aristotle

**Eric Engle**

## 15.1 Introduction

This article proposes a brief history of the concept of proportionality in law in order to understand the rule, found around the world, that state action must be a rational means to a permissible end which does not invade protected human rights unless strictly compelled by necessity. Although the concept of fundamental human rights is a very modern one, dating from the Scottish enlightenment (circa 1776)[1] the concept of proportionality, which is used to adjudicate conflicts between fundamental rights, is much older.

The idea of distributive justice as proportionality is first seen in Book V of Aristotle's *Nicomachean Ethics*.[2] Proportionality as commutative justice ("an eye for an eye" *lex talionis*) is even older and can be traced to the Code of Hammurabi.

[1] As legal concepts the idea of inalienable individual rights is first seen in the founding US constitutional documents such as the US Declaration of Independence (1776 "all men are created equal; they are endowed by their creator with certain inalienable rights") and the US Constitution. The first modern legal use of the exact term "human rights" is the French *Déclaration des Droits de l'Homme* of 1789 (lit.: *declaration of the rights of man*).

[2] Aristotle's idea is that distributive justice can be expressed as ratio, i.e. proportionality, is one form of justice. As one of my anonymous peer reviewers rightly pointed out, Aristotle also discusses justice in the sense of a personal virtue – a virtuous mean. (*EN* 1106b28) Thus, Aristotle's theory of justice also connects to his theory of moral virtue as the prudent mean between opposite extremes of vice. (*EN* 1107a2) As one of the reviewers noted, adjudication involves the judicial prudential determination of the mean between the opposing positions of the plaintiff and defendant. Adjudication is an act of practical reasoning – Aristotle's phronêsis. Aristotle's ideas of justice as ratio and virtue as mean explain the application of the proportionality to distributive and commutative justice – respectively, social justice (proportional shares in the constitution of the Polis, i.e. the State) on the one hand and proportional punishment of crimes on the other.

E. Engle (✉)
Humbolt Universität, Berlin, Germany
e-mail: eric.engle@ut.ee

L. Huppes-Cluysenaer and N.M.M.S. Coelho (eds.), *Aristotle and The Philosophy of Law: Theory, Practice and Justice*, Ius Gentium: Comparative Perspectives on Law and Justice 23,
DOI 10.1007/978-94-007-6031-8_15, 

The proportionality principle in law, though evolving, has shown remarkable continuity over centuries. The *theory* of distributive justice as geometric proportionality seems to spring forth from Aristotle's brain like Athena, fully formed and crying "Victory!"[3] However, the *practice* of proportionality in law, i.e. the implementation of the general concept in legal practice, occurred over several centuries. Athena's cry echoes to this day: a law must be a rational means to a permissible end; the punishment must fit the crime; the use of force in self-defence must be necessary, i.e. unavoidable, and limited to only that violence which is needed to extinguish the threat. The abstract proportionality concept became more precisely defined as a rule of law through historical experience.

The general principle of proportionality is now a world-wide principle of law[4] found in common law, civil law (Poto 2007) and international law (Andenas and Zleptnig 2007). The proportionality principle is a key organizing principle of contemporary legal thought, converging civil law and common law to a global uniform *ius commune* which hybridizes aspects of common law (binding inductive case law) *alongside* civil law (deductive general principles of civil law and common law fundamental rights). This hybridized *ius commune* in turn converges substantive rules of national law toward uniform global rules. Norm convergence arises due to intensified trade, vastly improved communication and transit, and machine translation. Base and superstructure here both reach toward the same goal, the withering of the state and its replacement by civil society through peaceful trade to replace war as the principal mode of state interactions. Contemporary global law is guided by a teleology of peace through interdependence and protection of human rights. This teleology seeks to replace conflict oriented negative and zero sum state interactions with positive sum private law interactions. War is to be replaced by law, and law in turn is to be replaced by voluntary market transactions. Thus, proportionality as an adjudicative principle which determines the outcomes of conflicts between fundamental rights is central to contemporary law. Legal theories are also converging globally. At the broadest level, conceptual jurisprudence (*Begriffsjurisprudenz*) is linked to legal process interest balancing (*Interessenjurisprudenz*). Conceptual jurisprudence is then applied to natural law, and legal process interest balancing is applied to positive law.

Proportionality, first clearly elucidated by Aristotle, has become the rational principle which adjudicates conflicts between state power and fundamental individual rights as well as conflicts between competing fundamental human rights. Aristotle did not however analyse law in terms of conflicts involving fundamental individual or human rights. The idea of individual fundamental rights as abstract deductive general principles of law which are hierarchically superior to ordinary laws only arose with the Scottish enlightenment circa 1776, which recognized

[3] *Id.*

[4] "From German origins, proportionality analysis spread across Europe, into Commonwealth systems (Canada, New Zealand, South Africa), and Israel; it has also migrated to treaty-based regimes, including the European Union, the European Convention on Human Rights, and the World Trade Organization" (Sweet 2008).

individuals as having inalienable fundamental rights. However, the early modern idea of inalienable rights is linked to the pre-modern idea of natural right. Natural right is universal because it is an inevitable consequence of the nature of things. Modernity's inalienable human rights are also universal, and universality links them to the pre-modern concept of international law; antiquity saw international law as those laws which were true in all countries due to the nature of things. Modernity's fundamental rights are also natural in that they are inevitably linked to the attainment of Aristotle's idea of the good life in political society. Because fundamental rights are universal and hierarchically superior to ordinary laws they are deductive general principles of the law of reason (*Vernunftrecht*). The idea of inalienable individual rights which are inviolable and hierarchically superior to ordinary laws is clearly an early modern idea, yet the early modern idea of human rights is intellectually coherent with pre-modern thought. The modern idea that inalienable individual rights are hierarchically superior to ordinary law also coheres with the idea of antiquity that law is subordinate to justice – that an unjust law is not law (*lex mala, lex nulla*). As fundamental rights were increasingly recognized, they also became relativized against each other in late modernity. Thus, the idea of the just as *recta ratio* and law as *recta ratio naturae congruens*, pre-modern concepts, became applied in late modernity to adjudicate conflicts between inalienable fundamental human rights, an early modern concept. Unlike inalienable fundamental human rights, which inhere by nature and are universal, economic rights are alienable, positive, and vary with time and place. Market rights may be bought and sold. Thus, they are positive and subject to economic analyses (interest evaluation and balancing) and are hierarchically subordinate to inalienable human rights.

To understand the global rise and success as well as the contours of this general principle of (constitutional) law and how proportionality serves the constitutionalization and globalization of law, we examine the history of the concept and then its legal practice.

## 15.2 Proportionality in Antiquity

The idea of justice as proportionality appears first and clearly in Aristotle's *Nicomachean Ethics* Book V, and also Book III Ch. 10–12. To understand Aristotle's idea of proportionality in law, and to see how it is the root of contemporary proportionality analysis, we must first understand Aristotle's theory of law and justice.

### *15.2.1 Aristotle's Theory of Law*

Aristotle's rightly distinguishes two types of law: nomos and dike (*EN* V.7). *Nomoi*, positive laws, are established by convention and like all man-made things (*technê* – *EN* 1140 a 1) vary from place to place. *Dikê* in contrast is the idea of universal right which is natural (*physis*) – that which is natural cannot be otherwise (*EN* 1139b20–23). Aristotle saw the co-existence of a universal natural law, valid in all places and times, *alongside* positive national laws which would hold true in one land, but not in another (*EN* 1134b18–20). Aristotle rightly regarded what we call positive law (*nomoi*) and natural justice (*dikê*) as complementary, (*EN* V III.4) not conflicting (*EN* V.7).

### *15.2.2 Aristotle's Theory of Justice*

Aristotle identifies several forms of justice (*EN* 1129a27). As a moral virtue, justice addresses the idea of men who are just and acts which are just (*EN* 1129a31–1129b7, 1136a25-3, 1135b25, 1135a16, 1133b29–1134a1). As a theory of law, Aristotle distinguishes commutative justice and social distributive justice (*EN* 1131b9–20). Distributive justice is positive, not natural, and may vary from one land to another and follows a geometric proportion. Commutative justice in contrast follows arithmetic proportionality (*EN* 1131b32–1132a1). Equity, in Aristotle's thinking, acting *ex aequo et bono*, is the means whereby Judges correct errors which result in applying laws ex post to situations legislated *ex ante* (*EN* 1137b10–14).

### *15.2.3 Aristotle's Theory of Proportionality*

Proportionately measures distributive justice (Hanau 2004) in Aristotle's schema, which appears to be the earliest known historical source of the contemporary general principle of proportionality in law (Radbruch 2003, 122, nt.).[5] Distributive justice addresses public law.[6] It is conventional: different societies have different distributive principles.[7]

In Aristotle, proportionality is justice as the right ratio – the relationship between a distributive principle and the shares apportioned thereby. Aristotle's idea has since evolved in theory and practice to become more precisely defined. However, the essence of the principle of proportionality in law is clearly seen in Aristotle's idea of distributive justice as a rational principle (*recta ratio*) which determines the attribution of rights between State and citizen as well as the rule of law.[8] The idea of

---

[5] *See also* Bydlinski (1991, 339); Engisch (1971, 162, 222, 229).

[6] Aristotle clearly intends his concept of distributive justice (the just as ratio, i.e. proportionality) as governing public-private relations. "Of particular justice and that which is just in the corresponding sense, (A) one kind is that which is manifested in distributions of honour or money or the other things that fall to be divided among those who have a share in the *constitution*" (*EN*, Book V Ch. II v. 8 (emphasis added).

[7] *EN* 1131 a 24–28.

[8] "We do not allow a man to rule, but rational principle, because a man behaves thus in his own interests and becomes a tyrant" (*EN* V, VI, 5).

proportionality as a specific rule of law emerged obliquely from Aristotle's thought as an abstract general principle and became an increasingly concrete and definite proposition of the law of self-defense in Cicero,[9] Justinian (Digest 43.16.3.9),[10] Augustine (City of God, 1998, ch. 7),[11] and Aquinas (Sum. Theol. 2a2ae 90–97 esp. 95/3, 96/1).[12]

### 15.2.4 Other Pre-modern Theorists on Proportionality

The well-defined abstract theoretical principle became concretized and realized by Cicero into positive law (Jolowicz and Nichols 1972, 104–5; Schiller 1978, 374–5; Jackson 1915). Cicero (De Republica) describes law as the *recta ratio naturae congruens* the right ratio, i.e. the proper proportion. This concretization (realization) was further refined by Aquinas in the law of self-defence of states (Sum. Theol. 2a2ae 40). Aquinas then presented the first decomposition of Aristotle's concept into the now known multi-step proportionality procedure (Sum. Theol. 2a2ae 90–97). In the law of self-defence, Aquinas argued that there are conditions which must exist for the use of force to be just (necessity); that force, when used, must not be excessive (proportionality), and that must be exercised by the sovereign according to rules. Aquinas' theory on proportional self-defence, in turn, became seen as a general principle of law by Grotius (War and Peace, 1901, introd. par. 62).[13] The principle would apply not just to states in their mutual relations but also to individuals

---

[9] "No war can be undertaken by a just and wise state, unless for faith or self-defence. This self-defense of the state is enough to ensure its perpetuity, and this perpetuity is what all patriots desire. Those afflictions which even the hardiest spirits smart under poverty, exile, prison, and torment private individuals seek to escape from by an instantaneous death. But for states, the greatest calamity of all is that death, which to individuals appears a refuge. A state should be so constituted as to live for ever. For a commonwealth, there is no natural dissolution, as there is for a man, to whom death not only becomes necessary, but often desirable. And when a state once decays and falls, it is so utterly revolutionized, that if we may compare great things with small, it resembles the final wreck of the universe. All wars, undertaken without a proper motive, are unjust. And no war can be reputed just, unless it be duly announced and proclaimed, and if it be not preceded by a rational demand for restitution.

*Our Roman Commonwealth, by defending its allies, has got possession of the world*" (*Treatise on the Commonwealth*, trans. Barham 1841–42, emphasis supplied).

[10] "Those who do damage because they cannot otherwise defend themselves are blameless...It is permitted only to use force against an attacker and even then only so far as is necessary for self-defense."

[11] Augustine discusses just war theory but doesn't use the term proportionality (between force and threat). He does however use the term "just war". This seems to be the first use of the signifier "just war" (certainly one of the earliest).

[12] *Also see* Aquinas Sum. Theol. 2a2ae 40.

[13] "The Law of Nations does not consist, therefore, of a mere body of deductions derived from general principles of justice, for there is also a body of doctrine" ( trans. Campbell).

in their mutual relations. Grotius transitions the concept into modernity and links the idea of *justice as proportion* (ratio) to the idea of *interest balancing* as a method for dispute resolution. Grotius unites the ancient concept of justice as ratio, the medieval concept of proportional self-defence, and the modern concept of balancing interests.[14] Modern proportionality thus emerged as a general principle of law. This legal principle of proportional self defence, first articulated in the law of nations (also known as *jus gentium* – public international law) was increasingly applied in cases of self-defence (Totten 2007, nt. 36) not only of states, but also of the person, and then in national police and then administrative law: the right to self defence must be exercised in proportion to the threat; punishments should be proportional to crimes; the administration must not act excessively. Rational principle rules – not caprice nor *Diktat*; and it rules with laws – laws which serve justice.

## 15.3 Proportionality in Early Modern Law

### *15.3.1 German Law*

This proportionality principle, brought into national law as the right to proportional individual self defence and the duty of the state to punish crime only proportionally, became a heavily litigated aspect of German national administrative law (Sweet and Mathews 2008). In German law, the principle, as Wieacker (1979) noted, is rooted in antiquity. The concept evolved from a prohibition of disproportionality (*Uebermassverbot*)[15] (the state must not act too broadly) toward a more clearly defined and restrictive principle that the state must use proportional means to legitimate ends (*Verhaeltnismaessigkeit*) in the post-war era. This became a key principle of German constitutional law (Baer 1999). The idea is so popular that it even found expression in *East* German law, (Mampel 1982, 731–743) evidence of a broader thesis that Socialist law is a variant of Western law, albeit organized by the general principle of *equality* rather than the general principle of *liberty*.

### *15.3.2 Proportionality in Anglo-American Common Law*

Proportionality, as a general principle of international law likewise found its way into Anglo-American common law (Gray 1763, 844–847). Proportionality in common

[14] Chapter XXIV: Precautions Against Rashly Engaging in War, Even Upon Just Grounds. "In all cases of deliberation, not only the ultimate but the intermediate objects leading to the principal ends are to be considered. The final object is always some good, or at least the evasion of some evil, which amounts to the same. The means are never to be considered by themselves, but only as they have a tendency to the proposed end. Wherefore in all cases of deliberation, the proportion, which the means and the end bear to each other, is to be duly weighed, by comparing them."

[15] Remmert (1995).

law is rooted in Magna Charta, which can be seen as the legal source of the principle of proportionality in British[16] and U.S. common law.[17] Thus, the Eighth amendment of the U.S. Constitution permits only proportional punishments.[18] Just as in German law, in common law the principle of proportionality found its earliest expression in the areas of police powers[19] Despite controversy (Pillai 2002), it is clear that punishment must be proportional to the crime.[20] The parallel evolution of the proportionality principle in German law and common law is likely due to the common connected conceptual roots of the principle in the thought of Aristotle, Cicero, Aquinas, and Grotius.

Early modernity developed two concepts in parallel:

1. Interest balancing (often seen as political and thus non-justiciable)[21]
2. Proportionality (legal and thus justiciable)

Grotius links interest balancing and proportionality (Ch XXIV). He appears to be the source of the fusion of interest balancing and the inquiry into whether law is proportional, i.e. a rational means to a permissible end. However, interest balancing and means-end review address two different types of rights: positive economic rights (interest balancing) and natural human rights (means-end review). The fusion of these two ideas does not appear in the works of the pre-modern primary sources surveyed here.

As we see first in Grotius, the general principle of proportionality (means-end review) and economic interest balancing are sometimes joined together. Their fusion is an error because they address two different categories of rights and are mathematically distinct. That which in Europe is referred to as proportionality analysis is known in

---

[16] Hodges v. Humkin (1615). "By the seventeenth century, England had extended this principle to punishments that called for incarceration. In one case, the King's Court ruled that "imprisonment ought always to be according to the quality of the offence".

[17] *See* Note *The Eighth Amendment, Proportionality, And The Changing Meaning Of Punishments* (2009). "In Solem v. Helm, Justice Powell traced the history of the Cruel and Unusual Punishments Clause back to the Magna Carta and the English Bill of Rights of 1689, which he found to have embodied a strong principle of proportional punishment." *See also Harmelin v. Michigan* (1991).

[18] *Coker v.* Georgia (1977), *Eberheart v.* Georgia (1977). "The Eighth Amendment requires that every punishment imposed by the government be commensurate with the offense committed by the defendant. Punishments that are disproportionately harsh will be overturned on appeal. Examples of punishments that have been overturned for being unreasonable are two Georgia statutes that prescribed the death penalty for rape and kidnapping".

[19] U.S. Const, Amdt. VIII. *See also Solem v. Helm* (1983), *overruled by Harmelin v. Michigan* (1991). The proportionality of punishment to crime is, however, differently tested than the proportionality of means to ends. In Solem, the Court determined that objective criteria should guide the proportionality analysis. (463 U.S. at 292) The objective criteria considered by the Court were "(i) the gravity of the offense and the harshness of the penalty; (ii) the sentences imposed on other criminals in the same jurisdiction; and (iii) the sentences imposed for commission of the same crime in other jurisdictions" (*Id.* at 292).

[20] *Coker v. Georgia* (1977); *Eberheart v. Georgia* (1977) (State death penalty for rape and kidnapping unconstitutional as disproportionate).

[21] *See* the *Federalist Papers* (1961); also Montesquieu's (1914), *Spirit of the Laws*.

U.S. courts, unequivocally, as means-end rational review with strict scrutiny for fundamental rights and suspect classes. Proportionality in Aristotle is expressed as a geometric ratio (A:B::C:D) which may be continuous or discontinuous (Aristotle). Interest-balancing, in contrast, is economic cost/benefit analysis – it is a simple inequality (A>B). To avoid confusing cost/benefit analyses of alienable *economic* interests with proportionality analysis of conflicting inalienable *constitutional* rights I advise avoiding the use of the term "balancing" entirely. Interest analysis is an unequivocal alternative term for interest balancing (*Interessenjurisprudenz*). It is especially important to avoid confusing interest balancing and proportionality because early modernity regarded interests as political and non-justiciable. The dichotomy of non-justiciable political interests versus justiciable legal rights subsists to this day in some views of international law and in the U.S. constitutional "political question" doctrine.

### *15.3.3 Proportionality in Contemporary Law (Late Modernity)*

In EU law the proportionality test (means-end rational review) is very well worked out. Most recently it was reiterated in *Viking* and *Laval* before the ECJ (Case C-438/05). According to the EU Treaty, an interference with a fundamental right "is warranted only if [1] it pursues a legitimate objective compatible with the Treaty [permissible end] and [2] is justified by overriding reasons of public interest [necessary means]; if this is the case, it must be [3] suitable for securing the attainment of the objective which it pursues [rational means to the legitimate end] and [4] not go beyond what is necessary in order to attain it" [least restrictive means; braces supplied by author] (Case C-341/05). "In simplest terms, the proportionality principle requires some articulable relationship between means and ends, specifically that the means chosen by an administration be suitable or appropriate, and no more restrictive than necessary to achieve a lawful end. The principle operates in each of the Member-State's domestic jurisprudence, though not always in hic verba, and it qualifies as a peremptory norm of international law." (Steinhardt 1994)

As already mentioned, people sometimes wrongly equate the last step in proportionality – the scrutiny as to whether the invasion of the fundamental right is as non-invasive as possible – with (economic) interest balancing. The proportionality inquiry goes to the determination of the right relationship between the means and ends of state action with respect to private rights in accord with the nature of things – it is expressed by Cicero's as *recta ratio naturae congruens* in De Republica (Commonwealth). Distributive proportionality does not concern the evaluation and comparison of costs and benefits or alienable economic interests. The last step in proportionality analysis is correctly referred to as proportionality *strictu sensu* (EU) or as "strict scrutiny" (US), and it applies respectively to fundamental rights and to "suspect classes". "Suspect classes" are discrete and insular politically powerless minorities who have suffered a history of invidious discrimination. Proportionality *strictu sensu* in the EU is also called proportionality in the narrower sense. Strict scrutiny in the US is also called "least restrictive means analysis".[22] One divergence in proportionality analysis is that the EU does not apply proportionality *strictu sensu* to govern minority/majority relations (suspect classes). It does however require strict necessity when the state invades a fundamental right. Regardless of this minor transatlantic rift, proportionality *strictu sensu* (strict scrutiny) is decidedly *not* the balancing of competing (economic) interests. Proportionality *strictu sensu* is the inquiry into whether state power is exercised in the least invasive manner possible. Any divergence between the ideas of strict scrutiny for suspect classes, least restrictive means analysis, and proportionality *strictu sensu* – which are all rooted in the same concept, necessity – would be found in the question of how majorities and minorities relate to each other. Proportionality *strictu sensu* (strict scrutiny) is not the question of identifying the interested parties, defining their competing interests, assigning weights to the interest, and then deciding the issue in favor of maximizing social wealth – which is the method known as legal process interest balancing (*Interessenjurisprudenz*).

The term "balancing" in law is used to indicate several different things, namely:

1. Commutative justice – after application, the scales of justice are restored to their *ex ante* balance, e.g., *lex talionis*.
2. Cost-benefit analysis (CBA) – we weigh the costs and benefits of a policy against the costs and benefits of another policy and decide for the one that generates the most social wealth. Cost-benefit analysis is the proper contemporary use of the term balancing, but that use rightly applies only to alienable economic rights, not to inalienable (fundamental, universal, human) rights. CBA is positive law, not natural justice.

---

[22] Aristotle (*EN* V.3.) "The just, then, is a species of the proportionate (proportion being not a property only of the kind of number which consists of abstract units, but of number in general). For proportion is equality of ratios, and involves four terms at least (that discrete proportion involves four terms is plain, but so does continuous proportion, for it uses one term as two and mentions it twice; e.g. 'as the line A is to the line B, so is the line B to the line C'; the line B, then, has been mentioned twice, so that if the line B be assumed twice, the proportional terms will be four); and the just, too, involves at least four terms, and the ratio between one pair is the same as that between the other pair; for there is a similar distinction between the persons and between the things. As the term A, then, is to B, so will C be to D, and therefore, alternando, as A is to C, B will be to D. Therefore also the whole is in the same ratio to the whole; and this coupling the distribution effects, and, if the terms are so combined, effects justly. The conjunction, then, of the term A with C and of B with D is what is just in distribution, and this species of the just is intermediate, and the unjust is what violates the proportion; for the proportional is *intermediate*, and the just is proportional. (Mathematicians call this kind of proportion geometrical; for it is in geometrical proportion that it follows that the whole is to the whole as either part is to the corresponding part.) This proportion is not continuous; for we cannot get a single term standing for a person and a thing." (trans. Thomson, emphasis added: Geometric justice is *distributive* it is the even handed application of a positive general principle to apportion shares of social goods to the citizens who then may interchange their goods according to the principles of commutative, i.e. arithmetic, transactional justice).

3. Proportionality in the narrower sense i.e. proportionality *strictu sensu*.

   The use of the term "balancing" to indicate proportionality in any sense should simply be avoided and allowed to fall into disuse to prevent confusion. "Proportionality in the narrower sense", "proportionality *strictu sensu*" or even "strict scrutiny".are all more exact terms for the last step in proportionality analysis and avoid confusing different concepts by referring to them with the same term. Multi-factor Interest balancing is the determination of interested parties, of their competing interests, and their evaluation against each other: its roots are in the free-law school (*freie Rechtslehre*), a forerunner of critical legal studies, and its methodology is known as *Interessenjurisprudenz* – interest analysis. Interest balancing is the comparison of costs and benefits associated with competing interests and is decided by the principle of social wealth maximization. Thus, interest balancing logically refers to alienable economic rights rather than inalienable fundamental human rights.

## 15.4 Conclusions

This historical overview shows that proportionality as a principle of law arose out of the Aristotelian concept of justice. This general theoretical fact partly explains the worldwide success of the concept, since it has deep, global common roots. After Aristotle, the principle was refined by Cicero, Aquinas and Grotius and instantiated into law by Justinian, Magna Charta, the US Constitution, and the Prussian Civil Code (ALR). Proportionality in law first appeared as a principle of the law of war, then in national police law and then administrative law from whence it evolved into a principle of constitutional law as a tool for the adjudication of conflicting constitutional rights. The principle has become a method for the formation of a transnational *ius commune*, a hybrid of common law (inductive binding case law) and civil law (deductive general principles). Because the proportionality principle is one of the most important methods of converging global norms, future developments of the general principle of proportionality should seek:

1. to develop a universally coherent terminology which avoids confusion. Means-end review with strict scrutiny for suspect classes and proportionality are methodologically synonymous and unequivocal. Interest balancing, in contrast, is a much broader term and is equivocal or at least polysemious.
2. to clearly delineate the positive law (economic interest balancing) versus the natural right aspects (means-end rational review) of proportionality discourse. Economic interest balancing through cost/benefit analysis and similar economic tests are inappropriate for adjudication of fundamental inalienable rights.

Aristotle's thought about law and justice structures jural relations to this day.[23] Proportionality, like causation (*Met.* 1.3.11) in tort (Engle 2009) is one more example

[23] One can for example trace the dichotomies of universal international law / particular national law, public law / private law and voluntary obligations (contract) / involuntary obligations (tort) to Aristotle (See, e.g. *EN* 1130 b 30–1131 a 8).

of Aristotle's influence on contemporary law. True, Aristotle was sexist (*Pol.* 1254b 10–15,1259b 2) racist (*Pol.* 1252 b 7, 1237 b 23–32) and homophobic. However, his contributions to justice through law are the greater and better part of his work. Aristotle's schema of justice, brilliantly and pithily exposed in Book V of *The Nicomachean Ethics* (see also *EN* III.10–12) holds true to this day (Engle 2008) as seen in concepts such as equity jurisprudence (ex aequo et bono), proportionality (law as right ratio) and causation in tort. Aristotle's distinction between commutative and distributive justice is tenable and useful. Aristotle serves as a good common universal starting point for contemporary thought about proportionality. This brief historical overview provides a synopsis of the history of proportionality discourse so that contemporary jurists can continue to develop the rule of law in the most rational way to serve justice.

## Bibliography

Andenas, M., and S. Zleptnig. 2007. Proportionality: WTO law in comparative perspective. *Texas International Law Journal* 42: 371–423.
Aquinas. 1947. *Summa Theologica*. New York: Benziger Bros.
Aristotle. 1955. *Nicomachean EN*. Trans. J.A.K. Thomson, and H. Tredennick and ed. J. Barnes. Harmondsworth, Middlesex: Penguin Books.
Aristotle. 1984. Metaphysics. Trans. W.D. Ross. In *The complete works of Aristotle*: *The revised Oxford Translation,* ed. J. Barnes. Princeton: Princeton University Press.
Augustine. 1998. *The city of God*. Trans. R.W. Dyson. New York: Cambridge University Press.
Baer, Susanne. 1999. Equality: The jurisprudence of the German Constitutional Court. *Columbia Journal of European Law* 5: 249–279.
Bydlinski, Franz. 1991. *Juristische Methodenlehre*. New York: Springer.
C-341/05, *Laval un Partneri Ltd v. Svenska Byggnadsarbetareförbundet* and Others.
C-438/05, *International Transport Workers' Federation and Finnish Seamen's Union v. Viking Line ABP and OÜ Viking Line Eesti*.
Cicero, Marcus Tullius. 1841–42. Treatise on the commonwealth. In *The political works of Marcus Tullius Cicero: Comprising his treatise on the commonwealth; and his treatise on the laws,* vol. 1. Trans. Francis Barham. London: Edmund Spettigue.
*Coker v. Georgia*. 433 U.S. 584 (1977).
*Eberheart v. Georgia*. 433 U.S. 917 (1977).
Engisch, Karl. 1971. *Auf der Suche nach Gerechtigkeit*. München: Piper.
Engle, Eric. 2008. Aristotle, law and justice: The tragic hero. *Northern Kentucky Law Review* 35: 1–17.
Engle, Eric Allen. 2009. Aristotelian theory and causation: The globalization of tort law. *Gujarat National Law Review* 2: 1–18.
France. 1789. *Déclaration des Droits de l'Homme* (lit.: declaration of the rights of man).
Gray, Anchitell. 1763. *Debates of the House of Commons, from the year 1667 to the year 1694*. London.
Grotius, Hugo. 1901. *The rights of war and peace, including the law of nature and of nations*. Trans. A.C. Campbell. New York: M. Walter Dunne.
Hamilton, Alexander, James Madison, and John Jay. 1961. Federalist papers. In *The federalist*, ed. Jacob E. Cooke. Middletown: Wesleyan University Press.
Hanau, Hans. 2004. *Der Grundsatz der Verhältnismaessigkeit als Schranke privater Gestaltungsmacht: Zu Herleitung und Struktur einer Angemessenheitskontrolle von Verfassungswegen*. Tübingen: Mohr-Siebecke.
*Harmelin v. Michigan*. 501 U.S. 957 (1991).

*Hodges v. Humkin*. 80 English Reports 1015, Kings Bench 1615, Coke, Justice.
Jackson, Hilton. 1915. *Latin for lawyers*. London: Sweet and Maxwell.
Jolowicz, H.F., and Barry Nichols. 1972. *Historical introduction to the study of Roman law*, 3rd ed. Cambridge: Cambridge University Press.
Justinian. *The digest of Justinian*. Alan Watson edition 1985.
*Magna Charta*, 1215.
Mampel, Siegfried. 1982. *Die Sozialistische Verfassung der Deutschen Demokratischen Republik*. Frankfurt: Alfred Metzner Verlag.
Montesquieu. 1914. *Spirit of the laws*. Trans. Thomas Nugent. London: G. Bell & Sons.
Note. 2009. The eighth amendment, proportionality, and the changing meaning of punishments. *Harvard Law Review* 122: 960–981.
Pillai, K.G.J. 2002. Incongruent disproportionality. *Hastings Const. L.Q* 29: 645–720.
Poto, Margherita. 2007. The principle of proportionality in comparative perspective. *German Law Journal* 8: 835–870.
Radbruch, Gustav. 2003. *Rechtsphilosophie*, 2nd ed. Heidelberg: C. F. Müller.
Remmert, Barbara. 1995. *Verfassungs und Verwaltungsrechtsgeschichtliche Grundlagen des Übermassverbotes*. Heidelberg: C.F. Mueller.
Schiller, A.Arthur. 1978. *Roman, law: mechanisms of development*. New York: Mouton.
*Solem v. Helm*. 463 U.S. 277 (1983).
Steinhardt, Ralph G. 1994. Book review, 'European administrative law'. *George Washington Journal of International Law and Economics* 28: 225–241.
Sweet, Alec Stone, and Jud Mathews. 2008. Proportionality balancing and global constitutionalism. *Columbia Journal of Transnational Law* 47: 72–164.
Totten, Mark. 2007. Using force first: Moral tradition and the case for revision. *Stanford Journal of International Law* 43: 95–126.
United States Constitution. 1792.
United States Declaration of Independence. 1776.
Wieacker, Franz. 1979. *Festschrift fuer Robert Fischer*. Berlin: De Gruyter.

CPSIA information can be obtained at www.ICGtesting.com
Printed in the USA
LVOW010021070313

323093LV00011B/135/P

9 789400 760301